# Biofuels Engineering Process Technology

Caye M. Drapcho, Ph.D.
Nghiem Phu Nhuan, Ph.D.
Terry H. Walker, Ph.D.

New York   Chicago   San Francisco
Lisbon   London   Madrid   Mexico City
Milan   New Delhi   San Juan
Seoul   Singapore   Sydney   Toronto

*The McGraw·Hill Companies*

Copyright © 2008 by The McGraw-Hill Companies, Inc. All rights reserved. Printed in the United States of America. Except as permitted under the United States Copyright Act of 1976, no part of this publication may be reproduced or distributed in any form or by any means, or stored in a data base or retrieval system, without the prior written permission of the publisher.

1 2 3 4 5 6 7 8 9 0    DOC/DOC    0 1 4 3 2 1 0 9 8

ISBN 978-0-07-148749-8
MHID 0-07-148749-2

**Sponsoring Editor**
Larry S. Hager

**Production Supervisor**
Pamela A. Pelton

**Editorial Supervisor**
Stephen M. Smith

**Project Manager**
Vastavikta Sharma, International Typesetting and Composition

**Copy Editor**
Megha RC, International Typesetting and Composition

**Proofreader**
Upendra Prasad, International Typesetting and Composition

**Indexer**
Broccoli Information Management

**Art Director, Cover**
Jeff Weeks

**Composition**
International Typesetting and Composition

Printed and bound by RR Donnelley.

McGraw-Hill books are available at special quantity discounts to use as premiums and sales promotions, or for use in corporate training programs. To contact a special sales representative, please visit the Contact Us page at www.mhprofessional.com.

This book is printed on acid-free paper.

Information contained in this work has been obtained by The McGraw-Hill Companies, Inc. ("McGraw-Hill") from sources believed to be reliable. However, neither McGraw-Hill nor its authors guarantee the accuracy or completeness of any information published herein, and neither McGraw-Hill nor its authors shall be responsible for any errors, omissions, or damages arising out of use of this information. This work is published with the understanding that McGraw-Hill and its authors are supplying information but are not attempting to render engineering or other professional services. If such services are required, the assistance of an appropriate professional should be sought.

To our parents Catherine and Cyril Drapcho and Pam and George Walker. They would have been proud of their children for trying to improve the world.
—Caye and Terry

To my wife, Minh Dzung, and to all the children of tomorrow with love and hope.
—Nhuan

## About the Authors

**Caye M. Drapcho, Ph.D.,** is an Associate Professor and the Graduate Coordinator in the Biosystems Engineering program at Clemson University. She has over 13 years of teaching and research experience in bioprocess and bioreactor design.

**Nghiem Phu Nhuan, Ph.D.,** is a Senior Research Biochemical Engineer in the Crop Conversion Science and Engineering Research Unit at the Eastern Regional Research Center, Agricultural Research Service, U.S. Department of Agriculture, and also an Adjunct Professor in the Biosystems Engineering program at Clemson University. He has more than 20 years of experience in bioprocess engineering in industrial and federal research laboratories.

**Terry H. Walker, Ph.D.,** is a Professor in the Biosystems Engineering program at Clemson University. He has over 10 years of experience in bioprocess engineering, specializing in fungal fermentation, bioproduct separations, and bioavailability studies.

# Contents

Preface .................................... xi

## Part 1  The Basics

### 1  Introduction ............................... 3
1.1  Biorefinery ............................ 3
1.2  Description of Biofuels ................ 5
1.3  Energy Use ............................. 6
1.4  Efficiency of Energy Use ............... 8
1.5  Biofuels Production and Use ........... 10
1.6  Alternative Energies .................. 12
1.7  Environmental Impact .................. 13
1.8  Book Overview ......................... 14
References ................................ 15

### 2  Harvesting Energy from Biochemical Reactions ... 17
2.1  Introduction and Basic Definitions .......... 17
2.2  Biochemical Pathways Review for Organoheterotrophic Metabolism .......... 19
    2.2.1  Aerobic Respiration ............. 19
    2.2.2  Anaerobic Respiration ........... 23
    2.2.3  Fermentation .................... 25
2.3  Biochemical Pathways Overview for Lithotrophic Growth .................. 30
2.4  Biochemical Pathways Overview for Phototrophic Metabolism .............. 31
    2.4.1  Light Reactions ................. 32
    2.4.2  Anabolic (Dark) Reactions ....... 33
2.5  Definition and Importance of Chemical Oxygen Demand ....................... 33
Acknowledgments ............................ 35
References ................................ 36

### 3  Microbial Modeling of Biofuel Production ...... 37
3.1  Introduction .......................... 37
3.2  Summary of Microbial Growth Models .... 37
    3.2.1  Unstructured, Single Limiting Nutrient Models ................. 38
    3.2.2  Inhibition Models ............... 39

|      |       | 3.2.3 | Models for Multiple Limiting Substrates | 42 |
|---|---|---|---|---|
|      |       | 3.2.4 | Yield Parameters | 44 |
|      | 3.3   | Kinetic Rate Expressions | | 45 |
|      |       | 3.3.1 | Temperature Effects | 47 |
|      | 3.4   | Bioreactor Operation and Design for Biofuel Production | | 48 |
|      |       | 3.4.1 | Batch Reactors | 50 |
|      |       | 3.4.2 | Continuous Stirred Tank Reactors | 50 |
|      |       | 3.4.3 | CSTR with Cell Recycle | 52 |
|      |       | 3.4.4 | Fed-Batch Systems | 54 |
|      |       | 3.4.5 | Plug Flow Systems | 55 |
|      | 3.5   | Bioreactor Design Strategies | | 57 |
|      | 3.6   | Modeling of Glucose Utilization and Hydrogen Production | | 58 |
|      |       | 3.6.1 | Batch Fermentations and Simulations | 59 |
|      |       | 3.6.2 | CSTR Fermentations and Simulations | 61 |
|      | Summary | | | 64 |
|      | References | | | 65 |

## Part 2  Biofuels

| 4 | **Biofuel Feedstocks** | | | **69** |
|---|---|---|---|---|
|   | 4.1 | Starch Feedstocks | | 69 |
|   |     | 4.1.1 | Cereal Grains | 69 |
|   |     | 4.1.2 | Other Grains | 78 |
|   |     | 4.1.3 | Tubers and Roots | 78 |
|   | 4.2 | Sugar Feedstocks | | 79 |
|   |     | 4.2.1 | Sugarcane | 79 |
|   |     | 4.2.2 | Sugar Beet | 80 |
|   | 4.3 | Lignocellulosic Feedstocks | | 80 |
|   |     | 4.3.1 | Forest Products and Residues | 81 |
|   |     | 4.3.2 | Agricultural Residues | 82 |
|   |     | 4.3.3 | Agricultural Processing By-Products | 84 |
|   |     | 4.3.4 | Dedicated Energy Crops | 84 |
|   | 4.4 | Plant Oils and Animal Fats | | 88 |
|   | 4.5 | Miscellaneous Feedstocks | | 91 |
|   |     | 4.5.1 | Animal Wastes | 91 |
|   |     | 4.5.2 | Municipal Solid Waste | 94 |
|   | References | | | 94 |

## Contents

**5  Ethanol Production** .......................... 105
- 5.1  Ethanol Production from Sugar and Starch Feedstocks .............................. 105
  - 5.1.1  Microorganisms ................. 105
  - 5.1.2  Process Technology ............. 111
- 5.2  Ethanol Production from Lignocellulosic Feedstocks .............................. 133
  - 5.2.1  Basic Concept .................. 133
  - 5.2.2  The Sugar Platform ............. 134
  - 5.2.3  The Syngas Platform ............ 158
- Acknowledgments......................... 174
- References .............................. 174

**6  Biodiesel** .................................. 197
- 6.1  Introduction .......................... 197
  - 6.1.1  Environmental Considerations ..... 199
- 6.2  Biodiesel Production Chemistry and Thermodynamic Aspects ................... 201
  - 6.2.1  Transesterification .............. 202
  - 6.2.2  Esterification ................... 202
  - 6.2.3  Lipase-Catalyzed Interesterification and Transesterification ........... 203
  - 6.2.4  Side Reactions: Saponification and Hydrolysis .................. 203
  - 6.2.5  Alcohol Effect .................. 204
  - 6.2.6  Base or Alkali Catalysis .......... 204
  - 6.2.7  Acid Catalysis .................. 206
  - 6.2.8  Enzyme Catalysis ................ 208
  - 6.2.9  Supercritical Esterification and Transesterification ........... 208
  - 6.2.10  Thermodynamics and Reaction Kinetics ................ 210
- 6.3  Oil Sources and Production ................ 219
  - 6.3.1  Plant Oils ...................... 219
  - 6.3.2  Microbial and Algal Oils .......... 223
  - 6.3.3  Used Cooking Oils ............... 233
  - 6.3.4  Straight Vegetable Oil ........... 233
  - 6.3.5  Biosynthesis of Oils and Modification ................. 234
- 6.4  Coproducts ............................ 236
- 6.5  Methods of Biodiesel Production ........... 238
  - 6.5.1  General Biodiesel Production Procedures ..................... 239
  - 6.5.2  Pilot and Commercial Scale ....... 245
  - 6.5.3  Quality Control Analytical Technique ..................... 247

| | | | |
|---|---|---|---|
| 6.6 | Economics | | 250 |
| | 6.6.1 Feedstock Cost | | 252 |
| | 6.6.2 Manufacturing Cost | | 255 |
| | 6.6.3 Capital Cost | | 255 |
| | 6.6.4 Operating Cost | | 257 |
| 6.7 | Summary and Conclusions | | 258 |
| | Acknowledgments | | 259 |
| | Problems | | 260 |
| | References | | 262 |

## 7 Biological Production of Hydrogen ... 269

- 7.1 Introduction ... 269
  - 7.1.1 Important Enzymes ... 269
  - 7.1.2 Abiotic $H_2$ Production ... 271
- 7.2 Photobiological $H_2$ Production ... 271
  - 7.2.1 Direct Biophotolysis ... 272
  - 7.2.2 Indirect Biophotolysis ... 273
  - 7.2.3 Photofermentation ... 273
  - 7.2.4 Photobiological $H_2$ Production Potential ... 274
- 7.3 Hydrogen Production by Fermentation ... 274
  - 7.3.1 Overview ... 274
  - 7.3.2 Energetics ... 275
  - 7.3.3 Thermotogales ... 276
  - 7.3.4 Biochemical Pathway for Fermentative $H_2$ Production by *Thermotoga* ... 276
  - 7.3.5 Hydrogen Production by Other Bacteria ... 277
  - 7.3.6 Coproduct Formation ... 279
  - 7.3.7 Batch Fermentation ... 280
  - 7.3.8 Hydrogen Inhibition ... 281
  - 7.3.9 Role of Sulfur—Sulfidogenesis ... 281
  - 7.3.10 Use of Other Carbon Sources Obtained from Agricultural Residues ... 284
  - 7.3.11 Process and Culture Parameters ... 287
- 7.4 Hydrogen Detection, Quantification, and Reporting ... 290
  - 7.4.1 Hydrogen Detection ... 291
  - 7.4.2 Total Gas Pressure ... 292
  - 7.4.3 Water Vapor Pressure ... 292
  - 7.4.4 Hydrogen Partial Pressure ... 292
  - 7.4.5 Hydrogen Gas Concentration ... 293
  - 7.4.6 Hydrogen Concentration Expressed as mol $H_2$/L Media ... 294

|  |  | 7.4.7 | Hydrogen Production Rate | 294 |
|  |  | 7.4.8 | Dissolved $H_2$ Concentration in Liquid | 294 |

7.5 Fermentation Bioreactor Sizing for PEM Fuel Cell Use ... 297
Acknowledgment ... 299
References ... 299

## 8 Microbial Fuel Cells ... 303
8.1 Overview ... 303
8.2 Biochemical Basis ... 303
8.3 Past Work Summary ... 305
8.4 Fuel Cell Design ... 308
    8.4.1 Anode Compartment ... 308
    8.4.2 Microbial Cultures ... 309
    8.4.3 Redox Mediators ... 310
    8.4.4 Cathode Compartment ... 311
    8.4.5 Exchange Membrane ... 312
    8.4.6 Power Density as Function of Circuit Resistance ... 313
8.5 MFC Performance Methods ... 314
    8.5.1 Substrate and Biomass Measurements ... 314
    8.5.2 Basic Power Calculations ... 315
    8.5.3 Calculation Example ... 317
8.6 MFC Performance ... 318
    8.6.1 Power Density as Function of Substrate ... 318
    8.6.2 Single-Chamber Versus Two-Chamber Designs ... 320
    8.6.3 Single-Chamber Designs ... 320
    8.6.4 Wastewater Treatment Effectiveness ... 321
8.7 Fabrication Example ... 322
8.8 Future Directions ... 323
References ... 325

## 9 Methane ... 329
9.1 Introduction ... 329
9.2 Microbiology of Methane Production ... 329
    9.2.1 Methanogenic Environments ... 329
    9.2.2 Methane Process Description ... 330
    9.2.3 Microbial Communities ... 332
9.3 Biomass Sources for Methane Generation ... 334

|     |       |                           |     |
|-----|-------|---------------------------|-----|
| 9.4 | Systems | ...................................... | 338 |
|     | 9.4.1 | Reactor Conditions ................ | 339 |
|     | 9.4.2 | Process Design .................. | 340 |
| 9.5 | Biogas Composition and Use ............... | | 343 |
|     | References .............................. | | 344 |

**Appendix: Conversion Factors and Constants** .... 347

**Index** ....................................... 351

# Preface

The development of renewable energy has attracted a great deal of interest not only because of the steady rise in oil prices, but also because of the limit of fossil fuel reserves. One day not very far into the future, refineries and coal-fire power plants may be closed forever because their reserves have been depleted. It took nature a very long time to create gas, oil, and coal, but it takes us just a blink of an eye within the geological time scale to burn them all.

There are many sources of renewable energy. Biofuels are just one source, but a very important one. Biofuels can be defined as fuels that are derived from biological sources. Among them, methane produced by anaerobic digestion has been used by the human race for hundreds, if not thousands, of years. More recently, ethanol produced from sugar- and starch-based feedstocks has become another important biofuel. Other biofuels such as lignocellulosic ethanol, biodiesel, biohydrogen, and bioelectricity have been the focus of vigorous research, and the technologies for their production are being developed, although most of these are not quite ready for commercialization.

This book is written with two objectives. First, it may be a reference book for those who are interested in biofuels. Second, it may be used as a textbook to teach biofuel technologies to science and engineering students who want to contribute to the development and implementation of processes for production of these important renewable energy sources. In this book, readers will find the fundamental concepts of important biofuels and the current state-of-the-art technology for their production.

We hope our book will serve our readers well. We will be very grateful to receive comments and suggestions for improvement from our colleagues in this field and also from the students who will use this book in their educational endeavors.

*Caye M. Drapcho, Ph.D.*
*Nghiem Phu Nhuan, Ph.D.*
*Terry H. Walker, Ph.D.*

# Preface

# PART 1
# The Basics

**CHAPTER 1**
Introduction

**CHAPTER 2**
Harvesting Energy from Biochemical Reactions

**CHAPTER 3**
Microbial Modeling of Biofuel Production

# CHAPTER 1

# Introduction

## 1.1 Biorefinery

Renewable energy deriving from solar, wind, and biomass sources has great potential for growth to meet our future energy needs. Fuels such as ethanol, methane, and hydrogen are characterized as biofuels because they can be produced by the activity of biological organisms. Which of these fuels will play a major role in our future? The answer is not clear, as factors such as land availability, future technical innovation, environmental policy regulating greenhouse gas emissions, governmental subsidies for fossil fuel extraction/processing, implementation of net metering, and public support for alternative fuels will all affect the outcome. A critical point is that as research and development continue to improve the efficiency of biofuel production processes, economic feasibility will continue to improve.

Biofuel production is best evaluated in the context of a biorefinery (Fig. 1.1). In a biorefinery, agricultural feedstocks and by-products are processed through a series of biological, chemical, and physical processes to recover biofuels, biomaterials, nutraceuticals, polymers, and specialty chemical compounds.[2,3] This concept can be compared to a petroleum refinery in which oil is processed to produce fuels, plastics, and petrochemicals. The recoverable products in a biorefinery range from basic food ingredients to complex pharmaceutical compounds and from simple building materials to complex industrial composites and polymers. Biofuels, such as ethanol, hydrogen, or biodiesel, and biochemicals, such as xylitol, glycerol, citric acid, lactic acid, isopropanol, or vitamins, can be produced for use in the energy, food, and nutraceutical/pharmaceutical industries. Fibers, adhesives, biodegradable plastics such as polylactic acid, degradable surfactants, detergents, and enzymes can be recovered for industrial use. Many biofuel compounds may only be economically feasible to produce when valuable coproducts are also recovered and when energy-efficient processing is employed. One advantage of microbial conversion processes over chemical processes is that microbes are able to select their substrate among a complex mixture of compounds, minimizing the need for isolation and purification of substrate prior to processing. This can translate to more complete use of substrate and lower chemical requirements for processing.

Early proponents of the biorefinery concept emphasized the *zero-emissions* goal inherent in the plan—waste streams, water, and heat from one process are utilized as feed streams or energy to another, to fully recover all possible products and reduce waste with maximized efficiency.[2,3] Ethanol and biodiesel production can be linked effectively in this way. In ethanol fermentation, 0.96 kg of $CO_2$ is produced per kilogram of ethanol formed. The $CO_2$ can be fed to algal bioreactors to produce oils used for biodiesel production. Approximately 1.3 kg $CO_2$ is consumed per kilogram of algae grown, or 0.5 kg algal oil produced by oleaginous strains. Another example is the potential application of microbial fuel cells to generate electricity by utilizing waste organic compounds in spent fermentation media from biofuel production processes.

Also encompassed in a sustainable biorefinery is the use of "green" processing technologies to replace traditional chemical processing. For example, supercritical $CO_2$ can be used to extract oils and nutraceutical compounds from biomass instead of using toxic organic solvents such as hexane.[4] Ethanol can be used in biodiesel production from biological oils in place of toxic petroleum-based methanol traditionally used. Widespread application of biorefineries

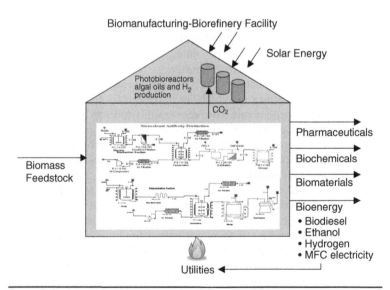

**FIGURE 1.1** Integrated biorefinery showing example bioprocesses of monoclonal antibody and ethanol production. (*Adapted from Walker, 2005.*)

would allow for replacement of petroleum-derived products with sustainable, carbon-neutral, low-polluting alternatives.

In addition to environmental benefits of biorefining, there are economic benefits as new industries grow in response to need.[2,3] A thorough economic analysis, including ecosystem and environmental impact, harvest, transport, processing, and storage costs must be considered. The R&D Act of 2000 and the Energy Policy Act of 2005 recommend increasing biofuel production from 0.5 to 20 percent and biobased chemicals and materials from 5 to 25 percent,[5] a goal that may best be reached through a biorefinery model.

## 1.2 Description of Biofuels

The origin of all fuel and biofuel compounds is ultimately the sun, as solar energy is captured and stored as organic compounds through photosynthetic processes. Certain biofuels, such as oils produced by plants and algae, are direct products of photosynthesis. These oils can be used directly as fuel or chemically transesterified to biodiesel. Other biofuels such as ethanol and methane are produced as organic substrates are fermented by microbes under anaerobic conditions. Hydrogen gas can be produced by both routes, that is, by photosynthetic algae and cyanobacteria under certain nutrient- or oxygen-depleted conditions, and by bacteria and archae utilizing organic substrates under anaerobic conditions. Electrical energy produced by microbial fuel cells—specialized biological reactors that intercept electron flow from microbial metabolism—can fall into either category, depending on whether electron harvest occurs from organic substrates oxidized by organotrophic cultures or from photosynthetic cultures.

A comparison of biofuel energy contents reveals that hydrogen gas has the highest energy density of common fuels expressed on a mass basis (Table 1.1). For liquid fuels, biodiesel, gasoline, and diesel have energy densities in the 40 to 46 kJ/g range. Biodiesel fuel contains 13 percent lower energy density than petroleum diesel fuel, but combusts more completely and has greater lubricity.[7] The infrastructure for transportation, storage, and distribution of hydrogen is lacking, which is a significant advantage for adoption of biodiesel.

Another measure of energy content is energy yield ($Y_E$), the energy produced per unit of fossil fuel energy consumed. $Y_E$ for biodiesel from soybean oil is 3.2 compared to 1.5 for ethanol from corn and 0.84 and 0.81 for petroleum diesel and gasoline, respectively.[8] Even greater $Y_E$ values are achievable for biodiesel created from algal sources or for ethanol from cellulosic sources.[9] The high net energy gain for biofuels is attributed to the solar energy captured compared to an overall net energy loss for fossil fuels.

| Fuel source | Energy density (kJ/g) | Density (kg/m³) | Energy content (GJ/m³) |
|---|---|---|---|
| Hydrogen | 143.0 | 0.0898 | 0.0128 |
| Methane (natural gas) | 54.0 | 0.7167 | 0.0387 |
| No. 2 diesel | 46.0 | 850 | 39.1 |
| Gasoline | 44.0 | 740 | 32.6 |
| Soybean oil | 42.0 | 914 | 38.3 |
| Soybean biodiesel | 40.2 | 885 | 35.6 |
| Coal | 35.0 | 800 | 28.0 |
| Ethanol | 29.6 | 794 | 23.5 |
| Methanol | 22.3 | 790 | 17.6 |
| Softwood | 20.4 | 270 | 5.5 |
| Hardwood | 18.4 | 380 | 7.0 |
| Rapeseed oil | 18.0 | 912 | 16.4 |
| Bagasse | 17.5 | 160 | 2.8 |
| Rice hulls | 16.2 | 130 | 2.1 |
| Pyrolysis oil | 8.3 | 1280 | 10.6 |

*Values reported at standard temperature and pressure
Source: Adapted from Brown, 2003.

**TABLE 1.1** Energy Density Values* for Common Fuels

## 1.3 Energy Use

The motivation for development and use of alternative fuels include (1) diminishing reserves of readily recoverable oil, (2) concern over global climate change,[10] (3) increasing fuel prices, and (4) the desire for energy independence and security. The U.S. Energy Information Administration determined that total world energy consumption in 2005 was 488 EJ (exajoule, $10^{18}$ J) or 463 Quad (quadrillion Btu, $10^{15}$ Btu), with U.S. consumption of 106 EJ (100.6 Quad) or 22 percent of the world total.[11] World consumption is expected to surpass 650 EJ by 2025.[11] The rates of increase in energy usage vary greatly by nation. Between 1985 and 2005, annual energy consumption increased 31 percent in the United States, while only 18 percent in Europe, and an overwhelming 250 percent in China and India,

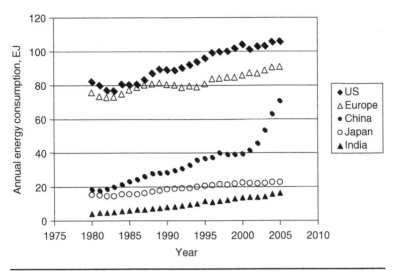

**FIGURE 1.2** Annual energy consumption values for selected countries. (*Adapted from Energy Information Agency, 2007.*)

although India's total consumption is small at only 3 percent of the world total (Fig. 1.2). These values reflect a host of factors, including degree of industrialization, gross domestic product, relative efficiency of primary energy source used, and energy conservation. In the United States, fossil fuels accounted for 86 percent of our total energy consumption in 2004. Petroleum fuels, natural gas, and coal accounted for 40, 23, and 23 percent, respectively, with an additional 8 percent from nuclear power and only 6 percent from renewable sources, including hydroelectric (2.7 percent), biomass/biofuels (2.7 percent), and 0.6 percent from solar, wind, and geothermal energy sources combined.[11,12] Currently available fossil fuel sources are estimated to become nearly depleted within the next century, with petroleum fuel reserves depleted within 40 years.[11,13] The United States imports 10 million barrels of oil per day of the existing world reserves (1.3 trillion barrels) (Table 1.2). Peak oil, the maximum rate of oil production, is expected to occur between 2010 and 2020.[11] Even with increasing attention on hydrogen as an alternative fuel, 95 percent of worldwide production of hydrogen gas is from fossil fuel sources, primarily the thermocatalytic reformation of natural gas.[14]

Approximately 50 percent of the U.S. trade deficit is attributed to the import of crude oil. Crude oil prices have risen from less than $20/barrel in the 1990s to nearly $100/barrel in 2007. Accounting for military aid and subsidies to protect and maintain an uninterrupted flow of crude oil from unstable regions of the world, the true cost of oil[15] has been estimated as greater than $100/barrel since 2004.

## The Basics

| Country | Oil reserves (billion barrels) | U.S. oil imports (million barrels/day) |
|---|---|---|
| Saudi Arabia | 267 | 1.50 |
| Canada | 179 | 1.62 |
| Iran | 132 | — |
| Iraq | 115 | 0.66 |
| Kuwait | 104 | 0.24 |
| United Arab Emirates | 98 | — |
| Venezuela | 80 | 1.30 |
| Russia | 60 | — |
| Libya | 39 | — |
| Nigeria | 36 | 1.08 |
| United States | 21 | — |
| China | 18 | — |
| Qatar | 15 | — |
| Mexico | 13 | 1.60 |
| Algeria | 11 | 0.22 |
| Brazil | 11 | — |
| Other | 91 | 1.84 |
| **Total** | **1290** | **10.06 (60%)** |

*Source*: Adapted from Energy Information Agency, 2007.

**TABLE 1.2** World Oil Reserves and U.S. Imports Based on Leading Producers

## 1.4 Efficiency of Energy Use

The main fossil fuels (coal, natural gas, and oil) are about 33 percent efficient when used for energy generation, and emit high levels of $CO_2$ (Fig. 1.3) and nitrogen oxides. Geothermal and solar energy are less than 20 percent efficient with current technology, but are nearly zero-emission energy sources. Wind power has both high efficiency and zero-emissions, but is restricted to certain regions. Home heating by natural gas has a high efficiency, with lower emissions than other fossil fuels.

Spark-ignition (SI) gasoline engines, the most commonly used for transportation in the United States, are the most inefficient of current technologies, with an average efficiency of 16 percent (Fig. 1.4) compared to biodiesel in diesel engines (29 percent efficiency).[15] The most efficient engines—hybrid diesel and hybrid hydrogen fuel cell—achieve nearly 50 percent efficiency. Further, emissions for hybrid hydrogen fuel cell (390 g $CO_2$/mile) are substantially less than diesel (475 g $CO_2$/mile) and SI gasoline engines (525 g $CO_2$/mile).[16]

Introduction 9

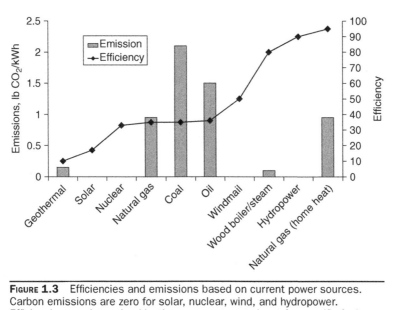

FIGURE 1.3 Efficiencies and emissions based on current power sources. Carbon emissions are zero for solar, nuclear, wind, and hydropower. Efficiencies are determined by the power output to input for specific fuel or energy sources. (*Adapted from http://www.memagazine.org/supparch/mepower03/gauging/gauging.html.*)

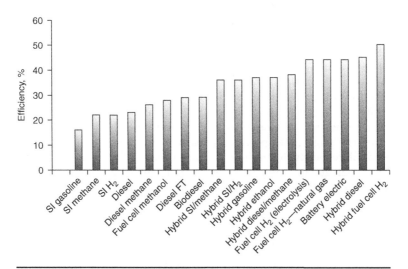

FIGURE 1.4 Tank to wheel efficiencies of existing engine technologies (SI: spark-ignition; FT: Fischer-Tropse; $H_2$: hydrogen). (*Adapted from http://www.memagazine.org/supparch/mepower03/gauging/gauging.html.*)

Hybrid diesel technology has already been demonstrated and commercialized in Germany by General Motors, and adoption in the United States could be initiated quickly due to existing diesel fuel storage and distribution infrastructure. Hydrogen fuel cell technology has been commercialized primarily for vehicles that are fueled by a centralized source, such as fork-lift vehicles used in warehouses and factories, and is under development for stationary power generation.

Transportation fuels account for nearly 25 percent of the energy consumed in the United States, of which more than half comes from foreign oil.[6] To displace the 120 billion gallons of gasoline and 60 billion gallons of diesel fuel used for transportation each year in the United States, 140 billion gallons of biodiesel would be required due to the 35 percent greater efficiency of biodiesel engines compared to SI gasoline engines.

Other inefficiencies of energy use, including waste of electricity due to insufficient insulation in homes and workplaces, lack of daylighting in buildings, use of inefficient incandescent lighting, and "vampire losses"—electricity consumption due to electronic devices such as TVs when off—increase our energy consumption unnecessarily. For example, vampire losses account for about 4 percent of total electricity consumed in the United States.[17] Fortunately, LEED-certified designs and increased car fuel efficiency (CAFE) standards are gaining acceptance. In 2007, the CAFE standards were raised to 35 mpg by 2020, an increase from 27.5 mpg for automobiles and a low 22.5 mpg average for trucks and SUVs.

## 1.5 Biofuels Production and Use

Ethanol and biodiesel production have increased 10 percent per year worldwide over the past decade (Fig. 1.5). Major world producers of ethanol include Brazil (primarily from sugarcane feedstock) and the United States (primarily from corn), with 10 and 13 billion gallons per year, respectively.[18] China (from corn and wheat) and India (using primarily sugarcane) produced nearly 1 and 0.5 billion gallons of ethanol, respectively. In the European Union (EU), France produces more than 200 million gallons of ethanol primarily from sugar beets and wheat feedstocks.[14] Further, 10 billion gallons of biodiesel are produced in the EU, far exceeding production from any other region. However, insufficient sources of sugarcane and corn for ethanol production and rapeseed, palm, sunflower, and soybean oils for biodiesel are expected to limit further increases in production of these biofuels[4] by 2020, unless more promising sources such as algal oils for biodiesel and cellulosic biomass sources for ethanol become prevalent.

Biomass feedstocks, including dedicated crops and agricultural and forestry by-products, can be converted into usable fuel by biological processing, thermal processing, and direct oil extraction. In general, biomass can be broken into two main categories—carbohydrate

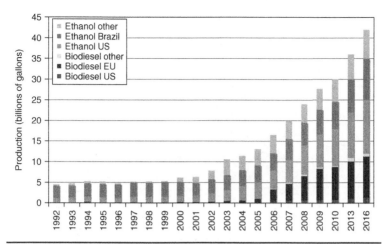

**FIGURE 1.5** World biodiesel and ethanol historical production and projections. (*Adapted from www.ers.usda.gov/publications/oce071/oce20071b.pdf.*)

materials containing sugars and starches and heterogeneous woody materials collectively termed *lignocellulosics*.[19] Corn stover (corn crop wastes) is a biomass source that has been heavily utilized for energy conversion. Thermal conversion processes include direct combustion, gasification, liquefaction, and pyrolysis. The greatest potential of biomass and biofuel resources for energy production in the United States (Fig. 1.6) occurs along the west coast, upper Midwest, Maine,

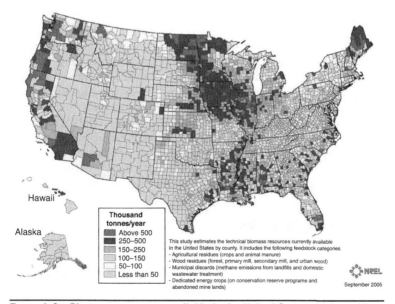

**FIGURE 1.6** Biomass resources available in the United States. (*Adapted from www.ers.usda.gov/publications/oce071/oce20071b.pdf.*)

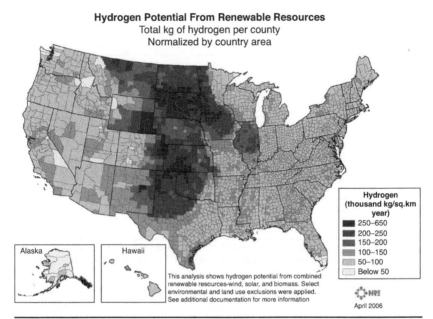

**FIGURE 1.7** Hydrogen production potential in the United States from biomass, wind, and solar energy sources. (*Adapted from www.ers.usda.gov/publications/oce071/oce20071b.pdf.*)

and the southeast centered around agricultural and forestry lands. The Midwest shows great energy production primarily through grasses and grain crops. The plains also have high solar and wind potential, which, combined with biomass sources, could increase hydrogen production potential in the United States (Fig. 1.7) as hydrogen gas is suggested by many as an effective means of storage of energy produced from wind and solar.[20] The distribution of biomass for food production must be accounted for in potential energy production as utilization of land not typically suitable for food production should be used for energy crops. For example, arid regions of the southwest are low terrestrial biomass production areas, but high solar radiation provides great potential for solar energy capture and production of algal biomass for production of oils.[21]

## 1.6 Alternative Energies

Alternative energy sources and "fuels of opportunity" are being developed to make up our energy deficit. Solar photovoltaic technology has improved dramatically since the 1970s, with a 95 percent decline in costs to $0.25/kWh that will continue with improved technology.[19] Solar energy hitting the entire earth (1000 W/m$^2$ at full sunlight) may be captured by solar photovoltaic cells, passive solar water heaters, and by photosynthetic capture and biochemical storage in the form of protein (4 cal/g), carbohydrates (4 cal/g), and fat (9 cal/g). The use of solar energy has increased by 15 percent per year over the past several years.[19] The use of wind energy has increased by 25 percent over the past decade, with a decline in cost from $0.3/kWh to less than $0.05/kWh, making wind energy competitive with even the cheapest sources of energy.[19]

"Opportunity fuels" are combustible resources that are outside of the mainstream fuels produced commercially, but that can be used productively to generate electricity.[19] The most common types of opportunity fuels are residues or low-value by-products produced from manufacturing of consumer goods and agricultural products. Examples include petroleum coke, sawdust, spent pulping liquor, rice hulls, oat hulls, wheat and other straws, and other solid materials.[19] A second category of opportunity fuels includes reclaimed wastes from energy industries such as from coal mining and coal processing wastes. For example, once-discarded coal fines are now utilized for energy in coal-water slurries. A third category includes waste commercial products such as vehicle tires. As our energy deficit grows, techniques for recycling and reuse of these by-products and wastes are being explored to recover all usable energy sources.

## 1.7 Environmental Impact

Biofuel production, if approached in a sustainable manner, can be more environmentally benign than fossil fuel technologies for several major reasons. First, biofuel production from biomass is largely carbon neutral—that is, the $CO_2$ produced as the fuel is combusted, is offset by the carbon absorbed as the biomass is grown.[22] Second, bioconversion processes in general do not produce hazardous compounds, and if toxic solvents and chemicals are avoided in the processing stages, then fewer environmental pollutants are produced. Third, biomass production and microbial conversion processes can be developed and used in a more distributed manner, avoiding the need for transport of fuels via cargo ships or pipelines for long distances.

However, just as extraction of fossil fuels places a heavy toll on the environment, large-scale biomass production and harvesting may burden the environment if not planned and executed using sustainable practices. Further, growing dedicated crops specifically for biofuel production may displace acreage used to produce food crops. Since 2005, nonfood use of rapeseed oil for biodiesel production has exceeded food uses in the EU, requiring the importation of nearly 0.6 million MT.[23] In the United States, animal feed/fuel corn prices have approached $4/bu due to rapidly increasing demands for ethanol. These large price increases have benefited agricultural producers and the corn-based economy, but have caused increased meat prices and decreased exports to countries such as Mexico. Food corn prices have been less impacted since it is different than corn grown for ethanol or feed. Also, $N_2O$ emissions from biofuel produced through nitrogen-intensive crops such as corn for ethanol and rapeseed could potentially eliminate the carbon savings effect on global warming.[24] Sustainable biomass production practices that limit nitrogen use are imperative to curtail the overall effects on global warming.

Carbon dioxide emissions due to combustion of fossil fuels have become a major environmental concern. The tipping point for atmospheric carbon dioxide has been estimated as 450 ppm with current levels reaching 370 ppm.[25] Carbon dioxide emissions in the United States have not met levels set by the Energy Policy Acts of 1992, 2000, and 2005. The Clean Air Act mandated cleaner fuel emissions in the largest U.S. cities and the use of oxygenated fuels, which can be accomplished through addition of ethanol or methyl tert-butyl ether (MTBE). California, which accounts for 13 percent of the U.S. economy, banned the use of MTBE because of environmental effects, thus increasing ethanol addition to gasoline and biodiesel. Ethanol addition to gasoline in gasohol varies from 10 percent (E10) in the United States to 22 percent in Brazil.[26] E-biodiesel is a blend of 20 percent ethanol with biodiesel produced from biological oils and ethanol, which also increases lubricity over traditional biodiesel.[26]

## 1.8 Book Overview

This book focuses on the biological processes that convert biomass feedstocks to fuels such as ethanol or hydrogen and those that capture/convert solar energy directly into fuels. The basic biochemical pathways for production of biofuels are reviewed in Chap. 2 for a complete understanding of these bioconversion processes. Chapter 3 provides an in-depth review of microbial modeling with examples given for biological hydrogen production. Chapter 4 reviews the main biomass feedstocks used for biofuels. Chapters 5, 6, and 7 cover the microbial process descriptions and systems design and engineering analysis needed for ethanol, biodiesel, and biohydrogen production. Chapter 8 covers microbial fuel cells, novel biological reactors used to generate electricity through microbial activity. Finally, Chap. 9 provides an overview of the bioproduction of methane.

To what extent will biofuels displace petroleum-based fuels in the twenty-first century and lessen the environmental impact of energy production and use? By adopting the biorefinery model, integrating green processing technologies, and applying sustainable biomass production practices, we may be able to shift from centralized energy production using coal, oil, nuclear, and natural gas to distributed energy sources such as solar and wind energy and biorefineries producing a diversified list of biofuels such as ethanol, hydrogen, and biodiesel. The aim of this book is to describe the process technologies and relative merits and value of the major biofuels to stimulate and facilitate that process.

## References

1. Walker, T. H. 2005. "White Paper on Biomanufacturing Capabilities." Clemson University, BioEnergy Research Committee. Available at: http://people.clemson.edu/~walker4/.
2. Gravitis, J. 1998. "A Biochemical Approach to Attributing Value to Biodiversity—The Concept of the Zero Emissions Biorefinery." *Presented at the 4th Annual World Congress on Zero Emissions in Windhoek*, Namibia, October 16.
3. Gravitis, J., and M. Suzuki. 1999. "Biomass Refinery—A Way to Produce Value-Added Products and Base for Agricultural Zero Emissions System." *Proceedings of 1999 International Conference on Agricultural Engineering*, Beijing, China.
4. Walker, T. H., H. D. Cochran, and G. J. Hulbert. 1999. "Supercritical carbon dioxide extraction of lipids from *Pythium irregulare*." *J Am Oil Chem Soc*. 76(5):595–602.
5. Biomass Technical Advisory Committee. 2002. "Vision for Bioenergy and Biobased Products in the U.S.." Available at: http://www.climatevision.gov/sectors/electricpower/pdfs/bioenergy_vision.pdf.
6. Brown, R. C. 2003. *Biorenewable Resources*. Iowa State Press, Blackwell Publishing Co., Iowa.
7. UNH Biodiesel Group. Available at: www.unh.edu/p2/biodiesel/article_alge.html.
8. http://www.mda.state.mn.us/ethanol/balance.html.
9. Christi, Y. 2007. "Biodiesel from microalgae." *Biotechnol Adv*. 25:294–306.
10. Intergovernmental Panel on Climate Change (IPCC). 2007. "Climate Change 2007: Synthesis Report." Available at: http://www.ipcc.ch/ipccreports/assessments-reports.htm.
11. Energy Information Agency. 2007. "Official Energy Statistics from the U.S. Government." Energy Information Administration. Available at: http://eia.doe.gov/.
12. U.S. DOE. 2004. "U.S. Energy Consumption by Energy Sources, 2000–2004." Available at: http://www.eia.doe.gov/cneaf/solar.renewables/page/trends/table01.pdf.
13. BP. 2005. "Putting Energy in the Spotlight. British Petroleum Statistical Review of World Energy." Available at: http://www.bp.com/liveassets/bp_internet/switzerland/corporate_switzerland/STAGING/local_assets/downloads_pdfs/s/statistical_review_of_world_energy_2005.pdf.
14. Sperling, D., and J. S. Cannon (eds.). 2004. *The Hydrogen Energy Transition: Moving Toward the Post Petroleum Age in Transportation*. p. 80.
15. Kinney, A. J, and T. E. Clemente. 2005. "Modifying soybean oil for enhanced performance in biodiesel blends." *Fuel Process Technol*. 86:1137–1147.
16. http://www.memagazine.org/supparch/mepower03/gauging/gauging.html.
17. Tidwell, M. 2006. *The Ravaging Tide: Strange Weather, Future Katrinas, and the Coming Death of America's Coastal Cities*. Free Press, New York.
18. Licht, F. O. 2005. "Homegrown for the Homeland: Industry Outlook 2005." Renewable Fuels Association, Washington, D.C., p. 14. Available at: http://www.earth-policy.org/Updates/2005/Update49_data.htm.
19. Murray, D. 2005. "Ethanol's potential: Looking beyond corn." Earth Policy Institute. Available at: http://www.earth-policy.org/Updates/2005/Update49.htm.
20. U.S. Department of Agriculture, Office of the Chief Economist. 2007. "USDA Agricultural Projections to 2016, OCE-2007-1." Available at: www.ers.usda.gov/publications/oce071/oce20071b.pdf.
21. Tillman, D., and N. S. Harding. 2004. *Fuels of Opportunity: Characteristics and Uses in Combustion Systems*. N.S. Harding & Associates, Salt Lake City, Utah.

22. Tillman, D., J. Hill, and C. Lehman. 2006. "Carbon-negative biofuels from low-input high-diversity grassland biomass." *Science*. 314:1598–1600.
23. USDA NREL. "Dynamic Maps, GIS Data and Analysis Tools." Available at: http://www.nrel.gov/gis/biomass.html.
24. Crutzen, P. J, et al. 2007. "$N_2O$ release from agro-biofuel production negates global warming reduction by replacing fossil fuels." *Atmos Chem Phys Discuss*. 7:11191–11205.
25. Jacquet, F., et al. 2007. "Recent developments and prospects for the production of biofuels in the EU: Can they really be 'part of the solution?'" Food & Feed Tradeoffs. http://www.farmfoundation.org/projects/documents/Jacquet.pdf.
26. Fernando, S., and M. Hanna. 2004. "Development of a novel biofuel blend using ethanol-biodiesel-diesel microemulsions: EB-Diesel." *Energy Fuels*, 18:1695–1703.

# CHAPTER 2
# Harvesting Energy from Biochemical Reactions

## 2.1 Introduction and Basic Definitions

Bioconversion processes are carried out by microorganisms of all three domains of life (Fig. 2.1)—the *archae*, the simplest and most ancient of the life forms; the *bacteria*; and the *eucarya*. Bacteria and archae are prokaryotic, lacking true nuclei, yet archae share more RNA with eucaryotes than bacteria. Eucarya, specifically yeast, and bacteria ferment carbohydrates to ethanol; archae and bacteria convert carbohydrates to hydrogen gas; photosynthetic bacteria and eucarya (algae and plants) convert solar energy directly into stored lipids that can be extracted and chemically converted to biodiesel; and all three domains can be used in microbial fuel cells to convert organic substrates to electricity.

Microorganisms can be classified as chemotrophs (those that gain energy through oxidation of chemical electron donor compounds) or phototrophs (those that gain energy for growth by utilizing solar radiation). Chemoorganotrophs obtain energy by oxidizing an organic compound, and chemolithotrophs obtain energy by oxidizing an inorganic compound. The amount of energy that a chemotroph can obtain from oxidizing the compound is proportional to the oxidation state of the compound. The more reduced the compound, the more electrons available for capture and use by the microbe.

The range of oxidation states of elements in common biological compounds is given in Table 2.1. Rules for assigning oxidation numbers include the following: (1) Compounds in their elemental form, such as $O_2$, $H_2$, or $N_2$, have a valence of zero; (2) the oxidation number for any single ion is equal to the charge on the ion; and (3) the sum of the oxidation numbers of all atoms in a compound must equal the overall charge of the compound. In common compounds involved in biological reactions, the oxidation number for hydrogen is +1 and for oxygen is −2.

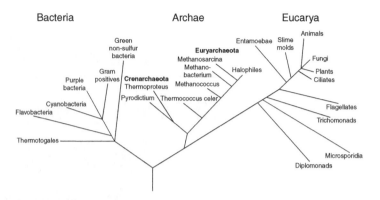

**FIGURE 2.1** Domains of life. (*Adapted from Madigan et al., 1997, and Woese et al., 1990.*)

Metabolism refers to two phases of biochemical processes that occur in a microorganism—catabolism and anabolism. *Catabolism* refers to the degradation reactions that occur as a chemotroph oxidizes the electron donor to yield energy in the form of adenosine triphosphate (ATP) and intermediate compounds that are used in the anabolic reactions. The series of catabolic reactions involved in the oxidation of a compound to yield energy is called the *biochemical pathway*. Anabolism refers to the biosynthesis reactions that use the energy and intermediate compounds produced in catabolic reactions to produce new cellular compounds, such as proteins, carbohydrates, and lipids, that are required for new cell growth.

The electrons removed from the electron donor as it is oxidized are transferred to intermediate electron acceptor compounds. In biological reactions, intermediate electron carrier molecules, such as nicotinamide adenine dinucleotide ($NAD^+$), nicotinamide adenine dinucleotide phosphate ($NADP^+$), and flavin adenine dinucleotide ($FAD^+$), carry 2 $e^-$ per molecule to other sites within the organism. As these compounds accept $e^-$, they are reduced to NADH, NADPH, and FADH, respectively, and subsequently transport the $e^-$ to other electron acceptor molecules and are reoxidized.

| Element | Reduced | | Oxidized | |
|---|---|---|---|---|
| | Oxidation state | Example | Oxidation state | Example |
| C | −4 | $CH_4$ | +4 | $CO_2$ |
| H | 0 | $H_2$ | +1 | $H_2O$, $C_6H_{12}O_6$ |
| O | −2 | $H_2O$, $C_6H_{12}O_6$ | 0 | $O_2$ |
| N | −3 | $NH_3$ | +5 | $NO_3^-$ |
| S | −2 | $H_2S$ | +6 | $SO_4^{2-}$ |

**TABLE 2.1** Range of Oxidation States for Common Elements

Knowledge of the basic biochemical pathways used by microorganisms is fundamental to process design for biofuel production. The main biochemical pathways for conversion of organic substrates are aerobic respiration, anaerobic respiration, and fermentation. Aerobic respiration commonly includes the Embden-Meyerhof pathway to convert sugars to pyruvate, the tricarboxylic acid cycle to convert pyruvate to carbon dioxide, and finally the electron transport system to process the electrons removed from the organic substrates to ATP production, with oxygen acting as the terminal electron acceptor. Anaerobic respiration is identical to aerobic respiration, except that inorganic compounds such as nitrate or sulfate are used as the terminal electron acceptor. Respiration processes result in the complete oxidation of the organic substrate to carbon dioxide, so that all usable energy is harvested from the organic compound. True fermentation, in the classical sense, is the biochemical pathway that occurs under anaerobic conditions in which no electron transport system is utilized. In fermentation, a portion of the organic compound is oxidized to carbon dioxide while the remaining portion becomes reduced as it accepts the electrons from the oxidized portion. However, in most bioprocessing industries, the term fermentation has come to mean any bioconversion process. The biological processing of organic substrates to biofuel compounds will often entail an anaerobic process, since common biofuels (i.e., ethanol, hydrogen, and methane) are reduced endproducts that are formed by fermentation. Oils produced by algae or fungi that may be converted to biodiesel represent cell storage products that are typically the result of an aerobic pathway.

The corresponding environments for the above biochemical pathways are described as aerobic, anoxic, and anaerobic. Aerobic environments are those that contain oxygen available for use as terminal electron acceptor. An anoxic environment is defined as one in which oxygen is not present, but other inorganic electron acceptors such as nitrate or sulfate are present for use.[3] Anaerobic environments do not contain usable inorganic electron acceptor compounds. Thus, aerobic respiration occurs in an aerobic environment, anaerobic respiration occurs in an anoxic environment, and fermentation occurs in an anaerobic environment. The net energy yield from the oxidation of 1 mol of glucose by procaryotic organisms is typically ~38 ATP for aerobic respiration, ~26 ATP for anaerobic respiration when $NO_3^-$ is used as electron acceptor, and ~2 ATP for most anaerobic fermentations.

## 2.2 Biochemical Pathways Review for Organoheterotrophic Metabolism

### 2.2.1 Aerobic Respiration

The reactions involved with oxidizing a simple sugar such as glucose under aerobic conditions can be broken down broadly into three phases: glucose oxidation, pyruvate oxidation, and electron transport system. If a complex carbohydrate is present instead of glucose, hydrolysis reactions must first break down the complex compound to simple sugars. Hydrolysis reactions do not result in oxidation or reduction of the substrate.

### 2.2.1.1 Glucose Oxidation

Glucose can be oxidized by several central pathways such as (1) the Embden-Meyerhof pathway (EMP), also called glycolysis; (2) the hexose monophosphate pathway; or (3) the Entner-Douderoff pathway. All of these pathways involve intracellular reactions in which glucose is first phosphorylated and then oxidized to pyruvate through a series of intermediate reactions. In glycolysis (Fig. 2.2) the

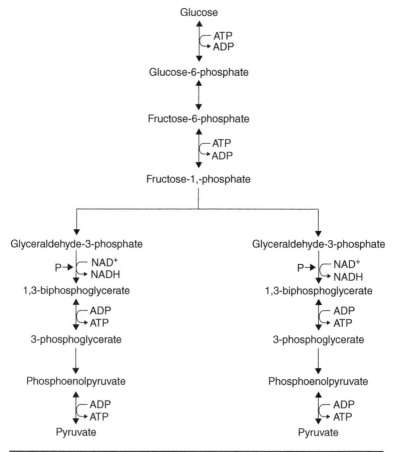

**FIGURE 2.2** Embden-Meyerhof pathway (glycolysis). (*Adapted from Madigan et al., 1997.*)

reactions yield 2 ATP, 2 pyruvate, and 2 NADH molecules. The overall reaction for glycolysis is

$$\text{Glucose} (+ 2\, \text{ADP} + 2\, \text{NAD}^+) \rightarrow 2\, \text{Pyruvate} + 2\, \text{ATP} + 2\text{NADH} \quad (2.1)$$

or

$$C_6H_{12}O_6 (+ 2\, \text{ADP} + 2\, \text{NAD}^+) \rightarrow 2\, CH_3COCOOH + 2\, \text{ATP} + 2\, \text{NADH} \quad (2.1a)$$

The C in glucose is at a 0 valence, while the C in pyruvate is at a $+2/3$ oxidation state. Thus, EMP represents a partial oxidation of glucose with 4 e$^-$ removed/mol glucose and transferred to NAD$^+$. EMP can occur under aerobic, anoxic, or anaerobic conditions.

### 2.2.1.2 Pyruvate Oxidation

In aerobic respiration, the pyruvate formed from the above pathways is processed through the tricarboxylic acid (TCA) (Fig. 2.3), also called

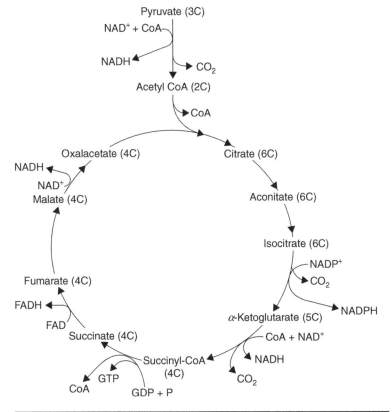

**Figure 2.3** The tricarboxylic acid (TCA) cycle. (*Adapted from Madigan et al., 1997.*)

the Krebs cycle. The oxidation of 2 mol of pyruvate in the TCA cycle yields 6 $CO_2$, 2 GTP, 8 NADH, and 2 FADH. The TCA cycle occurs under aerobic or anoxic conditions but does not occur under anaerobic conditions. The overall reaction for the TCA cycle is

$$\text{Pyruvate} (+ \text{GDP} + 4 \text{NAD}^+ + \text{FAD}) \rightarrow 3 CO_2 \\ + \text{GTP} + 4 \text{NADH} + \text{FADH} \qquad (2.2)$$

$$CH_3CH_2COOH (+ \text{GDP} + 4 \text{NAD}^+ + \text{FAD}) \rightarrow 3 CO_2 \\ + \text{GTP} + 4 \text{NADH} + \text{FADH} \qquad (2.2a)$$

In the TCA cycle, the carbon in pyruvate is completely oxidized to $CO_2$. Therefore, 10 $e^-$/mol pyruvate are removed and transferred to 4 mol $NAD^+$ and 1 mol FAD.

### 2.2.1.3 Electron Transport System

NADH, NADPH, and FADH carry electrons from glycolysis and the TCA cycle to the electron transport system (ETS), which is a series of membrane-bound proteins and enzymes that act as electron carriers. In eucaryotes, the processes of electron transport and ATP synthesis occur in the mitochondrial membrane; in procaryotes the ETS is embedded in the plasma membrane. The amount of energy produced through electron transport is proportional to the difference in redox potential between the reduced electron carrier molecule and the terminal electron acceptor. The redox potential difference between NADH/$NAD^+$ (−320 mV) and $O_2/H_2O$ (+820 mV) represents the largest change in potential, and therefore the largest ATP yield (Fig. 2.4). In general, if oxygen is used as the terminal electron acceptor, ~3 ATP are formed per electron pair transferred by NADH molecule while ~2 ATP are produced per electron pair transferred by FADH to the electron transport system.

The half reaction for the reduction of oxygen is given as

$$4 e^- + O_2 + 4 H^+ \rightarrow 2 H_2O \qquad (2.3)$$

In terms of ATP, the net result of the aerobic metabolism of 1 mol of glucose is 2 ATP produced directly in glycolysis, 2 GTP produced in the TCA cycle, and 34 ATP produced by ETS (6 ATP from 2 NADH produced in glycolysis; 24 ATP from 8 NADH produced in TCA cycle; and 4 ATP from 2 FADH produced in TCA cycle) for a total of 38 ATP/mol glucose oxidized.

The overall equation for the aerobic oxidation of an organic compound, not considering cell growth, is therefore

$$C_6H_{12}O_6 + 6 O_2 \rightarrow 6 CO_2 + 6 H_2O + 38 \text{ ATP} \qquad (2.4)$$

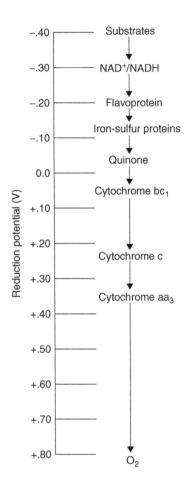

FIGURE 2.4
Redox potential of substrates and electron carriers in the electron transport system. (*Adapted from Madigan et al., 1997.*)

### 2.2.2 Anaerobic Respiration

The reactions involved with oxidizing a simple sugar such as glucose under anoxic conditions can be broken down into the same three phases: glucose oxidation, pyruvate oxidation, and electron transport system. The first two phases (glucose oxidation and pyruvate oxidation) are carried out in the same manner as they are in aerobic respiration. The final phase, the electron transport system, is the same as in aerobic respiration, except that oxygen does not serve as terminal electron acceptor. Instead, inorganic electron acceptors such as nitrate ($NO_3^-$) and sulfate ($SO_4^{2-}$) may be used. Because the redox potential values for alternative electron acceptors are lower than oxygen's (+0.433 V and −0.060 V for $NO_3^-$ and $SO_4^{2-}$, respectively), a smaller change in redox potential occurs, and less energy is gained by the cell (Fig. 2.5). On average, 2 ATP are produced per electron pair delivered to the electron transport system by NADH and accepted by $NO_3^-$, while 1 ATP is produced when $SO_4^{2-}$ is the electron acceptor (Table 2.2).

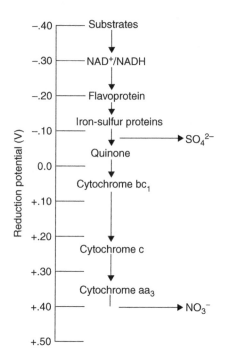

FIGURE 2.5 Electron transport system utilizing alternative inorganic electron acceptors.

The half reactions for use of nitrate and sulfate as electron acceptors are

$$5e^- + NO_3^- + 6H^+ \rightarrow \tfrac{1}{2} N_2 + 3H_2O \tag{2.5}$$

$$8e^- + SO_4^{2-} + 10H^+ \rightarrow H_2S + 4H_2O \tag{2.6}$$

It should be evident that aerobic and anaerobic respiration pathways are very similar. In both, the organic carbon compound is completely oxidized to $CO_2$, while the electrons removed from the organic compound are processed through an electron transport system. For facultative organisms, aerobic respiration will preferentially be carried out if oxygen is present due to the greater ATP yield, but will shift to use of alternative electron acceptors if oxygen becomes depleted from the environment.

| Electron carrier | mol ATP produced | | |
| --- | --- | --- | --- |
| | Electron acceptor | | |
| | $O_2$ | $NO_3^-$ | $SO_4^{2-}$ |
| NADH (or NADPH) | 3 | 2 | 1 |
| FADH | 2 | 1 | — |

TABLE 2.2 Approximate ATP Production per Mol of Electron Pairs Transferred

## 2.2.3 Fermentation

*Fermentation* is the internally balanced oxidation/reduction of organic compounds that takes place in the absence of external electron acceptors ($O_2$, $NO_3^-$, $SO_4^{2-}$, etc.). In fermentation, the oxidation of an intermediate organic compound is coupled to the reduction of another organic intermediate. Fermentation typically results in the formation of a carbon compound that is more reduced than the original organic electron donor, and a carbon compound that is more oxidized. Therefore, only a partial amount of the potential energy in the compound is captured by the microorganism.

### 2.2.3.1 Fermentation of Simple Sugars by Yeast

In many fermentation processes, the reactions involved with oxidizing a simple sugar under anaerobic conditions involve two phases: glucose oxidation and pyruvate metabolism. Glucose metabolism often occurs through glycolysis (EMP) in the same manner as in aerobic or anaerobic respiration. However, because oxygen is unavailable for use as an external electron acceptor or the microorganism does not have the capacity to use alternative inorganic compounds such as nitrate or sulfate, the electron carrier molecule $NAD^+$ must be regenerated by donating electrons to intermediate organic compounds. For example, in the fermentation of glucose by yeast such as *Saccharomyces cerevisiae*, glycolysis proceeds with the e$^-$ transferred to $NAD^+$. Pyruvate is split into acetaldehyde and $CO_2$. To regenerate $NAD^+$, acetaldehyde is reduced to ethanol by accepting the 2 e$^-$ (Fig. 2.6). Thus, the net energy yield for the yeast cell is 2 ATP/mol glucose formed by substrate level phosphorylation. The oxidation states of C in glucose, ethanol, and carbon dioxide are 0, –2, and +4, respectively, demonstrating the internal oxidation/reduction that is characteristic of fermentation processes.

The summary reactions for the fermentation of glucose by yeast are:

Glycolysis:

$$\text{Glucose} (+ 2\text{ ADP} + 2\text{ NAD}^+) \rightarrow 2\text{ Pyruvate} + 2\text{ ATP} + 2\text{ NADH} \quad (2.7)$$

Fermentation:

$$2\text{ Pyruvate} \rightarrow 2\text{ Acetaldehyde} + 2\text{ CO}_2 \quad (2.8)$$

$$2\text{ Acetaldehyde} + 2\text{ NADH} \rightarrow 2\text{ Ethanol} + 2\text{ NAD}^+ \quad (2.9)$$

The summary reaction of fermentation of sugar by yeast, not including cell growth, is

$$\text{Glucose} (+ 2\text{ ADP}) \rightarrow 2\text{ Ethanol} + 2\text{ CO}_2 + 2\text{ ATP} \quad (2.10)$$

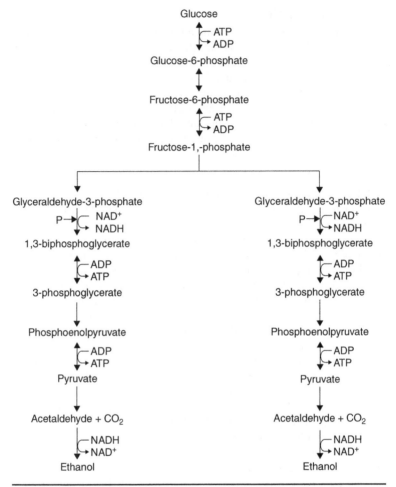

**Figure 2.6** Fermentation of glucose to ethanol by yeast. (*Adapted from Madigan et al., 1997.*)

or

$$C_6H_{12}O_6 \;(+ \text{ADP}) \rightarrow 2\, C_2H_5OH + 2\, CO_2 + 2\, \text{ATP} \quad (2.10a)$$

### 2.2.3.2 Fermentation of Simple Sugars by Bacteria

Fermentation of simple sugars such as glucose by various bacteria results in the formation of a wide array of alcohols and acids as reduced products and carbon dioxide (Fig. 2.7). These include many alcohols that are of use for industrial and bioenergy applications, such as ethanol, butanol, and isopropyl alcohol.

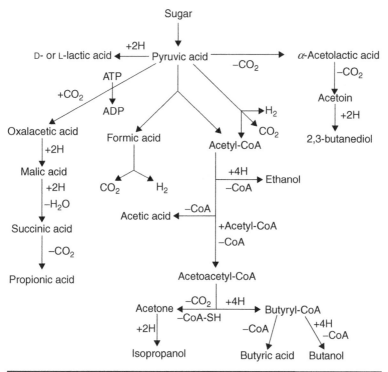

FIGURE 2.7  Fermentation of glucose by bacteria. (*Adapted from Shuler and Kargi, 1992.*)

### 2.2.3.3 Fermentation of Complex Organic Compounds by Mixed Cultures of Bacteria and Archae

When complex organic compounds containing proteins, carbohydrates, and lipids are fermented under anaerobic conditions by a consortium of bacteria, methane (a reduced product) and carbon dioxide (an oxidized product) are formed. The overall process, shown in Fig. 2.8, involves four primary processes: hydrolysis, fermentation, acetogenesis, and methanogenesis. Hydrolysis reactions are catalyzed by extracellular reactions that convert macromolecules to soluble compounds. Complex macromolecules are cleaved to their constituent protein, carbohydrate, and lipid fractions, which are further hydrolyzed to amino acid, sugar, and fatty acid monomers. Under anaerobic conditions, the amine and sulfide groups of amino acids are cleaved to release ammonia and sulfide. Hydrolysis reactions do not result in a net oxidation or reduction of substrate.

In the fermentation reactions, amino acids and sugars are fermented to a variety of volatile organic acids such as butyric and propionic acids, acetic acid, hydrogen gas, and carbon dioxide. In the acetogenesis reactions, the volatile organic acids and fatty acids are

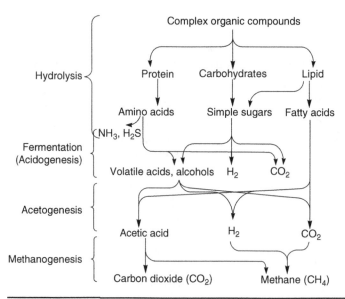

**FIGURE 2.8**  Anaerobic digestion of complex organic compounds by microbial consortia. (*Adapted from Grady et al., 1999.*)

fermented to acetic acid, hydrogen gas, and carbon dioxide by syntrophic organisms. These microbes rely on the subsequent oxidation of $H_2$ by the methanogenic organisms to lower the $H_2$ concentration and prevent end-product inhibition. This production of $H_2$ is also called *anaerobic oxidation* because electrons removed from organic substrate are transferred to inorganic electron acceptor, $H^+$, rather than to organic carbon intermediates. Finally, two groups of organisms are responsible for methane production. The first, the acetoclastic methanogens split acetic acid into methane and carbon dioxide and the second, the $H_2$-oxidizing methanogens oxidize $H_2$ and reduce $CO_2$ to produce methane. Approximately two-thirds of methane derives from the acetoclastic methanogens.

### 2.2.3.4 Anabolic Reactions for Organoheterotrophic Microorganisms

In the anabolic reactions, the energy and intermediate compounds produced during the catabolic reactions are used to form new cellular constituents, such as proteins, nucleic acids, polysaccharides, and lipids, needed for formation of new cell mass. For heterotrophic organotrophic microbes, intermediate compounds formed during glycolysis and the TCA cycle are diverted from the catabolic reactions to the biosynthesis of cellular constituents. Some key intermediates in glycolysis and the TCA cycle that are used for amino acid production are listed in Table 2.3.

| Intermediate compound | Process | | Amino acid |
|---|---|---|---|
| 3-Phosphoglycerate | Glycolysis | → | Serine → glycine, cysteine |
| Pyruvate | Glycolysis | → | Alanine |
| | | → | Valine, leucine |
| α-Ketoglutarate | TCA | → | Glutamine |
| | | → | Glutamate → proline, arginine |
| Oxaloacetate | TCA | → | Aspartate → lysine, asparagine, methionine |
| | | → | Threonine → isoleucine |

**TABLE 2.3** Organic Intermediate Compounds Used in Amino Acid Biosynthesis

The carbon contained in these intermediate compounds that are used for biosynthesis is roughly at a zero oxidation state. These compounds are removed from glycolysis or the TCA cycle before they are oxidized to $CO_2$ and are used to form new cellular constituents. Bacterial biomass is often assigned the representative formula of $C_5H_7O_2N$, so that the carbon in bacterial biomass is also at a neutral oxidation state.[3] Therefore, there is no change in oxidation state of C in these intermediates when they are used to form new cell mass. This is a significant difference between heterotrophic growth and autotrophic growth, since autotrophs must assimilate inorganic carbon to form new cellular compounds in a process that requires substantial energy and reducing power.

Therefore, when we sum the catabolic reactions (oxidation of organic electron donor to yield energy) and anabolic reactions (biosynthesis) for organoheterotrophic growth, we can develop the equation that describes the total metabolic reaction (catabolic and anabolic) for the specific growth reaction.

For the oxidation of glucose by heterotrophic bacteria under aerobic conditions, using $C_5H_7O_2N$ as a representative formula for bacterial biomass[3]:

$$1000\ C_6H_{12}O_6 + 1740\ O_2 + 852\ NH_3 \rightarrow 852\ C_5H_7O_2N$$
$$+ 1740\ CO_2 + 4296\ H_2O \quad (2.11)$$

For the oxidation of glucose by heterotrophic bacteria under anoxic conditions, using $C_5H_7O_2N$ as a representative formula for bacterial biomass[3]:

$$1000\ C_6H_{12}O_6 + 1392\ NO_3^- + 852\ NH_3 + 852\ HCO_3^-$$
$$+ 2244\ H^+ \rightarrow 852\ C_5H_7O_2N + 2592\ CO_2$$
$$+ 696\ N_2 + 5844\ H_2O \quad (2.12)$$

For the fermentation of glucose by yeast under anaerobic conditions, using $C_{100}H_{174}O_{45}N_{20}$ as a representative formula for yeast biomass[4]:

$$1000\ C_6H_{12}O_6 + 118\ NH_3 \rightarrow 5.9\ C_{100}H_{174}O_{45}N_{20} + 1300\ C_2H_5OH$$
$$+ 1540\ CO_2 + 430\ C_3H_8O_3 + 36\ H_2O \qquad (2.13)$$

In each of the above cases, a portion of the carbon is oxidized as it is used as electron donor, and a portion is used as carbon source.

## 2.3 Biochemical Pathways Overview for Lithotrophic Growth

*Chemolithotrophs* are organisms that derive energy from the oxidation of inorganic compounds. Many compounds can be used as inorganic electron donors for bacteria, and a complete review of all pathways of chemolithotrophic growth is beyond the scope of this chapter. Common inorganic electron donor compounds include ammonium ($NH_4^+$), nitrite ($NO_2^-$), hydrogen sulfide ($H_2S$), ferrous iron ($Fe^{+2}$), and hydrogen gas ($H_2$). Catabolism of any of these compounds involves the oxidation of the compound for the microorganism to capture the electrons contained in the compound. For example, in nitrification ammonia is oxidized to nitrite by one group of chemolithotroph bacteria (the genus *Nitrosomonas* and others), and then nitrite is oxidized to nitrate by a second group (*Nitrobacter* and other genera) as shown here.

$$NH_4^+ \rightarrow NO_2^- \qquad (2.14)$$

$$NO_2^- \rightarrow NO_3^- \qquad (2.15)$$

In nitrification, 6 e⁻ are removed per mol of N as $NH_4^+$ is oxidized to $NO_2^-$ and 2 e⁻ are removed per mol of N as $NO_2^-$ is oxidized to $NO_3^-$. Nitrification is an aerobic process; therefore, the electrons removed from ammonia and nitrite are carried by electron carrier molecules such as NADH to the electron transport system for ATP generation.

Many lithotrophs are autotrophs—they obtain their carbon for cell synthesis from inorganic carbon. The carbon in $CO_2$ is at oxidation state of +4; therefore, the cells must use some of the reducing power (i.e., electrons) obtained from the oxidation of the inorganic compound to reduce the carbon in order to use it for cell synthesis. The Calvin cycle (Fig. 2.9) is utilized to reduce carbon dioxide to carbohydrate, in a process also referred to as carbon fixation. The overall reaction for the Calvin cycle is

$$6\ CO_2 + 12\ NADPH + 18\ ATP \rightarrow C_6H_{12}O_6$$
$$+ 12\ NADP^+ + 18\ ADP \qquad (2.16)$$

The process of carbon fixation requires ATP and NADPH, so less energy is available for new cell synthesis. The electrons carried in NADPH must be diverted from the electron transport system to

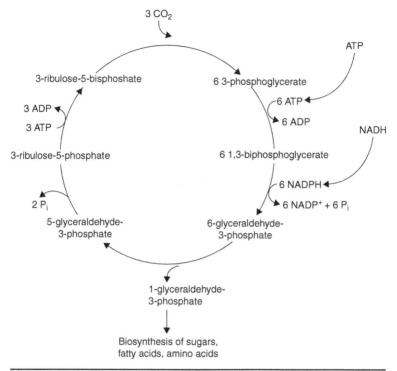

**FIGURE 2.9** Calvin cycle.

reduce the carbon in $CO_2$, so less energy is formed in the ETS. The end result is that much less energy is obtained for growth when inorganic carbon is used as C source. The organic compounds produced in the Calvin cycle are then used to produce new cell biomass in reactions similar to those for heterotrophs.

When we sum the catabolic and anabolic reactions for the utilization of an inorganic electron donor, the overall reaction for the metabolism of a compound can be obtained. For example, the oxidation of ammonia to nitrite by nitrifying bacteria with the formation of new cell mass, using $C_5H_7O_2N$ as a representative formula for bacterial biomass, is represented as[3]

$$55\ NH_4^+ + 76\ O_2 + 5\ HCO_3^- \rightarrow C_5H_7O_2N$$
$$+ 54\ NO_2^- + 104\ H^+ + 57\ H_2O \tag{2.17}$$

## 2.4 Biochemical Pathways Overview for Phototrophic Metabolism

Phototrophic metabolism is typically described as two phases: (1) light reactions, in which light energy is captured and converted to chemical energy and reducing power; and (2) anabolic (sometimes called dark) reactions, in which the chemical energy and reducing power are used to fix atmospheric $CO_2$.

### 2.4.1 Light Reactions

Phototrophic microorganisms gain energy for growth by utilizing solar radiation. Solar energy primarily in the wavelength range of 400 to 700 nm is captured by chlorophylls within the cells of green algae and cyanobacteria (blue-green algae). This wavelength range is called the photosynthetically active radiation (PAR) range, and is approximately 50 percent of the complete solar spectrum radiation.[5] When solar radiation values are reported, it is important to note whether the values are full-spectrum or PAR. Solar radiation measurements are commonly expressed in units of $W/m^2$ or $\mu E/m^2\text{-s}$ where 1E (Einstein) = 1 mol of photons. For conversion, $1\ W/m^2 \approx 4.6\ \mu E/m^2\text{-s}$.

To study energy production in photosynthesis, the flow of electrons will again be followed (Fig. 2.10). In oxygenic photosynthesis

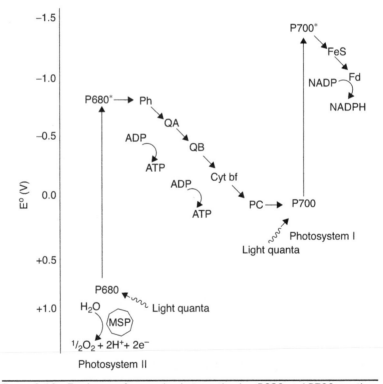

**FIGURE 2.10** Z-scheme of oxygenic photosynthesis. *P680 and P700 are the chlorophylls of photosystem II and I, respectively. (MSP: manganese stabilizing protein; Ph: pheophytin; Q: quinone; Cyt: cytochrome; PC: plastocyanin; FeS: nonheme iron-sulfur protein; Fd: ferredoxin.) (*Adapted from Madigan et al., 1997.*)

(oxygen producing photosynthesis), two photosystems are used. In photosystem II (PSII), light energy at 680-nm wavelength is used to split water molecules (Eq. 2.18).

$$2 H_2O \xrightarrow{\text{light energy}} O_2 + 4 H^+ + 4e^- \qquad (2.18)$$

The electrons are accepted by the photosystem II (PSII) chlorophyll and reduce it from a reduction potential of +1 V to approximately −0.8 V. The electrons then are transferred through a series of membrane-bound electron carrier molecules to photosystem I (PSI). ATP is produced as the electrons are transferred due to a proton-motive force that develops as protons are pumped across the thykaloid membrane. Acceptance of the $e^-$ reduces the potential of PSI to approximately −1.4 V. The reduction potential of PSI is then sufficient to reduce ferredoxin, which in turn reduces $NADP^+$ to NADPH. This NADPH is then used to reduce inorganic carbon for new cell synthesis.

Anoxygenic photosynthesis (nonoxygen producing photosynthesis) is carried out by purple and green bacteria. In this type of photosynthesis, light energy is used to split $H_2S$ to produce the $e^-$ needed for transfer to the photosystem. Since water is not split, oxygen is not produced through this type of photosynthesis. Only PSII is utilized.

### 2.4.2 Anabolic (Dark) Reactions

Most photosynthetic microbes are autotrophic, which means they derive their carbon from inorganic carbon sources. The anabolic reactions for photoautotrophic microorganisms are the same as the anabolic reactions for other autotrophs—the NADPH and ATP produced in the light reactions are used to reduce carbon dioxide to organic compounds through the Calvin cycle.

An important consideration of phototrophic metabolism is that phototrophs also carry out aerobic respiration in the absence of light. Glucose that has been produced from the Calvin cycle is oxidized to pyruvate, then carbon dioxide, using oxygen as terminal electron acceptor. The ATP formed through respiration is used to continue the anabolic reactions.

## 2.5 Definition and Importance of Chemical Oxygen Demand

Chemical oxygen demand (COD) is a useful parameter for measuring and expressing the electron donor strength of organic substrates used by chemoorganotrophic cultures for biofuel production. COD is a measure of the amount of oxygen required to completely oxidize an organic compound to $CO_2$ by chemical oxidation. COD is useful in biofuel applications because mixtures of organic compounds derived from agricultural feedstocks or wastewaters are often used as substrates for bioenergy, and analysis of individual compounds in the mixture is costly and time-consuming.

The COD content may be measured with the "closed reflux," colorimetric method (method 5220 D) of *Standard Methods*.[6] The closed reflux COD test uses concentrated sulfuric acid to chemically hydrolyze complex organic compounds to soluble compounds and a strong chemical oxidant (potassium dichromate, $K_2Cr_2O_7$) to chemically oxidize the compound in the solution. The sample and reagents are digested at 150°C for 2 h. After digestion, the organic compounds are oxidized and the chromate is reduced. The oxidized chromate ($Cr^{+6}$) has an orange/red color, and the reduced form of the chromium ($Cr^{+3}$) has a blue/green color. The intensity of the blue/green color is proportional to the concentration of organic compounds in the original sample, and is measured with a spectrophotometer at 600 nm. Oxidation of 95 to 100 percent of organic compounds is achieved with this method. However, this procedure should not be used for samples containing a chloride content of >2000 mg/L.

The COD test is an easier, quicker, and more reliable test than the BOD test for many bioprocessing and environmental engineering applications. Since microorganisms will not be able to oxidize substrates as completely as the chemical oxidant used in the COD test, the $BOD_5$ value of a solution will be less than the COD value. The more easily degradable the organic compound (as in the case of simple sugars), the closer the $BOD_5$ and COD values will be.

The advantages of utilizing the COD test for analysis and modeling of bioreactor systems include: (1) the electron donor strength in heterogeneous mixtures can be measured by using the COD test; (2) mass balances calculations and interpretation can be performed more easily by expressing biomass, substrate, product, and electron acceptor on a COD basis; (3) COD can be used to represent the number of electrons available in a substrate regardless of the electron acceptor used; and (4) inorganic electron donor compounds may also be represented as having an equivalent COD.

In bioreactor applications, the COD represents the number of electrons contained in a compound, expressed as the amount of oxygen required to accept the electrons as the compound is completely oxidized. The COD value therefore represents the energy available in a substrate for microbial growth. The growth of the organism in a biofuel bioconversion process and the power output from a microbial fuel cell are directly related to electrons captured from substrate.

In biotechnology applications, the "degree of reduction" is also used to represent the oxidation state of the substrate and the energy available, and is therefore analogous to the COD value of a substrate. The degree of reduction is calculated based on the number of O, H, C, and N in the compound. The disadvantage of the degree of reduction is that it can only be calculated when the compounds in the media are known. If an undefined medium (broths, extracts, hydrolysates, or wastewaters) is used, the individual compounds must be determined before the degree of reduction can be calculated.

The COD value of a known compound can be calculated by balancing the equation for oxidation of the compound with respect to the amount of electron acceptor (oxygen) needed. For example, the theoretical COD calculation for glucose involves expressing the oxidation as two half reactions as

$$C_6H_{12}O_6 + 6\,H_2O \rightarrow 6\,CO_2 + 24\,H^+ + 24\,e^- \quad (2.19)$$

$$24\,H^+ + 24\,e^- + 6\,O_2 \rightarrow 12\,H_2O \quad (2.20)$$

Each mol of $O_2$ accepts 4 $e^-$, so 6 mol $O_2$ are required to accept 24 mol $e^-$ as 1 mol glucose is oxidized to 6 mol $CO_2$. Therefore, 1 mol glucose (180 g) has an equivalent oxygen demand of 6 mol $O_2$ (192 g) or 1.067g COD/g glucose.

Solutions containing biomass can also be represented on a COD basis. All biomass, regardless of microorganism cultured (i.e., bacterial, fungal, yeast, or algal) is composed of organic compounds, so the equivalent COD can be calculated or measured. Using the empirical formula $C_5H_7O_2N$ to represent bacterial biomass,[3] the equivalent COD of biomass can be calculated as

$$C_5H_7O_2N + 8\,H_2O \rightarrow 5\,CO_2 + 20\,e^- + 20\,H^+ + NH_3 \quad (2.21)$$

$$20\,H^+ + 20\,e^- + (^{20}/_4)\,O_2 \rightarrow 10\,H_2O \quad (2.22)$$

Therefore, 1 mol of biomass (113 g) requires 5 mol $O_2$ (160 g) to accept the 20 mol $e^-$ as biomass is oxidized to 5 mol $CO_2$, for an equivalent ratio of 1.42 g COD/g biomass. This relationship is obtained experimentally when biomass concentration is measured as ash free dry weight (i.e., volatile suspended solids). If biomass is measured simply as dry weight using total suspended solids test, then the empirical ratio of COD to biomass is empirically 1.20 g COD/g biomass.[3]

## Acknowledgments

The authors would like to thank Biosystems Engineering students Keelin Cassidy and Derek Little for their assistance with this chapter.

## References

1. Madigan, M. T., J. M. Martinko, and J. Parker. 1997. *Brock Biology of Microorganisms*. 8th ed. Prentice Hall: Upper Saddle River, NJ.
2. Woese, C. R., O. Kandler, and M. L. Wheelis. 1990. "Towards a natural system of organisms: Proposal for the domains Archae, Bacteria, and Eucarya." *Proc Natl Acad Sci USA*. 87:4576–4579.
3. Grady, C. P. L., G. T. Daigger, and H. C. Lim. 1999. *Biological Wastewater Treatment*. 2d ed. Marcel Dekker, Inc.: New York, NY.
4. Shuler, M. L., and F. Kargi. 1992. *Bioprocess Engineering*. Prentice Hall PTR: Englewood Cliffs, NJ.
5. Monteith, J. L. 1973. *Principles of Environmental Physics*. Edward Arnold: London. p. 25.
6. APHA. 1995. *Standard Methods for the Examination of Water and Wastewater*. 19th ed. American Public Health Association: Washington, D.C.

# CHAPTER 3
# Microbial Modeling of Biofuel Production

## 3.1 Introduction

A wide variety of biofuels can be produced through the bioconversion of substrates contained in agricultural crops and residues. Bioconversion of the sugars, starches, and other organic substrates contained in agricultural residues can be converted to ethanol by a variety of yeasts, to hydrogen by a variety of fermentative bacteria and archae, to methane by a consortium of bacteria and archae, and to oils for biodiesel production by fungi and algae. All of these microbial processes can be described mathematically to simulate the bioprocess.

Bioreactor modeling and design based on microbial growth and product formation kinetics may be used to optimize production of high-value biofuels or maximize utilization of feedstock nutrients. A range of wild-type microorganisms such as fungi, bacteria, yeasts, and algae in addition to recombinant organisms may be used. Steps to the design based on microbial kinetics include expression of growth, decay, and product formation rates; development of mass-balance models of the bioreactor; simulation of substrate, biomass, and product concentrations and rates of production for various reactor configurations; selection of reactor configuration; and design optimization to maximize product formation rate or product concentration based on simulation. The focus of this chapter is on microbial modeling and bioreactor design for suspended growth of shear-tolerant, wild-type strains for biofuel production.

## 3.2 Summary of Microbial Growth Models

Basic microbial growth models for suspended growth, shear-resistant cells should consider the potential impact of multiple growth-limiting substrates and inhibitory substrates, products, and/or xenobiotic compounds. Structured models consider individual reactions occurring within the cell involving specific components, such as DNA or proteins. Unstructured models view the cell as an entity and model growth and death of the microorganism. Growth rates for unstructured models are expressed as $v$ (cell number specific growth rate) or $\mu$ (biomass specific growth rate), both in units of inverse time. Balanced growth occurs when $v = \mu$ and cell composition is constant with time. Unbalanced growth occurs as cell composition changes with time, resulting in a change in cell mass per cell; therefore $v \neq \mu$.[1]

### 3.2.1 Unstructured, Single Limiting Nutrient Models

The Monod model[2] is a widely applied model used to describe microbial growth. The model was developed for the growth of a single microorganism (*Escherichia coli*) growing on a medium with a single limiting organic substrate (glucose) as

$$\mu = \frac{\hat{\mu}\, S_S}{K_S + S_S} \qquad (3.1)$$

where  $\mu$ = specific growth rate coefficient, h$^{-1}$
$\hat{\mu}$ = maximum specific growth rate, h$^{-1}$
$S_S$ = soluble substrate concentration, mg/L
$K_S$ = half-saturation constant, mg/L

The $K_S$ value is the substrate concentration at which the growth rate $\mu$ is equal to half of $\hat{\mu}$.

The Monod model has been used successfully to model the growth of many pure cultures of heterotrophic and autotrophic organisms growing on single substrates and mixed microbial cultures using mixed substrates, such as wastewater treatment applications.[3,4] Estimates of $\hat{\mu}$ and $K_S$ parameter values may be obtained by collecting data of specific growth rate values as a function of soluble substrate concentration during the exponential phase of batch growth, and applying a linearization technique, such as Lineweaver-Burk or Hanes (Hofstee) equations.[1,3,5,6]

Zero- and first-order approximations of the Monod model [Eqs. (3.2) and (3.3)] may be applied when the substrate concentration is high and low relative to the $K_S$ value, respectively.

$$\mu \approx \hat{\mu} \qquad \text{Zero-order appropriation when } S_S \gg K_S \qquad (3.2)$$

$$\mu \approx \frac{\hat{\mu} S_S}{K_S} \qquad \text{First-order appropriation when } S_S \ll K_S \qquad (3.3)$$

Due to high substrate concentrations typically found in biofuel production applications, many researchers report $\hat{\mu}$ values and not $K_S$.

Maximum specific growth rate values for biofuel production vary greatly, depending on the type of microorganism, nature of the substrate, and biochemical environment. In general, maximum specific growth rate values are greater for heterotrophic growth than autotrophic and greater for aerobic processes than anaerobic. Most of the common biofuel production processes occur under anaerobic conditions. Use of reduced substrates under aerobic conditions results in greater specific growth rates than for less reduced substrates, yet the more reduced substrates often cannot be used under anaerobic conditions, resulting in relatively low maximum specific growth rate values for most biofuel production processes.

### 3.2.2 Inhibition Models

Expansions of the Monod model to include inhibition by substrate, product, or xenobiotic compounds have been developed. Many inhibition models for microbial growth are empirical in nature, while others apply enzyme kinetics concepts.[1] In general, *competitive* inhibitors compete with substrate for binding sites of the enzyme, and may be xenobiotic compounds or products that accumulate in the reactor. *Uncompetitive* inhibition occurs when a compound binds to the enzyme-substrate (ES) complex. A special form of uncompetitive inhibition is substrate inhibition, where substrate binds to an alternative site on the enzyme leading to a nonreactive ES complex. *Noncompetitive* inhibitors can bind to either the free enzyme or the ES complex. Product inhibition may be modeled as noncompetitive inhibition.[1]

#### 3.2.2.1 Substrate Inhibition

High substrate concentrations relative to the $K_S$ value may limit growth. Modifications of the Monod model, such as the Andrews equation,[7] may be used to describe substrate inhibition.[3]

$$\mu = \frac{\hat{\mu} S_S}{K_S + S_S + \frac{S_S^2}{K_I}} \tag{3.4}$$

where $K_I$ = inhibition coefficient, mg/L.

As the value of $K_I$ increases, the above equation simplifies to the Monod model. In many cases of inhibitory substrate, the value of $\hat{\mu}$ may never be observed and therefore $K_S$ cannot be calculated. In these cases, $\mu^*$, the maximum observed growth rate, and $S_S^*$, the substrate concentration at which $\mu^*$ occurs, are determined and used to calculate $\hat{\mu}$ and $K_S$.[3]

$$\mu^* = \frac{\hat{\mu}}{2(K_S / K_I)^{0.5} + 1} \text{ and } S_S^* = (K_S / K_I)^{0.5} \tag{3.4a}$$

An alternative model [Eq. (3.5)] was proposed by Han and Levenspiel[8] to describe substrate inhibition.

$$\mu = \frac{\hat{\mu} S}{K_S + S}\left(1 - \frac{S}{S_{crit}}\right)^n \qquad (3.5)$$

where $S_{crit}$ = critical concentration of substrate causing complete growth inhibition
$n$ = inhibition factor (unitless)

Values of $n$ and $S_{crit}$ can be determined using nonlinear least squares analysis.

Van Neil[9] used Eq. (3.5) to model (sucrose) inhibition on the batch growth of the thermophilic anaerobic bacterium *Caldicellulosiruptor saccharolyticus*. They determined $\hat{\mu}$ = 0.073 h$^{-1}$ at 70°C, $K_S$ = 2.09 mM sucrose, $S_{crit}$ = 292 mM sucrose, and $n$ = 1.39.

For phototrophic microorganisms, Steele's model[10] is often used to describe growth considering potentially inhibitory light intensity.[10–12]

$$\mu = \hat{\mu}\frac{I_L}{I_{opt}} e^{\left(1 - \frac{I_L}{I_{opt}}\right)} \qquad (3.6)$$

where $I_L$ and $I_{opt}$ = light and optimal light intensity, respectively, μmol/m$^2$-s.

Heterotrophic algal growth may be modeled as a function of organic substrate as other heterotrophic microorganisms.

### 3.2.2.2 Product Inhibition

High product concentrations may inhibit microbial growth. Product inhibition may be modeled as

$$\mu = \frac{\hat{\mu}}{\left(1 + \frac{K_S}{S}\right)}\left(1 - \frac{P}{P_{crit}}\right)^n \quad \text{or} \quad \mu = \frac{\hat{\mu} S}{K_S + S}\left(1 - \frac{P}{P_{crit}}\right)^n \qquad (3.7)$$

where $P$ = product concentration, mg/L
$P_{crit}$ = critical product concentration causing inhibition of growth
$n$ = inhibition factor, unitless

For example, fermentation of organic substrates under anaerobic conditions by organotrophic bacteria of the genera *Enterobacter, Bacillus, Clostridium, Thermotoga,* and *Caldicellulosiruptor* produce hydrogen gas as hydrogen ions are used as electron acceptor. The production of hydrogen is inhibited by high $H_2$ concentrations in the liquid phase that occurs if $H_2$ is allowed to accumulate in the headspace above the reactor. Hydrogen synthesis pathways are sensitive to $H_2$

concentration and subject to end-product inhibition. As the $H_2$ concentration increases, $H_2$ synthesis decreases and metabolic pathways shift to production of more reduced substrates such as lactate, ethanol, acetone, butanol, or alanine. A $P_{crit}$ value of 27 mM $H_2$ (mmol gas in headspace/L of reactor volume) was determined for *Thermotoga neapolitana* in batch fermentation.[13]

Van Niel[9] used Eq. (3.7) to model the impact of $H_2$ inhibition of *Caldicellulosiruptor saccharolyticus* growth at 70°C. They found that the inhibitory level decreased as the stage of batch growth proceeded: the value of $P_{crit}$ for $H_2$ was 27.7 mM during lag phase, 25.1 mM during early exponential phase, and 17.3 mM during late exponential phase. This indicates that product inhibition is more critical at later stages of microbial growth.

Further, the degree of inhibition may be a function of reactor environment. The growth of the yeast *Candida shehatae* was completely inhibited by ethanol concentrations of 25 g/L under microaerobic conditons[14] and 37.5 g/L under aerobic conditions.[15]

Another equation to describe noncompetitive product inhibition is represented in several different forms as:[1,16,17]

$$\mu = \frac{\hat{\mu}}{\left(1+\frac{K_S}{S_S}\right)\left(1+\frac{P}{K_p}\right)} \quad \text{or} \quad \mu = \frac{\hat{\mu} S_S}{K_S + S_S}\left(\frac{K_P}{K_P + P}\right) \quad \text{or}$$

$$\mu = \hat{\mu}\left(\frac{S}{K_S + S}\right)\frac{1}{1+\frac{P}{K_p}}$$

(3.8)

where $K_p$ = product inhibition constant, mg/L.

### 3.2.2.3 Inhibition by Xenobiotic Compounds

A general form of uncompetitive inhibition of enzymes can be used to model inhibition by other compounds (rearranged from Blanch and Clark[1]).

$$\mu = \frac{\hat{\mu} S_S}{S_S\left(1+\frac{I}{K_I}\right) + K_S}$$

(3.9)

where $I$ = inhibitor concentration, mg/L
$K_I$ = inhibition constant, mg/L

Another equation used to model general inhibition is given as

$$\mu = \hat{\mu}\left(\frac{S_S}{K_S + S_S}\right)\left(\frac{K_I}{I + K_I}\right)$$

(3.10)

For example, sugarcane bagasse hydrolysate produced from acid hydrolysis contains compounds such as furfural and 5-hydroxymethylfurfural (HMF) that may inhibit yeast growth at high concentrations. Inhibition of cell growth occurred at furfural and HMF concentrations of 1.0 and 1.5 g/L, respectively, for the yeast *Candida guilliermondi*.[18] A $K_I$ value of 0.01 g/L in Eq. (3.10) would result in a predicted 99 percent rate reduction of the specific growth rate at furfural concentration of 1.0 g/L.

### 3.2.3 Models for Multiple Limiting Substrates

Substrate or nutrients can be thought of in a broad sense as complementary or substitutable. *Complementary* nutrients meet different needs for the microorganism; for example, oxygen may serve as an electron acceptor for yeast growth while glucose serves as an electron donor. *Substitutable* substrates meet the same need for the cell.

#### 3.2.3.1 Complementary Substrates

To model multiple complementary nutrients, an interactive form of the Monod model may be used.[3,4,16]

$$\mu = \hat{\mu} \left( \frac{S_1}{K_{S1} + S_1} \right) \left( \frac{S_2}{K_{S2} + S_2} \right) \quad (3.11)$$

The interactive model is based on the assumption that both substrates influence the growth rate. This model predicts a lower specific growth rate than the noninteractive model, particularly when substrate concentrations are small compared to their $K_S$ values.

An alternative approach to model complementary nutrients, called the noninteractive model, assumes one nutrient limits growth.[3]

$$\mu = \hat{\mu} * \text{minimum of} \left[ \left( \frac{S_1}{K_{S1} + S_1} \right), \left( \frac{S_2}{K_{S2} + S_S} \right) \right] \quad (3.12)$$

The predicted growth rate will be the lowest value based on each substrate concentration. This model is a discontinuous function at the transition from one nutrient limitation to another. Both models have advantages and have been used successfully to describe microbial growth.

#### 3.2.3.2 Substitutable Substrates

When multiple substitutable substrates are present in the media, simultaneous use of the substrates may occur. In this case, one compound may be preferred by the microorganism over the other. If $S_1$ is preferred, then its presence in the media will inhibit the use of $S_2$.

The growth rate of the microorganism using $S_2$ is modeled as a function of $S_2$ with inhibition by $S_1$ as.[3,4]

$$\mu_1 = \hat{\mu}\left(\frac{S_1}{K_{S1}+S_1}\right) \quad (3.13)$$

$$\mu_2 = \hat{\mu}\left(\frac{S_2}{K_{S2}+S_2}\right)\left(\frac{K_{S1}}{K_{S1}+S_1}\right) \quad (3.14)$$

For example, both glucose and xylose may be used by certain yeasts such as *Candida*. Glucose is the preferred substrate, so its presence in media suppresses the use of xylose by reversibly inactivating the enzyme xylose reductase,[19] while the presence of xylose in the media does not influence the use of glucose. The sequential use of glucose, followed by xylose, in mixed substrate media by *Candida* has been documented by many researchers.[19-21] Thus, the growth rate of *Candida* based on glucose use may be modeled as

$$\mu_{Glu} = \hat{\mu}\left(\frac{S_{Glu}}{K_{Glu}+S_{Glu}}\right) \quad (3.13a)$$

where  $S_{Glu}$ = glucose concentration, g/L
 $K_{Glu}$ = half-saturation constant for glucose, g/L
 $\mu_{Glu}$ = specific growth rate using glucose as organic substrate, h$^{-1}$

The growth rate of *Candida* based on use of xylose may be modeled as

$$\mu_{Xyl} = f * \hat{\mu}\left(\frac{S_{Xyl}}{K_{Xyl}+S_{Xyl}}\right)\left(\frac{K_{Glu}}{K_{Glu}+S_{Glu}}\right) \quad (3.14a)$$

where  $S_{Xyl}$ = xylose concentration, g/L
 $K_{Glu}$ = half-saturation constant for xylose, g/L
 $\mu_{Xyl}$ = specific growth rate using xylose as organic substrate, h$^{-1}$
 $f$ = factor for decreased maximum specific growth rate for nonpreferred substrate

These equations represent the growth of *Candida* as exclusively using glucose when the glucose concentration is high. As the glucose concentration declines, the growth rate based on glucose declines, with concomitant increase in the growth rate based on use of xylose.

However, not all fermentations utilizing multiple substrates display a clear preference by the organism for one substrate over another. In the hydrolysate of the feedstock *Miscanthus*, both glucose and xylose are present. Fermentation of *Miscanthus* hydrolysate[22] by the thermophilic bacterium *Thermotoga elfii* resulted in $H_2$ production rates equivalent to rates achieved with glucose media (approximately 1.3 and 1.2 mmol/L/h, respectively). Utilization of both glucose and xylose occurred simultaneously and at equal rates with *Thermotoga elfii*, with no clear preference by the organism for either substrate.[22]

Further, for other fermentations, the preference for a certain substrate or nutrient is more difficult to model. For the fermentative bacterium *Thermotoga* that uses organic substrates such as glucose for electron donor and protons ($H^+$) as electron acceptor, other electron acceptors are preferred. If elemental sulfur ($S^0$) is present in media, it is used preferentially by the organism as the electron acceptor, producing $H_2S$ and lowering the rate of $H_2$ production.[23] Therefore, $S^0$ is the preferred electron acceptor, and its presence in media will suppress the use of $H^+$, while the presence of $H^+$ in the media will not influence the use of $S^0$. Conceptually, this effect could be modeled as above, simulating the growth of *Thermotoga* as exclusively using $S^0$ when the sulfur concentration is high, and switching to the use of protons as the sulfur concentration declined. However, modeling of growth and electron acceptor utilization by *Thermotoga* would also need to consider the impact of pH as elevated $H^+$ concentration indicates low pH, and pH below optimum will decrease the organism's growth rate.

Inhibition terms can be combined with the complementary and substitutable substrate models. In the case of *Thermotoga*, combining Eq. (3.7) to describe product inhibition by hydrogen with Eq. (3.11) to describe complementary nutrients glucose and ammonia, results in the following growth model.

$$\mu = \hat{\mu} \left( \frac{S_{glu}}{K_{glu} + S_{glu}} \right) \left( \frac{S_{NH}}{K_{NH} + S_{NH}} \right) \left( 1 - \frac{P_{H_2}}{P_{H_2,crit}} \right) \qquad (3.15)$$

where $S_{NH}$ = ammonia-N concentration, g/L
$K_{NH}$ = half-saturation constant for ammonia-N, g/L

### 3.2.4 Yield Parameters

*Biomass yield* ($Y_{X/S}$) is defined as the mass of cells produced per mass of substrate utilized. *Product yield* ($Y_{P/S}$) is defined as the mass of product formed per mass of substrate utilized. True biomass and product yields may be determined from the stoichiometry of a balanced growth equation. For the balanced equation describing anaerobic fermentation of glucose by yeast[17] [Eq. (3.16)], $Y_{X/S}$ and $Y_{EtOH/S}$ values are 0.078 mg biomass/mg glucose and 0.33 mg ethanol/mg glucose, respectively.

$$1000\ C_6H_{12}O_6 + 118\ NH_3 \rightarrow 5.9\ C_{100}H_{174}O_{45}N_{20} + 1300\ C_2H_5OH \\ + 1540\ CO_2 + 430\ C_3H_8O_3 + 36\ H_2O \qquad (3.16)$$

Observed growth yields can be obtained by determining the amount of biomass formed per substrate utilized. Biomass and product yields are generally assumed to not vary with respect to substrate concentration, but will vary for different electron acceptor environments.

Hydrogen is formed as the reduced electron acceptor in the dark fermentation of organic substrates by organisms such as hyperthermophilic bacterium *Thermotoga*. When acetate is the main carbon end-product, the maximum yield of hydrogen is 4 mol $H_2$/mol glucose. Schroder et al. found that the observed biomass yield and $H_2$ product yield of *Thermotoga maritima* in batch culture at 80°C incubation were $Y_{X/S}$ = 0.25 g $X_B$/g glucose; $Y_{H_2/S}$ = 4 mol $H_2$/mol glucose (0.0448 g $H_2$/g glucose), 2 mol acetate/mol glucose, 2 mol $CO_2$/mol glucose, and 4 mol ATP/mol glucose.[23] These rates were determined for conditions when $H_2$ partial pressure was not inhibiting production of $H_2$.

## 3.3 Kinetic Rate Expressions

Once the appropriate relationship to describe the specific growth rate is determined, the rates of utilization or formation can be developed for important reactor compounds. The rate of biomass formation is given as

$$r_{X_B} = \mu\, X_B \tag{3.17}$$

where $r_{X_B}$ = rate of biomass formation, mg/L-h

For growth rate expressions developed for microorganisms using substitutable substrates, the rates of biomass formation using both substrates could be described as shown in Eq. (3.17a), following models proposed[3,4] for wastewater treatment.

$$r_{X_B} = \mu_1 X_B + \mu_2 X_B \tag{3.17a}$$

where $\mu_1$ and $\mu_2$ = specific growth rates based on substrates 1 and 2, respectively, $h^{-1}$.

Products formed from bioconversion processes were classified by Gaden[24] as one of three types: Type I—growth-associated products arising directly from metabolism of carbohydrate; Type II—products arising indirectly from carbohydrate metabolism that accumulate under abnormal metabolism; Type III—nongrowth-associated products formed as a result of processes other than growth/energy metabolism.

Using this classification, hydrogen produced by fermentation of glucose by *Thermotoga* or ethanol produced by fermentation of glucose by yeasts are Type I products, and oil produced by algae or fungi are Type II or Type III products.

The rate of growth-associated product formation can be expressed as a function of the rate of biomass formation.[25]

$$r_{pg} = k_{pg} r_{XB} = k_{pg} \mu X_B \qquad (3.18)$$

where $r_{pg}$ = rate of growth-associated product formation, mg/L-h
$k_{pg}$ = growth-associated product formation ratio, mg product/mg biomass

For example, the $k_{pg}$ value for ethanol based on the anaerobic fermentation of glucose by yeast can be determined by the stoichiometry of the balanced Eq. (3.16)[17] as 4.26 g ethanol/g biomass. The value of $k_{pg}$ may also be calculated as the ratio of $Y_{P/S}/Y_{X/S}$ or measured directly.

The observed $k_{pg}$ value for hydrogen production by *Thermotoga neapolitana* has been found to be 0.115 g $H_2$/g biomass.[13]

The rate of nongrowth-associated (NGA) product formation is modeled solely as a function of the concentration of biomass in the reactor, and not the rate of biomass formation.[25]

$$r_{pn} = k_{pn} X_B \qquad (3.19)$$

where $r_{pn}$ = rate of nongrowth-associated product formation, mg/L-h
$k_{pn}$ = nongrowth-associated product formation constant, mg product/mg biomass-h

Products formed from a combination of growth and nongrowth-associated means may be modeled as[25]

$$r_p = k_{pg} \mu X_B + k_{pn} X_B \qquad (3.20)$$

Oils produced by fungi such as *Pythium irregulare* typically accumulate in the fungal biomass as a nongrowth-associated or mixed growth product.[26]

Other reactions that occur as a result of growth, such as substrate and oxygen utilization, are represented also as functions of the rate of biomass formation. The rate of organic substrate utilization is therefore

$$r_S = \frac{r_{X_B}}{Y_{X/S}} = \frac{\mu X_B}{Y_{X/S}} \qquad (3.21)$$

The rate of substrate utilization that occurs for products formed during nongrowth processes can be modeled as

$$r_S = \frac{r_{pn}}{Y_{P/S}} \qquad (3.22)$$

The rate of $O_2$ utilization during growth can be expressed as a function of a stoichiometric factor based on the balanced growth equation or measured value. For example, in a representative equation describing organotrophic bacteria ($C_5H_7O_2N$) grown on glucose, 0.578 mg $O_2$ are required per mg $X_B$ formed;[3] therefore,

$$r_o = ratio * r_{X_B} = 0.578 * \mu X_B \qquad (3.23)$$

where $r_o$ = rate of oxygen utilization, mg/L-h $O_2$.

If the stoichiometry is not known, then the rate of oxygen utilization can be expressed as given in Eq. (3.24) if biomass and organic substrate are measured on a chemical oxygen demand (COD) basis.[3,4]

$$r_o = (1 - Y_{X/S}) * r_S = \frac{1 - Y_{X/S}}{Y_{X/S}} \mu X_B \qquad (3.24)$$

A final expression important to bioreactor modeling is the rate of microbial decay. This rate of natural death or decay may be modeled as first order with respect to biomass.

$$r_{X_B/d} = -b X_B \qquad (3.25)$$

where $r_{X_B/d}$ = rate of biomass decay, mg/L-h $X_B$
$b$ = decay constant, h$^{-1}$

Hydrolysis of particulate organic substrate can be modeled as[4]

$$r_{X_S} = k_h \left( \frac{X_S/X_B}{K_X + (X_S/X_B)} \right) \qquad (3.26)$$

where $r_{X_S}$ = rate of particulate organic substrate hydrolysis, mg/L-h
$k_h$ = maximum specific hydrolysis rate, mg $X_S$/mg $X_B$-h
$X_S$ = particulate organic substrate concentration, mg/L
$K_x$ = $K_s$ for hydrolysis of particulate organic substrate, mg $X_S$/mg $X_B$

### 3.3.1 Temperature Effects

Microbial growth models and design should consider impact of temperature on kinetic constant values. Biomass yield, product yield, half-saturation constant, and growth-associated product constant values are often considered independent of temperature.

Kinetic constant values, such as the maximum specific growth rate, decay constant, nongrowth-associated constant, and maximum hydrolysis rate values, may be corrected for temperature differences using the Arrhenius or modified Arrhenius equations.[3]

$$k = Ae^{-E_A/RT} \quad \text{Arrhenius equation} \quad (3.27)$$

where  $k$ = temperature-dependent reaction rate constant
 $A$ = Arrhenius constant
 $E_A$ = activation energy, kJ/mol
 $R$ = gas constant, kJ/mol K
 $T$ = absolute temperature, K

$$k_{T1} = k_{T2}\theta_T^{(T_1-T_2)} \quad \text{Modified Arrhenius equation} \quad (3.28)$$

where $k_{T1}$ and $k_{T2}$ = reaction rate constants at temperatures 1 and 2
 $\theta_T$ = temperature correction factor, dimensionless
 $T$ = temperature, °C or K

The Arrhenius expression is applicable over a wide range of temperatures. The modified Arrhenius expression should be applied only over a narrow range of temperatures.[3]

In addition, other temperature effects such as impact on solubility, chemical equilibria, and chemical reaction rates need to be considered. For example, gas solubility decreases as temperature increases. For gaseous products such as hydrogen that exert an inhibitory effect on its production, increasing culture temperature will lower dissolved gas concentration and lessen inhibitory conditions.

## 3.4 Bioreactor Operation and Design for Biofuel Production

The kinetic rate expressions developed above are used in mass balance equations with respect to substrate, biomass, and products for different bioreactor configurations to simulate performance. To develop the mass balance equations with respect to products in the reactor, characterization of the product as a growth-associated (GA), nongrowth-associated (NGA), or mixed metabolite is needed. Basic bioreactor designs for suspended growth cultures are batch, continuous (flow) stirred tank reactor (CSTR), and CSTR with external or internal biomass recycle (Fig. 3.1). Further characterization of the product as intracellular vs. extracellular, and "soluble" vs. "particulate" as compared to cell separation technique is needed for CSTR with recycle designs. If the means of cell separation (filtration, centrifugation, settling) removes the compound with the biomass, then the product is considered particulate. Hydrogen and ethanol are examples of soluble, extracellular products, while oils produced by

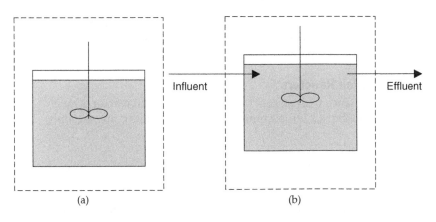
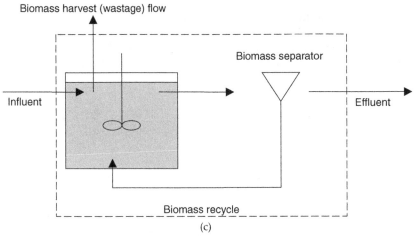
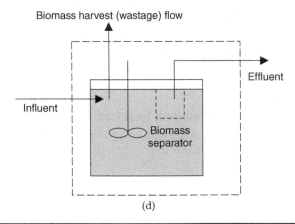

**FIGURE 3.1** Basic bioreactor types. (*a*) Batch. (*b*) Simple CSTR. (*c*) CSTR with external biomass recycle. (*d*) CSTR with internal biomass recycle. Dashed lines indicate system boundary used for developing mass balance equations.

filamentous fungi *Pythium* are intracellular products. Products such as monoclonal antibodies are extracellular "particulate" compounds since they can be separated by ultrafiltration.

### 3.4.1 Batch Reactors

The mass balance with respect to biomass for a batch reactor is given as shown in Eq. (3.29), assuming growth and decay are the only reactions.

$$\mu X_B - b X_B = \frac{dX_B}{dt} \quad (3.29)$$

Although there is a temporary balancing of the growth and decay terms during the stationary phase, batch reactors never reach true steady-state conditions, where concentrations in the reactor do not change with time. During exponential growth in a batch reactor, decay is often assumed to be negligible. Integration of Eq. (3.29) and substitution of the definition of biomass yield results in the logistic growth equation.[11,17]

$$(t - t_0)\hat{\mu} = \left( \frac{K_S Y_{X/S}}{X_{B0} + S_0 Y_{X/S}} + 1 \right) \ln\left( \frac{X_{Bt}}{X_{B0}} \right) + \frac{K_S Y_{X/S}}{X_{B0} + S_0 Y_{X/S}} \ln \frac{S_{S0}}{S_{St}} \quad (3.30)$$

where  $t$ and $t_0$ = time and initial time, h
 $X_{Bt}$ and $X_{B0}$ = biomass concentration at time $t$ and $t_0$
 $S_{St}$ and $S_{S0}$ = soluble substrate concentration at time $t$ and $t_0$

This equation describes the biomass increases while substrate decreases with time during exponential growth. The length of incubation in a batch reactor may vary depending on growth kinetics, accumulation of inhibitory products, and product type. Equation (3.30) can also be used to determine the Monod kinetic parameters for batch growth data, a technique that is primarily utilized in the wastewater treatment literature[27,28] but is being applied to bioprocessing applications.[29]

### 3.4.2 Continuous Stirred Tank Reactors

For a simple CSTR with a single influent and effluent flow, the mass balance with respect to biomass is

$$X_{Bi} \frac{Q}{V} - X_B \frac{Q}{V} + \mu X_B - b X_B = \frac{dX_B}{dt} \quad (3.31)$$

where  $Q$ = volumetric flow rate, L/h
 $V$ = reactor volume, L
 $X_{Bi}$ = biomass concentration in influent flow, mg/L

The hydraulic retention time for a CSTR reactor can be calculated as

$$\tau = \frac{V}{Q} \tag{3.32}$$

where $\tau$ = hydraulic retention time, h.

For cases with no biomass in the influent flow, growth based on a single limiting nutrient, and no substrate or product inhibition, the mass balance simplifies to the following at steady state.

$$\mu = \frac{1}{\tau} + b \tag{3.33}$$

where $\tau$ = hydraulic retention time in reactor, h.

The hydraulic retention time is the main design parameter for a CSTR, since it controls the microbial growth rate in the reactor. A simple CSTR can be operated at a short retention time to maintain a high specific growth rate to mimic a batch reactor during exponential growth, with sustained formation of growth-associated products. Alternatively, it may be operated at a long retention time to maintain a low specific growth rate to mimic a batch reactor during declining exponential or stationary phases, resulting in continuous formation of nongrowth-associated products.

Substitution of the single Monod model for $\mu$ results in

$$S_S = \frac{K_S(1/\tau + b)}{\hat{\mu} - (1/\tau + b)} \tag{3.34}$$

The effluent substrate concentration is a function of the hydraulic retention time and the kinetic parameter values, but not a function of the influent substrate concentration. Rearranging Eq. (3.34) gives another form of the expression, written as the hydraulic retention time needed to achieve a certain effluent substrate concentration that may be used for preliminary bioreactor design calculations.

$$\tau = \frac{K_S + S_S}{S_S(\hat{\mu} - b) - K_S b} \tag{3.34a}$$

Washout of microorganisms from a CSTR refers to the condition where the retention time is too short relative to the microbial growth rate to allow for growth to occur in the reactor. At washout conditions, the effluent substrate concentration would be the same as the influent concentration because no substrate utilization would occur. The retention time at which washout would occur can be calculated using Eq. (3.34a) with the $S_S$ value equal to the influent substrate value.

The mass balance with respect to substrate for a simple CSTR is

$$S_{Si}\frac{Q}{V} - S_S\frac{Q}{V} - \frac{\mu X_B}{Y_{X/S}} = \frac{dS_S}{dt} \tag{3.35}$$

where $S_{Si}$ = soluble organic substrate concentration in influent flow, mg/L.

After substitution of Eq. (3.32), the resulting steady-state concentration of biomass is given as

$$X_B = \frac{Y_{X/S}(S_{Si} - S_S)}{1 + \tau b} \tag{3.36}$$

The mass balance with respect to a general product is

$$P_i \frac{Q}{V} - P \frac{Q}{V} + k_{pg} \mu X_B + k_{pn} X_B = \frac{dP}{dt} \tag{3.37}$$

The steady-state solutions for GA and NGA products are[25]

$$P_{GA} = k_{pg} \mu X_B \tau \tag{3.38}$$

where $P_{GA}$ = growth-associated product concentration, mg/L;

$$P_{NGA} = k_{pn} X_B \tau \tag{3.39}$$

where $P_{NGA}$ = nongrowth-associated product concentration, mg/L.

At steady state, concentrations of substrate, biomass, and product will vary as a function of retention time. In general, the specific growth rate and effluent substrate concentration decline and biomass concentration increases with increase in hydraulic retention time. Thus, reactor design may be based on achieving the desired effluent concentrations.

To maximize rates of biomass or GA product formation or substrate utilization, both a high specific growth rate and a high biomass value must be achieved. The hydraulic retention time that maximizes these rates can be estimated from Eq. (3.40).[25]

$$\frac{1}{\tau} = \hat{\mu}\left[1 - \left(\frac{K_S}{K_S + S_{Si}}\right)^{0.5}\right] \tag{3.40}$$

For a NGA product, there is no single hydraulic retention time that maximizes the rate of production. The above equations may be used for preliminary analysis and design of a CSTR. For more in-depth analysis, and for systems where multiple substrates or inhibition occur, dynamic modeling may be used to determine the retention time that maximizes biomass or GA formation rates or the product concentration as a function of retention time.

### 3.4.3 CSTR with Cell Recycle

A CSTR with biomass recycle includes some means to remove biomass from the reactor effluent flow—through external or internal filter, settling device, or centrifugation—and return biomass flow to the reactor. By separating biomass from the effluent flow, the length of time that cells remain in the reactor, the cell retention time ($\theta$), can be controlled independently of the hydraulic retention time. To control $\theta$, a portion of biomass is harvested or "wasted" from the reactor. When the harvest flow is pumped directly from the reactor, the biomass concentration in the reactor is equal to the concentration in the harvest flow and $\theta$ is calculated as

$$\theta = \frac{\cancel{V} X_B}{Q_H X_{BH}} = \frac{\cancel{V}}{Q_H} \qquad (3.41)$$

where  $\theta$ = cell retention time, h
$X_B$ = biomass concentration in reactor, mg/L
$X_{BH}$ = biomass concentration in the harvest flow, mg/L
$Q_H$ = harvest flow rate, L/h

The mass balance equation with respect to biomass for a CSTR with recycle is

$$X_{Bi}\frac{Q}{\cancel{V}} - X_B\frac{Q_H}{\cancel{V}} - X_{Be}\frac{(Q-Q_H)}{\cancel{V}} + \mu X_B - bX_B = \frac{dX_B}{dt} \qquad (3.42)$$

where $X_{Be}$ = biomass concentration in final effluent flow, mg/L. $X_{Be} \sim 0$ if cell separation step is assumed 100 percent efficient.

At steady state, the mass balance with respect to biomass simplifies to

$$\mu = \frac{1}{\theta} + b \qquad (3.43)$$

Therefore, $\theta$ controls $\mu$ for a CSTR with recycle in the same manner as $\tau$ does for a simple CSTR. Substitution of the single Monod model for $\mu$ results in the following expression to describe the steady-state effluent substrate concentration.

$$S_S = \frac{K_S(1/\theta + b)}{\hat{\mu} - (1/\theta + b)} \qquad (3.44)$$

The mass balance with respect to soluble substrate for a CSTR with recycle is

$$S_{Si}\frac{Q}{\cancel{V}} - S_S\frac{Q_H}{\cancel{V}} - S_S\frac{(Q-Q_H)}{\cancel{V}} - \frac{\mu X_B}{Y_{X/S}} = \frac{dS_S}{dt} \qquad (3.45)$$

Because soluble organic substrate will not be removed by the cell separation technique, the same concentration will be present in the wastage and main effluent flow. The mass balance at steady state simplifies to Eq. (3.46), which describes the biomass concentration as a function of both the hydraulic and cell retention times. Influent organic substrate concentration and hydraulic retention time can be adjusted to achieve the desired cell concentration in the reactor.

$$X_B = \frac{\theta}{\tau} * \frac{Y_{X/S}(S_{Si} - S_S)}{1 + b\theta} \qquad (3.46)$$

The mass balance with respect to a general product is

$$P_i \frac{Q}{V} - P \frac{Q_H}{V} - P \frac{(Q - Q_H)}{V} + k_{pg} \mu X_B + k_{pn} X_B = \frac{dP}{dt} \qquad (3.47)$$

Product concentrations at steady state, assuming no product in influent, will vary according to whether the product is GA or NGA, and particulate or soluble, as

$$P = k_{pg}(1/\theta + b)X_B \tau \qquad \text{if product is GA, soluble} \qquad (3.48)$$

$$P = k_{pg}(1/\theta + b)X_B \theta \qquad \text{if product is GA, particulate} \qquad (3.49)$$

$$P = k_{pn} X_B \tau \qquad \text{if product is NGA, soluble} \qquad (3.50)$$

$$P = k_{pn} X_B \theta \qquad \text{if product is NGA, particulate} \qquad (3.51)$$

In general, the specific growth rate and effluent soluble substrate concentration will decline as cell retention time increases, and biomass and nongrowth-associated product concentrations, especially for particulate products, will increase with increase in cell retention time. In a CSTR with recycle, a single optimum cell retention time that maximizes the rate of product formation for all types of products does not exist. A model of the system based on mass balance equations may be used to run simulations to determine the product concentration and rate of product formation at different hydraulic and cell retention times. System optimization for a biological CSTR with recycle is a matter of determining the best combination of cell density and production rates for the system.

### 3.4.4 Fed-Batch Systems

The important difference between fed-batch systems and batch and CSTR systems is that the specific growth rate decreases with time due to increasing volume at constant feed rate[1] as shown by

$$\frac{d\mu}{dt} = \frac{d}{dt}\left[\frac{Q}{V_0 + Qt}\right] = \frac{-Q^2}{(V_0 + Qt)^2} \qquad (3.52)$$

where $V_0$ = initial volume, L.

However, steady-state operation may be achieved by exponentially adjusting the feed rate, $Q$, as shown by

$$\frac{dV}{dt} = Q = \lambda V \qquad (3.53)$$

Equation (3.53) may be integrated to obtain

$$Q = \lambda V_0 e^{\lambda t} \quad (3.54)$$

where $\lambda$ is the dilution rate or $1/\tau$.

At steady-state operation, the expression for substrate concentration mimics that developed for the CSTR system [Eq. (3.34)] as shown in Eq. (3.55) excluding the decay term.

$$\mu = \frac{1}{\tau} = \frac{\hat{\mu} S_S}{K_S + S_S} \quad \text{and} \quad S_S \cong \frac{K_S/\tau}{\hat{\mu} - (1/\tau)} \quad (3.55)$$

Fed-batch operation may be represented at steady state in terms of biomass and product formation excluding decay,[17] respectively, as

$$X^t = X_0^t + Q Y_B S_{Si} t \quad (3.56)$$

where $X^t$ and $X_0^t$ = mass of biomass at time $t$ and initial time, mg;

$$P = P_0 \frac{V_0}{V} + k_p X_B \left( \frac{V_0}{V} + \frac{t}{2\tau} \right) t \quad (3.57)$$

where $P$ and $P_0$ = product concentration at time $t$ and time $t_0$, mg/L
$V_0$ = initial volume, L

A fed-batch bioreactor system has inherent advantages of both batch and continuous systems since the operation is semi-continuous with only substrate feed and no effluent stream. For this reason, these systems are often employed in many commercial applications within the food and pharmaceutical industries. The system works well for NGA production typically accomplished by introduction of secondary metabolite inducers, specific nutrient deprivation, or substrate inhibition at specific times during the process. The fed-batch system is also applied to form GA and mixed-growth–associated (MGA) products, such as heterotrophic algal oils, by adjusting substrate feed rate to obtain maximum conversion to oil stored in high cell density culture.

### 3.4.5 Plug Flow Systems

In a plug flow reactor system, mass movement occurs by bulk fluid flow and by hydrodynamic dispersion. Hydrodynamic dispersion is the spreading of a liquid or gas from the bulk flow path due to mixing effects caused by both concentration gradients (diffusion) and velocity differences.

The governing equation for one-dimensional mass movement (bulk flow and diffusion) is[30]

$$E\frac{\partial^2 C}{\partial x^2} - V_x\frac{\partial C}{\partial x} \pm \Sigma r = \frac{\partial C}{\partial t} \qquad (3.58)$$

where  $E$ = hydrodynamic dispersion coefficient, m²/s
 $C$ = concentration of compound, g/m³
 $V_x$ = velocity of flow in the longitudinal direction, m/s

The hydrodynamic dispersion coefficient, $E$, comprises both diffusion in the longitudinal direction and longitudinal dispersion due to velocity differences (mixing).[30]

$$E = D_{AB} + E_L \qquad (3.59)$$

where $D_{AB}$ = binary diffusion coefficient of compound A in compound B, m²/s
 $E_L$ = longitudinal dispersion coefficient, m²/s

The longitudinal dispersion term is a function of velocity, whereas $D_{AB}$ is not, so that as the velocity increases, the longitudinal dispersion coefficient increases and the diffusion term becomes negligible.

For a continuous input of a compound into a plug flow system, the concentration in the effluent can be calculated as:

$$C_t = \frac{\dot{m}}{A\sqrt{V^2 + 4kE_L}} * \exp\left[\left(\frac{xV}{2E_L}\right) * \left(1 \pm \sqrt{1+(4kE_L/V^2)}\right)\right] \qquad (3.60)$$

where $C_t$ = concentration of compound at time $t$, g/m³
 $\dot{m}$ = mass flow rate of compound entering system, g/s
 $A$ = cross-sectional area, m²
 $k$ = first-order chemical reaction constant, s⁻¹
 $\pm$ = + for $x < 0$ and − for $x > 0$.

This solution can be used to determine the dispersion coefficient for a given plug flow system. For an insert compound, $K = 0$

An alternative way to model plug flow systems is to model the system as a series of completely mixed, continuous flow reactors.[3] An experiment can be conducted in which a tracer compound is added to a plug flow reactor, and the effluent concentration is measured. The solution to the mass balance equation for this system is[3]

$$\frac{C_t}{C_i} = 1 - \sum_1^n \left[\frac{(t/\tau_n)^{n-1}}{(n-1)!}\right] e^{-t/\tau_n} \qquad (3.61)$$

where $C_t$ and $C_i$ = concentration in the final effluent flow at time $t$ and in influent
 $\tau_n$ = hydraulic retention time for individual cell (= system hydraulic retention time/$n$)
 $n$ = number of individual CSTRs used to represent the plug flow reactor

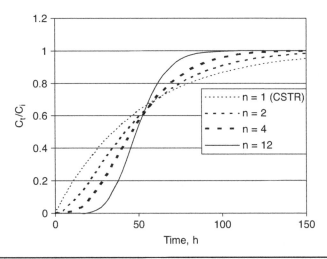

**FIGURE 3.2** Idealized effluent concentrations for plug flow reactor (total system $\tau = 50$ h). $C_t$ and $C_i$ = effluent and influent concentrations, respectively.

For an example of four reactors in series ($n = 4$),

$$\frac{C_t}{C_i} = 1 - \left[1 + \frac{t}{\tau_n} + \frac{(t/\tau_n)^2}{2!} + \frac{(t/\tau_n)^3}{(3)!}\right]e^{-t/\tau_n} \quad (3.62)$$

The equivalent number of CSTRs for a given plug flow reactor is determined by the $n$ value that best represents the data (Fig. 3.2).

## 3.5 Bioreactor Design Strategies

The strategy for bioreactor design may be based on maximizing the rate of product formation, biomass production, or substrate utilization, or on achieving target concentrations of product, biomass, or substrate. As revealed by the above rate expressions, rates of substrate/nutrient utilization and primary product formation will be high when the *rate* of biomass formation is high, while the rates of secondary product formation and decay will be high when the *concentration* of biomass is high. In general, rates of primary product formation in a batch reactor will be highest during the mid-exponential phase, while product concentration will be highest at the end of the exponential phase. Secondary product concentration in a batch reactor will peak at the end of the stationary phase. For a primary, extracellular, soluble product such as ethanol, a batch reactor or simple CSTR would achieve a high product formation rate. But for nongrowth-associated products or reactions that form inhibitory products, batch reactors may not be advantageous due to the low productivity rate or elevated product concentrations that may form. Selection of a simple CSTR allows for control of the specific growth rate through control of the hydraulic retention time to maximize the rate of a primary product. In a simple CSTR or CSTR with biomass recycle, the steady-state specific growth rate is inversely proportional to the hydraulic or cell retention time, respectively, while biomass concentration is, in general, directly proportional to the retention times. Therefore, these designs allow for efficient production of either primary or secondary products by achieving target-specific growth rate and biomass or product concentrations through control of the retention time(s). Dynamic modeling and simulation of microbial growth and product formation are needed to optimize the design based on predicted production rates and concentrations.

For secondary intracellular products such as certain lipids produced by fungi or for secondary extracellular, particulate products, rates of formation and product concentrations will be maximized by a CSTR with biomass recycle design with fairly long cell retention time—one that easily allows for high biomass concentration to be achieved without exposing cells to inhibitory substrate concentrations.

Alternating environments with respect to substrate, electron acceptor, or nutrient levels or types can be used to maximize product formation or to select for desired biomass cultures.[1,3] Biofuel bioprocesses that rely on anaerobic fermentations for product formation could be likely candidates for a two-stage CSTR system. For example, use of an aerobic environment for maximizing cell production in the first CSTR with anaerobic environment for ethanol production in the second CSTR could achieve greater ethanol production over single-stage systems.

## 3.6 Modeling of Glucose Utilization and Hydrogen Production

The fermentation of glucose under anaerobic conditions by *Thermotoga neapolitana* to produce hydrogen gas can be used to demonstrate the mass balance model development and reactor design. The first case will demonstrate $H_2$ production and cell growth in a batch reactor with Monod kinetics used to model single nutrient limitation. The second case will demonstrate $H_2$ production and growth in a batch reactor using single limiting nutrient limitation with product inhibition model. The third case will demonstrate growth and product formation in a CSTR with single limiting nutrient model with and without product inhibition.

For these simulations, BioMASS[31] (Biological Modeling and Simulation Software) will be used. BioMASS was developed at

Clemson University, Clemson, South Carolina, and is available for download at the web site www.clemson.edu/agbioeng/bio/Drapcho/BioMASS.htm. BioMASS incorporates forward finite difference to solve the simultaneous mass balance equations presented in this chapter to calculate the dynamic concentrations of substrate, biomass, and product. This user-friendly bioprocess simulator was developed using Microsoft Visual Basic 6.0 to model microbial growth, substrate utilization, and product formation for multiple batch or continuous reactors. The simulator provides extensive options for microbial kinetics such as multiple-limiting substrates (complementary and substitutable), and substrate, product, and/or xenobiotic compound inhibition models.

### 3.6.1 Batch Fermentations and Simulations

A batch growth curve for *Thermotoga neapolitana* cultured at 77°C utilizing glucose as the carbon source is shown in Fig. 3.3.

The peak concentrations were approximately 27 mM $H_2$ (mmol $H_2$ gas in headspace per L of reactor volume) and 0.3 g/L biomass. Growth and product formation ceased by approximately 10 h, despite only 50 percent glucose utilization. In this culture system, accumulated $H_2$ gas was not removed from the reactor headspace which may have caused product inhibition, and the media was not fully buffered, so that pH was not maintained within desired levels for *Thermotoga*. Estimates of the observed yield and kinetic parameters for *T. neapolitana* for these conditions were biomass yield ($Y_{X/S}$) = 0.248 g $X_B$/g glucose; product yield ($Y_{P/S}$) = 0.0286 g $H_2$/g glucose; maximum specific growth rate ($\mu_{max}$) with glucose as carbon source = 0.94 $h^{-1}$ at 77°C; and half-saturation constant ($K_S$) = 0.57 g/L.[13] From this data, $H_2$ gas can be

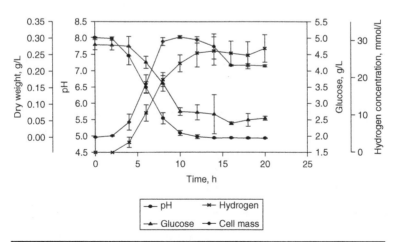

**FIGURE 3.3**  Batch growth curve for *Thermotoga neapolitana* at 77°C incubation. (*Adapted from Yu, 2007.*)

determined to be a primary (growth-associated) product with $k_{pg} = 0.115$ g $H_2$/g $X_B$. The decay constant (b) for this organism was not determined, but typical values may range from 0.01 to 0.05 $h^{-1}$.

Using BioMASS with these kinetic parameter values, specifying the single limiting nutrient Monod model to describe growth, and identifying $H_2$ gas as a primary product, the following simulation is obtained (Fig. 3.4).

As shown in Fig. 3.4, the predicted peak biomass (1.3 g/L) and $H_2$ (70 mmol/L) concentrations are three to four times greater than those obtained for the actual fermentation, and near complete utilization of glucose is predicted. This simulation indicates that the single limiting nutrient model to describe microbial growth is not sufficient to model growth and hydrogen production by *T. neapolitana* for these conditions where inhibition due to elevated $H_2$ concentrations or low pH may have occurred.

By incorporating the product inhibition term to modify the specific growth rate for inhibitory levels of $H_2$ [Eq. (3.7)], with a $P_{crit}$ value of 27 mmol $H_2/L^{13}$ and assumed $n = 1$, the following simulation results were obtained (Fig. 3.5).

These simulations result in predicted peak hydrogen (26.7 mmol/L) and biomass (0.50 g/L) concentrations, very near to those obtained for the actual fermentation. These results suggest that reactor designs including continuous removal of hydrogen gas could be used to

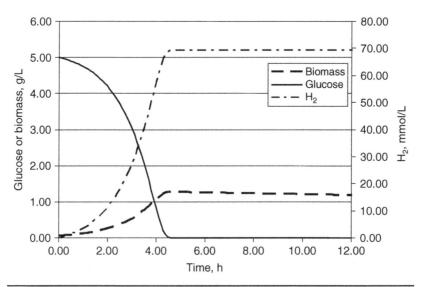

**Figure 3.4** Simulation of concentrations in batch reactor using single nutrient Monod model with no product inhibition. Kinetic parameters: $\hat{\mu} = 0.94$ $h^{-1}$; $K_s = 0.57$ g/L; $Y_{X/S} = 0.248$ g $X_B$/g $S_S$; $Y_{P/S} = 0.0286$ g $H_2$/g $S_S$; $k_{pg} = 0.115$ g $H_2$/g $X_B$; and b = 0.01 $h^{-1}$ (assumed). Initial conditions: $X_{B,0} = 0.05$ g/L, $S_{S,0} = 5$ g/L.

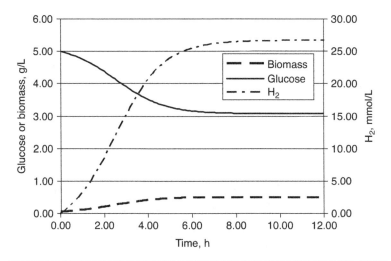

**FIGURE 3.5** Simulation of concentrations in batch reactor using single nutrient Monod with $H_2$ inhibition model. Kinetic parameters: $\hat{\mu} = 0.94$ h$^{-1}$; $K_s = 0.57$ g/L; $Y_{X/S} = 0.248$ g $X_B$/g $S_S$; $Y_{P/S} = 0.0286$ g $H_2$/g $S_S$; $k_{pg} = 0.115$ g $H_2$/g $X_B$; $P_{crit} = 27$ mmol $H_2$/L; and b = 0.01 h$^{-1}$(assumed). Initial conditions: $X_{B,0} = 0.05$ g/L, $S_{S,0} = 5$ g/L.

avoid end-product inhibition and increase the hydrogen production rate for production reactors.

### 3.6.2 CSTR Fermentations and Simulations

Simulation of biomass, substrate, and product concentrations in a CSTR with a 6-hour hydraulic retention using the same kinetic parameter values, with and without consideration of $H_2$ inhibition, results in the following predictions (Figs. 3.6 and 3.7).

Again, the inhibitory effect of $H_2$ concentration on $H_2$ production is clearly predicted. These results predict that sustained $H_2$ gas concentration of approximately 70 mmol/L may be achieved in a CSTR with $H_2$ gas removal to prevent end product inhibition. This concentration translates to 11.7 mmol $H_2$/L-h hydrogen production rate.

Finally, a simulation of $H_2$ concentrations and production rates as a function of hydraulic retention time in a CSTR using two different influent glucose concentrations (5 g/L and 20 g/L) is shown in Fig. 3.8. This simulation indicates that at a greater influent glucose concentration and with effects of inhibitory $H_2$ removed, $H_2$ production rates could potentially reach as high as 170 mmol $H_2$/L-h (for 20 g/L influent glucose case) or 33 mmol $H_2$/L-h (for the 5 g/L influent glucose case) at the optimum retention time of 1.6 h. The washout retention time is approximately 1.2 h.

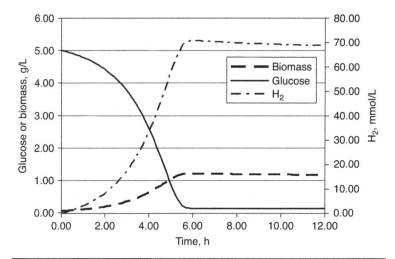

**FIGURE 3.6** Simulation of concentrations in CSTR using single nutrient Monod model with no product inhibition. Kinetic parameters: $\hat{\mu} = 0.94$ h$^{-1}$; $K_s = 0.57$ g/L; $Y_{X/S} = 0.248$ g $X_B$/g $S_S$; $Y_{P/S} = 0.0286$ g H$_2$/g $S_S$; $k_{pg} = 0.115$ g H$_2$/g $X_B$; and b = 0.01 h$^{-1}$ (assumed). Initial conditions: $X_{B,0} = 0.05$ g/L, $S_{S,0} = 5$ g/L. Hydraulic retention time = 6 h.

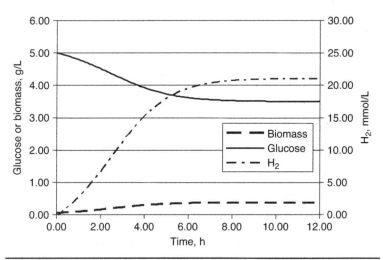

**FIGURE 3.7** Simulation of concentrations in CSTR using single nutrient Monod with H$_2$ inhibition model. Kinetic parameters: $\hat{\mu} = 0.94$ h$^{-1}$; $K_s = 0.57$ g/L; $Y_{X/S} = 0.248$ g $X_B$/g $S_S$; $Y_{P/S} = 0.0286$ g H$_2$/g $S_S$; $k_{pg} = 0.115$ g H$_2$/g $X_B$; $P_{crit} = 27$ mmol H$_2$/L; and b = 0.01 h$^{-1}$ (assumed). Initial conditions: $X_{B,0} = 0.05$ g/L, $S_{S,0} = 5$ g/L. Hydraulic retention time = 6 h.

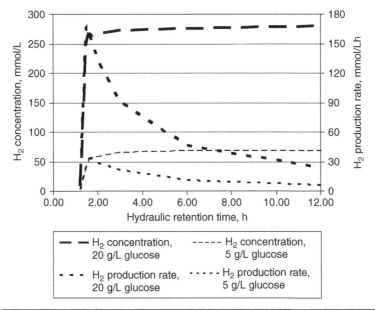

**Figure 3.8** Simulations of steady-state $H_2$ concentrations and $H_2$ production rates as function of hydraulic retention time in CSTR at influent glucose concentrations of 5 and 20 g/L, modeled using single nutrient Monod model with no product inhibition. Kinetic parameters: $\hat{\mu} = 0.94$ h$^{-1}$; $K_s = 0.57$ g/L; $Y_{X/S} = 0.248$ g $X_B$/g $S_S$; $Y_{P/S} = 0.0286$ g $H_2$/g $S_S$; $k_{pg} = 0.115$ g $H_2$/g $X_B$; and $b = 0.01$ h$^{-1}$ (assumed). Initial conditions: $X_{B,0} = 0.05$ g/L, $S_{S,0} = 5$ or 20 g/L.

**Example 3.1** Calculate the results summarized in Fig. 3.7 for $H_2$ production by *Thermotoga neapolitana* using 5 g/L influent glucose concentration. Assume all other nutrients are available in media, and no product inhibition. Use reactor volume of 100 L. Use the steady-state solutions to the mass balance equation for a simple CSTR to calculate (1) the minimum hydraulic retention time for a simple CSTR used to culture *T. neapolitana* at 77°C and (2) the retention time that will maximize $H_2$ production. Then, for that optimum retention time, calculate (3) the effluent glucose, biomass, and hydrogen gas concentrations; (4) the rate of $H_2$ production; and (5) the media flow rate required.

**Solution**

1. Use Eq. (3.34a) to calculate the washout retention time, using influent value for $S_S$:

$$\tau_{washout} = \frac{K_S + S_S}{S_S(\hat{\mu}-b) - K_S b} = \frac{0.57 \text{ g/L} + 5.0 \text{ g/L}}{5.0 \text{ g/L}(0.94 \text{h}^{-1})} = 1.18 \text{ h}$$

The minimum hydraulic retention time for reactor operation must be above 1.18 h. The exact value would depend on safety factor that your design would include to ensure prevention of washout of microorganisms from the reactor.

2. The hydraulic retention time that maximizes $H_2$ production is calculated using Eq. (3.40):

$$\frac{1}{\tau} = \hat{\mu}\left[1 - \left(\frac{K_S}{K_S + S_{Si}}\right)^{0.5}\right] = 0.94\,h^{-1}\left[1 - \left(\frac{0.57\,g/L}{0.57\,g/L + 5.0\,g/L}\right)^{0.5}\right] = 0.64\,h^{-1}$$

Therefore, the retention time to maximize $H_2$ production is 1.56 h.

3. The effluent substrate (glucose) and biomass concentrations at this retention time may be calculated using Eqs. (3.34), (3.36), and (3.38).

$$S_S = \frac{K_S(1/\tau + b)}{\hat{\mu} - (1/\tau + b)} = \frac{0.57\,g/L\left(\frac{1}{1.56\,h} + 0.01\,h^{-1}\right)}{0.94\,h^{-1} - \left(\frac{1}{1.56\,h} + 0.01\,h^{-1}\right)} = 1.28\,g/L$$

$$X_B = \frac{Y_{X/S}(S_{Si} - S_S)}{1 + \tau b} = \frac{0.248\,g\,X_B/g\,S_S(5\,g/L - 1.28\,g/L)}{1 + (1.56\,h * 0.01\,h^{-1})} = 0.908\,g/L\,X_B$$

$$H_2 = P_{GA} = k_{pg}\mu X_B \tau$$

$$= 0.115\,g\,H_2/g\,X_B * \left(\frac{1}{1.56\,h} + 0.01\,h^{-1}\right) * 0.908\,g/L\,X_B * 1.56\,h$$

$$= 0.106\,g/L\,H_2 = 52.5\,mmol/L\,H_2$$

4. The rate of hydrogen production at this retention time may be calculated using Eq. (3.18):

$$r_{pg} = k_{pg}\,\mu X_B = 0.115\,g\,H_2/g\,X_B * \left(\frac{1}{1.56\,h} + 0.01\,h^{-1}\right) * (0.908\,g/L\,X_B)$$

$$= 0.0680\,g\,H_2/L\text{-}h = 33.6\,mmol\,H_2/L\text{-}h$$

Using BioMASS, the washout retention time is calculated to be 1.2 h, and optimum retention time is 1.6 h. At this retention time effluent substrate, biomass, and hydrogen concentrations are 1.19, 0.93, and 0.109 g/L (54.0 mmol/L), respectively. The rate of $H_2$ production is predicted to be 33.7 mmol $H_2$/L-h. Notice that the hydraulic retention times, concentrations, and rate of $H_2$ production calculated using the steady-state equations are very similar to the values obtained using BioMASS.

# Summary

The kinetics of microbial growth, substrate utilization, and product formation can greatly influence the effectiveness of a bioconversion process. Modeling involves characterization of growth and product formation kinetics and development of mass balance models of system. Dynamic modeling and simulation of the bioprocess can be used to select, design, and optimize a bioreactor for production of biofuel products.

## References

1. Blanch, H. W., and D. S. Clark. 1997. *Biochemical Engineering.* Marcel Dekkar: New York, NY.
2. Monod, J. 1949. "The growth of bacterial cultures." *Ann Rev Microbiol.* 3:371.
3. Grady, C. P. L., G. T. Daigger, and H. C. Lim. 1999. *Biological Wastewater Treatment.* 2d ed. Marcel Dekker, Inc.: New York, NY.
4. Henze, M., et al. 1987. "A general model for single-sludge wastewater treatment systems." *Water Research.* 21:505.
5. Bailey, J., and D. Ollis. 1986. *Biochemical Engineering Fundamentals.* 2d ed. McGraw-Hill Book Co.: New York, NY.
6. Doran, P. M. 1995. *Bioprocess Engineering Principles.* Academic Press, Inc.: San Diego, CA.
7. Andrews, J. F. 1968. "A mathematical model for the continuous culture of microorganisms utilizing inhibitory substrates." *Biotechnol Bioeng.* 10:707.
8. Han, K., and O. Levenspiel. 1988. "Extended Monod kinetics for substrate, product, and cell inhibition." *Biotechnol Bioeng.* 32:430–437.
9. Van Niel, E. W. J., P. A. M. Claassen, and A. J. M. Stams. 2003. "Substrate and product inhibition of hydrogen production by the extreme thermophile, *Caldicellulosiruptor saccharolyticus.*" *Biotechnol Bioeng.* 81:255–262.
10. Steele, J. H. 1962. "Environmental control of photosynthesis in the sea." *Limnol Oceanogr.* 7:137.
11. Lee, J. M. 1992. *Biochemical Engineering.* Prentice Hall: Upper Saddle River, NJ.
12. Welch, E. B. 1980. *Ecological Effects of Wastewater.* Cambridge University Press: Cambridge, Great Britain.
13. Yu, X. 2007. "Biohydrogen production by the hyperthermophilic bacterium *Thermotoga neapolitana.*" PhD dissertation, Clemson University, Clemson, SC.
14. Kastner, J. R., et al. 1992. "Viability of *Candida shehatae* in D-xylose fermentations with added ethanol." *Biotechnol Bioeng.* 40:1282.
15. Du Preez, J. C., M. Bosch, and B. A. Prior. 1987. "Temperature profiles of growth and ethanol tolerance of the xylose-fermenting yeast *Candida shehatae* and *Pichia stipitis.*" *Appl Microbiol Biotechnol.* 25:521.
16. Doyle, A., J. B. Griffiths, and D. G. Newell (eds.). 1995. *Cell & Tissue Culture: Laboratory Procedures.* John Wiley & Sons: New York, NY.
17. Shuler, M. L., and F. Kargi. 1992. *Bioprocess Engineering.* Prentice Hall PTR: Englewood Cliffs, NJ.
18. Felipe, M. G. A., et al. 1997. "Environmental parameters affecting xylitol production form sugarcane bagasse hemicellulosic hydrolysate by *Candida guilliermondii.*" *J Indust Microbiol & Biotech.* 18:251.
19. Pfeifer, M. J., et al. 1996. "Effect of culture conditions on xylitol production by *Candida guilliermondii* FTI 20037." *Appl Biochem Biotechnol.* 57:423.
20. Oh, D.-K., and S.-K. Kim. 1998. "Increase of xylitol yield by feeding xylose and glucose in *Candida tropicalis.*" *Appl Microbiol Biotechnol.* 50:419.
21. Pessoa, A., I. M. Mancilha, and S. Sato. 1996. "Cultivation of *Candida tropicalis* in sugarcane hemicellulosic hydrolyzate for microbial protein production." *J Biotechnol.* 51:83.
22. De Vrije, T., et al. 2002. "Pretreatment of *Miscanthus* for hydrogen production by *Thermotoga elfii.*" *Int J Hydrogen Energy.* 27:1381–1390.
23. Schroder, C., M. Selig, and P. Schonheit. 1994. "Glucose fermentation to acetate, $CO_2$ and $H_2$ in the anaerobic hyperthermophilic eubacterium *Thermotoga maritima*—Involvement of the Embden-Meyerhof pathway." *Arch Microbiol.* 161:460–470.
24. Gaden, E. L. 1959. "Fermentation process kinetics." *J Biochemical Microbiol Technol Eng.* 1:413.

25. Van Dam-Mieras, M. C. E., et al. 1992. *Bioreactor Design and Product Yield.* Butterworth-Heinemann: Boston, MA.
26. Zhu, H., C. M. Drapcho, and T. H. Walker. 2001. "Bioconversion of rice bran to omega-3 fatty acids of *Pythium irregulare* submerged culture." *ASAE* Paper 01-7020.
27. Simkins, S., and M. Alexander. 1985. "Nonlinear estimation of the parameters of Monod kinetics that best describe mineralization of several substrate concentrations by dissimilar bacterial densities." *Appl Environ Microbiol.* 50:816.
28. Zhang, C., et al. 1999. "Aerobic biodegradation kinetics of four anionic and nonionic surfactants at sub- and supra-critical micelle concentrations (CMCs)." *Water Resources.* 33:115.
29. Reeves, E. G. M. 2004. "Kinetic analysis of *Kluyveromyces marxianus* yeast strain." MS Thesis, Louisiana State University, Baton Rouge, LA.
30. Tchobanoglous, G., and F. Burton. 1991. *Wastewater Engineering: Treatment, Disposal, and Reuse.* Metcalf & Eddy, Inc. 3d ed. McGraw-Hill Publishing Co.: New York, NY.
31. Deshpande, B. 2006. *Simulator for Microbial Growth Kinetics.* MS Thesis, Clemson University, Clemson, SC.

# PART 2
# Biofuels

**CHAPTER 4**
Biofuel Feedstocks

**CHAPTER 5**
Ethanol Production

**CHAPTER 6**
Biodiesel

**CHAPTER 7**
Biological Production of Hydrogen

**CHAPTER 8**
Microbial Fuel Cells

**CHAPTER 9**
Methane

# CHAPTER 4
## Biofuel Feedstocks

There are many forms of feedstocks for biofuel production. Animal manures and municipal solid wastes have been used to generate methane for on-farm and municipality energy uses. Fuel ethanol has been produced commercially using plant-derived starch and sugar feedstocks. Technologies for production of ethanol from lignocellulosic biomass and biodiesel from seed and algal oils have been developed and tested at various scales up to semicommercial demonstration plants. This chapter discusses the characteristics of feedstocks relevant to biofuel production and their availability in the United States and other regions of the world where data is available. Conversion technologies of these feedstocks for biofuel production will be discussed in the following chapters.

### 4.1 Starch Feedstocks

#### 4.1.1 Cereal Grains

Cereal grains are used mostly for food and feed. However, because of their high starch contents they are also good feedstocks for conversion to biofuels and other bio-based products. Ethanol is the only biofuel that has been produced commercially from these feedstocks in large quantities. Starch contents and theoretical ethanol yield of cereal grains which are relevant to ethanol production are summarized in Table 4.1.

Starch is a polymer of glucose and consists of two main structural components, amylose and amylopectin. *Amylose* is essentially a linear polymer in which the glucose residues are connected by the α-1,4 linkages (Fig. 4.1). The other main component of starch is *amylopectin*, which is a larger and branched molecule with both α-1,4 and α-1,6 linkages (Fig. 4.2). Most common starch contains about 25 percent amylose and 75 percent amylopectin. Starch in waxy grain varieties contains amylopectin almost exclusively (greater than 98 percent). Starch can be hydrolyzed by enzymes to produce the monomeric sugar glucose, which is readily metabolized by the yeast *Saccharomyces cerevisiae* to produce ethanol at high yields. Starch processing is a mature industry, and commercial enzymes required for starch

|  | | Theoretical ethanol yield* | |
| --- | --- | --- | --- |
| Feed grain | Starch (%) | (L/kg) | (gal/MT†) |
| Corn | 72 (65–76) | 0.52 | 125 |
| Wheat | 77 (66–82) | 0.55 | 132 |
| Barley | 57 (55–74) | 0.41 | 98 |
| Sorghum | 72 (68–80) | 0.52 | 125 |
| Oat | 58 (45–69) | 0.42 | 101 |
| Rice | 79 (74–85) | 0.57 | 137 |

*Theoretical yield of ethanol is 0.51 kg ethanol/kg glucose. Upon hydrolysis, 1 kg starch produces 1.11 kg glucose.
†MT: metric ton.
Source: Adapted from Dairy Research and Technology Centre (http://www.afns.ualberta.ca/Hosted/DRTC/Articles/Barley_Dairy.asp) and Juliano, 1993.

TABLE 4.1  Starch and Theoretical Ethanol Yield of Relevant Cereal Grains (Dry Basis)

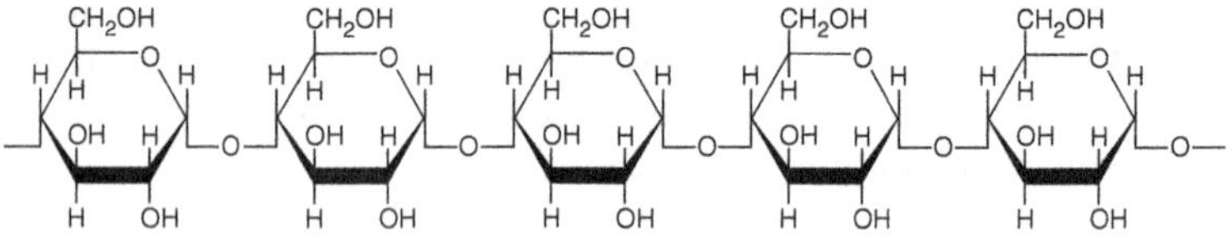

FIGURE 4.1  Amylose.

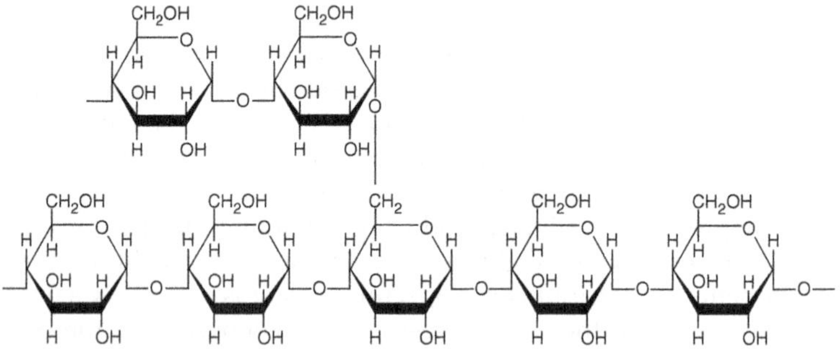

FIGURE 4.2  Amylopectin.

hydrolysis are available at low costs. Hydrolysis of starch in commercial practice will be discussed in the next chapter where ethanol production from grain feedstocks is considered.

In addition to starch cereal grains also contain other components, which can be recovered into value-added coproducts in the ethanol fermentation process.[3] Sales of these high-value coproducts play an important role in determining the process economics of fuel ethanol production.

### 4.1.1.1 Corn

Corn is the main feedstock for ethanol production in the United States and Canada. More than 95 percent of the ethanol currently produced in the United States comes from corn. In Canada, this figure is about 85 percent.[4] The United States is the largest producer of corn in the world. In 2005, 280 million MT of corn was produced in the United States. China is the second largest corn producer. In 2005, China produced 131 million MT of corn.[5] One-fifth of this amount was used for industrial applications and ethanol production accounted for 40 percent of total industrial corn use. The corn demand for ethanol production in China increased so fast that in December 2006 the government ordered suspension of all new development of the corn-based ethanol industry to avert further increase in corn price, which jumped 5 percent in the previous month.[6]

The total corn production and the corn usage for ethanol production in the United States from 1992 to 2007 are shown in Fig. 4.3. The corn farm price for the same period is shown in Fig. 4.4. Figure 4.5 compares the percentages of corn usage for ethanol production with respect to

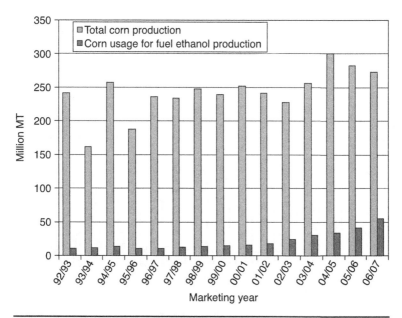

**FIGURE 4.3** Total corn production and corn usage for ethanol production in the United States. (*Adapted from USDA Feed Outlook, 2007.*)

*Note*: Marketing year for corn begins in September and ends in August the following year. Data in bushel is converted to metric ton assuming 56 lb/bu. The latest data is projected.

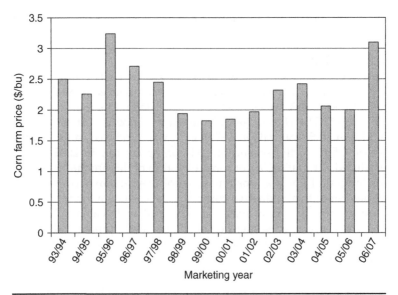

**FIGURE 4.4** Corn farm price in the United States from 1993 to 2007. (*Adapted from USDA Feed Outlook, 2007.*)

*Note*: The latest data is projected.

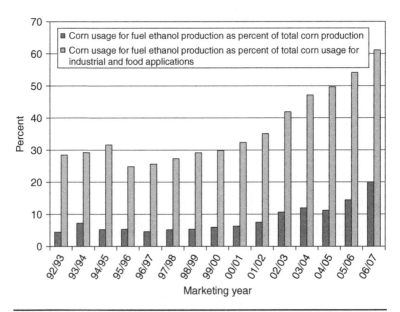

**FIGURE 4.5** Corn usage for ethanol production as percent of total corn production and total corn usage for industrial applications in the United States. (*Adapted from USDA Feed Outlook, 2007.*)

*Note*: The latest data is projected.

total corn production and total corn usage for all industrial and food applications (fuel ethanol, high fructose corn syrup (HFCS), starch, glucose and dextrose, alcoholic beverages, cereals, and other minor applications). The data clearly shows an increasing trend in corn usage for ethanol production, especially after the marketing year of 2000–2001. This increase did not seem to have an impact on total corn production and corn farm price. However, the corn farm price jumped to $3.12/bu in November 2006, which was 76 percent higher than that of the same period of the previous year ($1.77/bu). The U.S. Department of Agriculture (USDA) also predicted an increase in corn farm price for the marketing year of 2006–2007.[7] Corn production in the United States is concentrated in the "corn belt" in the Midwest. Illinois, Indiana, Iowa, Minnesota, and Nebraska are the states with largest corn production. In 2005 corn production in these five states accounted for 65 percent of the total U.S. production. Corn production by individual states in the United States is given in App. 4.1. The USDA predicted that corn production in the United States would be increased to meet future demand and could reach 400 million MT. About 25 percent of this could be used for ethanol production, which would produce 42 billion liters (11 billion gallons) fuel ethanol.[7]

In addition to starch the corn kernel also contains 10 percent protein, 4.5 percent oil, and 10 to 15 percent other materials such as fiber and ash. Protein, oil, and fiber can be recovered in the ethanol fermentation process and sold as coproducts. The rule of thumb used in the U.S. ethanol industry is that for 1 bu of corn (56 lb at 15.5 percent moisture), one can make 2.5 to 2.8 gal ethanol plus 17 lb distillers dried grains without or with solubles (DDG/S) in the dry-grind process or 1.6 lb corn oil, 2.6 lb gluten meal (60 percent protein), and 13.5 lb gluten feed (20 percent protein) in the wet-mill process.[8] DDGS is the key coproduct in the dry-grind ethanol process. Revenues brought in by selling this coproduct for animal feed help make the process economically viable. Corn oil, corn gluten meal, and corn gluten feed also are important coproducts in the wet-mill process.

### 4.1.1.2 Wheat

Wheat starch is currently used for ethanol production at only two plants in the United States. In other places, however, it is an important feedstock for production of this biofuel. In Canada, 15 percent of the ethanol produced by fermentation comes from wheat.[4] In Europe and Australia, wheat is the primary feedstock considered for expansion of the starch-based ethanol industry. The trend of wheat production in Canada and Australia from 1990 to 2004 is shown in Fig. 4.6. In the European Union (EU), the production of wheat in 2004 was totaled at 124 million MT.[5]

Currently production of ethanol from wheat in Canada is practiced on a relatively modest scale in the western part of the country. However, there have been proposals to significantly expand

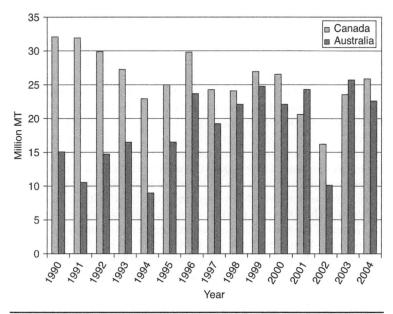

**FIGURE 4.6** Total wheat production in Canada and Australia from 1990 to 2004. (*Adapted from United Nations FAOSTAT, 2004.*)

ethanol production from wheat in Saskatchewan and Manitoba. Plans also are devised for construction of several wheat-based ethanol plants in Australia and the European Union. Wheat normally has higher protein contents than corn. Canadian wheat varieties and their protein contents are shown in Table 4.2. Wheat protein has better quality for food and feed applications than corn protein because it

| Wheat class | Typical protein level (% - dry basis) |
|---|---|
| Canadian Western Red Spring | 13.2 |
| Canadian Western Extra Strong | 12.2 |
| Canadian Prairie Spring Red | 11.5 |
| Canadian Western Red Winter | 11.3 |
| Canadian Prairie Spring White | 11.2 |
| Canadian Western Soft White | 10.5 |
| Canadian Western Amber Durum | 12.8 |

*Source*: Adapted from (S&T)² Consultants Inc., 2003.

**TABLE 4.2** Canadian Wheat Varieties and Their Protein Contents

contains the essential amino acid lysine. However, since wheat protein is insoluble in water it may cause problems in ethanol production, especially in downstream processing. Thus, only those with low protein contents such as Canadian Prairie Spring (Red and White), Canadian Western Red Winter, and Canadian Western Soft White are suitable for ethanol production. The use of low-protein wheat for ethanol production has the additional advantage of increased starch content. Coproducts of ethanol production include DDGS and gluten, which constitutes 80 to 90 percent of the wheat protein. Gluten is an important additive in bakery and breakfast foods and in processed meat and fish products. It is widely used in vegetarian cuisine, especially in East and Southeast Asia, where there are large Buddhist populations. Wheat also contains oil but the levels are low (less than 3 percent).

### 4.1.1.3 Barley

Barley is a potential alternative grain feedstock for ethanol production in the United States, especially in the Mid-Atlantic and some Midwest states, where it can be grown as a winter crop. It thus offers potential for double cropping with soybean. Barley production in the United States has declined steadily in the last 15 years, from 9.2 million MT in 1990 to 4.6 million MT in 2005.[7] However, this trend may be reversed if barley becomes a feedstock for ethanol production. If barley production is brought back to the 1990 level and all of the added barley is used for ethanol production, a theoretical ethanol yield of 1.7 billion liters (450 million gallons) can be realized. The Southeastern states potentially can produce 14.5 million MT (32 billion lb) of barley to be used solely for ethanol production.[10] This will add 5.4 billion liters (1.4 billion gallons) more ethanol. Despite its potential for being a strong feedstock candidate, production of ethanol from barley faces several problems. Barley has much lower starch content than corn (Table 4.1), thus ethanol yield per unit mass will be significantly lower. Barley hull is abrasive and can cause damage to milling and grain handling equipment. Lastly, barley contains significant levels of $\beta$-glucan, a linear unbranched polysaccharide of $\beta$-D-glucose like cellulose, but with one $\beta$-1,3 linkage for every three or four $\beta$-1,4 linkages (Fig. 4.7).

Beta-glucan is soluble in hot water. The soluble $\beta$-glucan causes the viscosity of the fermentation mash to rise to such levels that at 30 percent total dry solids, which is the typical condition used in industrial ethanol fermentation, it is practically impossible to provide adequate mixing to the mash. The presence of $\beta$-glucan in the coproduct DDGS makes it a poor feed for monogastric animals such as poultry and swine. Considerable research efforts have been made to alleviate the problems associated with using barley for ethanol production. Researchers at the Virginia Polytechnic Institute and State University have developed hulled barley varieties with higher

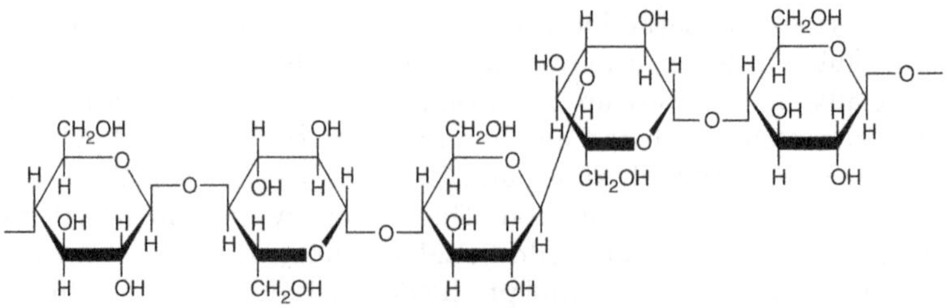

FIGURE 4.7  β-Glucan.

starch contents and hull-less barley varieties with lower fiber, no abrasiveness, and higher starch contents (approaching those of corn). Enzyme companies such as Novozymes and Danisco have developed commercial products to reduce the viscosity of the mash. These product lines include Viscozymes (Novozymes) and Optimash (Danisco). The Eastern Regional Research Center (ERRC) of the USDA's Agricultural Research Service (ARS) currently has a program with industrial collaboration to develop fermentation processes for ethanol production using both hulled and hull-less barley as feedstocks. Since the cost of ethanol production from barley in a conventional process is higher than that from corn due to lower ethanol yield, development of value-added coproducts or purchase of barley at a suitable cost relative to corn will be essential in a barley-based process. Barley protein contains lysine and hence is a valuable coproduct. Beta-glucan may be a second potential high-value coproduct due to its many health benefits that have recently been reported such as speeding of the immune response and reduction of the LDL cholesterol levels.

#### 4.1.1.4  Sorghum

Sorghum is another alternative grain feedstock for ethanol production. It has starch content similar to that of corn (Table 4.1). In 2006, 7.3 million MT of sorghum grains were produced in the United States. About 15 percent of this was used for ethanol production at eight plants in Kansas and Nebraska. Five more sorghum-based ethanol plants were proposed in these two states.[11] The total world production for 2006 was 41.2 million MT. Nigeria and India were ranked second and third behind the United States in sorghum production.[5] Sorghum and corn are interchangeable in a dry-grind ethanol plant. Several ethanol plants in the United States use both feedstocks. Ethanol yields from sorghum and from corn are equal. The composition of the DDGS coproducts in ethanol production from corn and sorghum also are very similar. In a proposed process using decortication (removal of the bran and outer layers of the grain) high-value coproducts such as phytochemicals can be recovered from the bran, and higher loadings of starch in the fermenter can be made possible.[12] One advantage of sorghum is that it is very drought resistant and heat tolerant, probably because of a waxy coating on its leaves and stems that keeps water in the plant for periods of intense heat. Thus, it is a suitable feedstock for ethanol production in arid regions such as Africa and South Asia.

### 4.1.1.5 Oat

Oat is a potential feedstock for ethanol production with total world production of 24 million MT in 2005.[5] It has not been used commercially for ethanol production because of a number of reasons. It has relatively low starch content (Table 4.1) and hull mass as high as 34 percent of the grain weight, which lowers starch loading in the fermenter. The grain also contains $\beta$-glucan and pentosans, which upon solubilization in hot water, cause significant rise in the mash viscosity. Commercial production of ethanol from oat grains can be made possible by making process modifications, which include removal of the hulls along with other suspended particles by decantation or filtration prior to starch hydrolysis to increase starch loading and addition of enzymes ($\beta$-glucanase and hemicellulase) to reduce the mash viscosity. Since the oat mash contains sufficient quantities of assimilable nitrogen, addition of urea or other nitrogen sources may not be necessary, and this may offset some of the costs of viscosity-reducing enzymes.[13] Development of high value-added coproducts will also help improve the process economics.

### 4.1.1.6 Rice

In terms of total production, rice is the third most important grain crop in the world behind wheat and corn. The total world production in 2005 was 622 million MT.[5] Rice has several characteristics that make it a potential feedstock for fuel ethanol production. It has high starch content (Table 4.1), and the starch can be readily fermented to ethanol as demonstrated in the preparation of many alcoholic beverages in Asia. Rice hull and bran can easily be separated and removed from the grain to allow high starch loading in a fermenter. Rice bran contains several high-value nutraceuticals that can be recovered as valuable coproducts. Despite these advantages rice has not been seriously considered as a feedstock for commercial production of fuel ethanol. Although rice can be grown almost anywhere, even on very steep hillsides, its cultivation requires very large quantities of water for irrigation and its planting is a very labor-intensive process. In regions of Asia where labor cost is sufficiently low, virtually all the rice produced is used for food or exported. Even the broken grains generated in rice processing are used for human consumption. Thus, there will be no extra rice for fuel ethanol production in these regions. However, in more developed countries like the United States, damaged rice grains, which normally are wasted or used for animal feed, can be used to produce ethanol. One report estimated that wasted rice in North America can generate 170 million liters (45 million gallons) of fuel ethanol annually.[14]

### 4.1.2 Other Grains

There are other grains that can be used to make ethanol albeit in smaller quantities. One example is pearl millet. The total world production of pearl millet in 2005 was 15 million MT, which accounted for about 50 percent of total millet production. Pearl millet is planted on about 1.5 million acres in the United States. It can grow in arid conditions with very low rainfall (less than 300 mm) and can survive in areas where corn and sorghum suffer even total crop failure.[15] Pearl millet contains about 70 percent starch, which gives it a theoretical ethanol yield of 0.43 L/kg. This grain is a good feedstock for ethanol production in areas where it is too hot or too dry for other grains to be cultivated.

### 4.1.3 Tubers and Roots

Root and tuber crops can be grown in the humid and subhumid tropics, which may not be suitable for cereal grains. Because of their high contents of starch they also are potential feedstocks for ethanol production. The two crops that have been given much attention are cassava and sweet potato.

#### 4.1.3.1 Cassava

Cassava, which is also known as manioc, sagu, yucca, and tapioca, is one of the most important tropical root crops. It can survive drought extremely well and is particularly suitable for regions with low nutrient availability. It also does not require a lot of attention and hence is suitable for regions with a small labor force such as Africa. Africa and Asia lead the world in cassava production. Cassava production in Africa in 2005 was 118 million MT and accounted for 56 percent of the total world production, whereas Asia contributed 57 million MT or 27 percent of the total world production. Nigeria is the world's largest cassava producer. In 2005 41.6 million MT of cassava was produced in this country. This was more than triple the production 20 years earlier (12.1 million MT in 1985).[5] Cassava used to be considered as a famine-reserve crop in Africa. It has recently emerged as the primary starch-based feedstock for future fuel ethanol production in Africa and Asia. Cassava typically contains 22 percent starch and has 70 percent moisture.[16] Thus, on a dry basis it contains about 73 percent starch, which gives it a theoretical ethanol yield of 0.45 L/kg. Cassava starch can be easily hydrolyzed to sugars for production of fermentation-based products. In China cassava is considered as the alternate starch-based feedstock to replace corn in ethanol production. The Guangxi Zhuangzu autonomous region in southern China planned to complete construction of a plant to produce 139 million liters ethanol in 2007 and expand the capacity to 1.27 billion liters in 2010 using cassava as the primary feedstock.[17] There also are proposals to build several ethanol plants using cassava in other countries, for example, Nigeria and Thailand. The government in Thailand recently approved a plan to construct 12 cassava ethanol plants with a total capacity of 1.24 billion liters/yr. These plants are expected to be in full operation in 2007 and 2008 and will use 35 percent of the total national cassava production.[18]

#### 4.1.3.2 Sweet Potato

The carbohydrate content of sweet potato normally is about 80 to 90 percent on a dry basis. Most of the carbohydrate is starch and the rest consists of sugars such as glucose, fructose, and sucrose,[19] which are all fermentable by yeast to produce ethanol. Sweet potato has been considered for ethanol production in the United States and other parts of the world. There were plans to construct several ethanol plants using sweet potato in North Carolina. An ethanol plant using sweet potato feedstock with a projected 2007 capacity of 127 million liters is under construction in Hebei in China.[18] Production of ethanol from sweet potato in Thailand is also reported.

## 4.2 Sugar Feedstocks

### 4.2.1 Sugarcane

Sugarcane is the primary feedstock for ethanol production in Brazil, the world's largest biofuel ethanol producer. In 1970 only 20 percent of total sugarcane produced in Brazil was used for ethanol production while the rest was used to produce sugar. Production of ethanol in Brazil started to increase in the late 1970s and early 1980s. It is projected that 53 percent of the total sugar produced in Brazil in the 2005–2006 season will be used for ethanol production.[4] Sugarcane production in Brazil in the last 25 years is shown in Fig. 4.8.

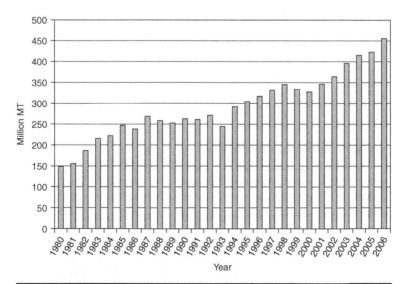

**FIGURE 4.8** Sugarcane production in Brazil from 1980 to 2006. (*Adapted from United Nations FAOSTAT, 2006.*)

Typically sugar cane contains 12 to 17 percent total sugars on a wet weight basis, with 68 to 72 percent moisture. The sugars comprise about 90 percent sucrose and 10 percent glucose plus fructose. All these three sugars are readily fermented by the yeast *Saccharomyces cerevisiae* to produce ethanol. The typical efficiency of extraction of the juice by crushing is very high, about 95 percent. The solid residues are called *bagasse*, which normally are burned to satisfy part of the energy requirements in the ethanol plant. An important by-product of sugarcane processing is molasses, which contains up to 65 percent wet weight sugars. Molasses also can be used for ethanol production after adjustment of the sugar concentrations. Removal of suspended solids prior to fermentation may also be needed. Both sugarcane juice and molasses normally have sufficient nutrients to support ethanol fermentation.[4] In addition to Brazil, ethanol is produced from sugarcane in Australia, Thailand, Vietnam, India, Cuba, El Salvador, Guatemala, Honduras, Nicaragua, Costa Rica, Peru, Colombia, Ethiopia, South Africa, and Zimbabwe.[20,21]

### 4.2.2 Sugar Beet

In addition to wheat, sugar beet is an important potential feedstock for ethanol production in the EU. In 2004, the EU members produced 181 million MT of sugar beet; only 1 million tons, or 0.6 percent, were used for ethanol production. However, the use of sugar beet as feedstock for ethanol production in the EU is expected to increase significantly. Due to a recent reform of the European Sugar Market Organization, as of July 2006, sugar beet production would qualify for both set-aside payments when grown as a nonfood crop and for energy crop aid of €45/ha on nonset-aside area. Furthermore, sugar used for fuel ethanol production would be excluded from sugar production quota. Sugar beet normally contains 16 to 18 percent sugar, which is slightly higher than sugarcane. It is estimated that in the EU ethanol can be produced from sugar beet with a yield of 86 L/MT of feedstock. The potential ethanol yields in the EU vary widely from one region to another. They are estimated to range from a low of 2964 L/ha in Lithuania to a high of 7980 L/ha in France.[22]

## 4.3 Lignocellulosic Feedstocks

The three main components of lignocellulosic biomass are cellulose, hemicellulose, and lignin. Cellulose and hemicellulose can be hydrolyzed with chemicals and/or enzymes to monomeric sugars, which can subsequently be converted biologically to biofuels. All three components also can be converted to synthesis gas or syngas by the gasification process. Syngas then can be converted to ethanol either biologically or catalytically. The first option, that is, hydrolysis of the two carbohydrate fractions followed by fermentation, has been investigated much more extensively and hence has much better chance of reaching commercialization first. The three main sources of lignocellulosic biomass are forest products and residues, agricultural residues, and dedicated energy crops. The availability of lignocellulosic feedstocks for biofuel production in the United States and other regions of the world are discussed in this section.

### 4.3.1 Forest Products and Residues

In a recent report[23] the currently available quantities of dry biomass residues from logging and other operations such as land clearing in the United States are estimated to be 37 million MT/yr. Fuel treatments to reduce fire hazard can contribute 54 million MT of biomass per year. Most of the wood residues generated in the forest products industry processing mills are recovered or burned. Out of 146 million MT generated annually, only 7 million MT are available for conversion to biofuels or other products. Pulp and paper mills generate 47 million MT black liquors per year, but all of these are burned for thermal energy generation. Even with all of the black liquors burned, the mills still cannot meet all their energy needs; they have to supplement with fossil and wood fuel burners. Therefore, there are essentially no excess black liquors available for biofuel production. Urban wood residues such as tree trimmings and construction and demolition debris are an important source of biomass. Of 57 million MT generated annually, 25 million MT are available for conversion to biofuels; the rest is burned for energy or is unrecoverable. In summary, a total amount of 123 million MT of forest-derived biomass residues is available for biofuel production. The available forest-derived biomass resources are not distributed evenly but rather concentrated in certain areas. Most of the available forest residues are concentrated in the Southeast, Northeast, Great Lakes, and Northwest regions, whereas the available urban wood residues are concentrated in states with large cities (App. 4.2).

Besides forest residues, wood could also be utilized for biofuels. Direct use of wood for production of biofuels has just begun to be developed but is expected to reach commercialization quickly as the infrastructure is already in place for the pulp and paper industry. The quantities of forest-derived biomass, which includes forest residues and surplus forest growth, that can be recovered in an economical way and still support ecological diversity (i.e., available for biofuel production), in various regions of the world[24] are projected for 2050 and are shown in Table 4.3.

In the United States it has been estimated that 334 million MT forest-derived biomass would potentially be available annually for biofuel production.[23] This figure, however, includes 67 million MT black liquors, which are currently burned for thermal energy and hence are not likely to be available for biofuel production. The actual amount of forest-derived biomass that would potentially be available annually in the United States therefore should be only 267 million MT.

| World region | Recoverable quantities (million dry MT/yr)* |
|---|---|
| North America | 595 |
| Oceania | 60 |
| Japan | 75 |
| Western Europe | 255 |
| Eastern Europe | 55 |
| CIS and Baltic States | 75 |
| Sub-Saharan Africa | 65 |
| Caribbean and Latin America | 320 |
| Middle East and North Africa | 20 |
| East Asia | 230 |
| South Asia | 40 |
| World total | 1780 |

*Data converted to dry MT using the conversion factor of 20 GJ/dry ton.
Source: Adapted from Smeets and Faaij, 2006.

**TABLE 4.3** Quantities of Economically and Ecologically Recoverable Forest-Derived Biomass Projected for 2050 in Various Regions of the World

## 4.3.2 Agricultural Residues

The currently available agricultural biomass residues in the United States are summarized in Table 4.4. The carbohydrate composition and theoretical ethanol yields of these potential feedstocks are shown in Table 4.5. These are the quantities that can be collected for biofuel production under current crop yields, sustainable tillage practices (20 to 40 percent for major crops), and residue collection technology (40 percent recovery). Small grain residues include sorghum, barley, oat, and rice straw. Other crop residues include cotton, other oil seeds, tobacco, sugar crops, potatoes, beans, miscellaneous root crops, and double crops (App. 4.3). At 75 million MT/yr corn stover is the largest available agricultural biomass resource that has not been used. To illustrate the potential of corn stover let us consider how much ethanol can be produced from this biomass resource. Corn stover typically contains 36 weight percent glucan, 19 weight percent xylan, 2.3 weight percent arabinan, 0.9 weight percent galactan, and 0.5 weight percent mannan;* thus the theoretical ethanol yield will be 112 gal (426 L)/dry

---

*These are averages calculated from data in Biomass Feedstock Composition and Property Database.

| Biomass | Available quantities (million dry MT/yr) |
|---|---|
| Corn stover | 75 |
| Wheat straw | 11 |
| Small grain residues | 6 |
| Other crop residues | 21 |
| Total | 113 |

Source: Adapted from Perlack et al., 2005.

**TABLE 4.4** Current Sustainable Availability of Agricultural Residues in the United States

| Residue | Cellulose (%) | Hemicellulose (%) | Theoretical ethanol yield* | |
|---|---|---|---|---|
| | | | (L/kg dry) | (gal/MT dry) |
| Barley straw | 38.9 | 20.9 | 0.43 | 104 |
| Corn stover | 36.9 | 21.3 | 0.43 | 101 |
| Oat straw | 38.2 | 30.0 | 0.50 | 119 |
| Rice straw | 38.6 | 19.7 | 0.42 | 101 |
| Sorghum straw | 34.0 | 20.0 | 0.39 | 94 |
| Wheat straw | 39.6 | 26.4 | 0.48 | 115 |
| Bagasse | 39.1 | 22.5 | 0.45 | 107 |

*It is assumed that the hemicellulose fractions are all polymers of xylose.
Source: Adapted from Biomass Feedstock Composition and Property Database; Thiago and Kellaway, 1982; Lequart et al., 1999; Linde et al., 2006; Herrera et al., 2004; and Zhu et al., 2005.

**TABLE 4.5** Chemical Composition of Agricultural Biomass Residues

MT of corn stover.[†] The total theoretical production of ethanol from corn stover therefore will be 8.4 billion gallons (31.8 billion liters)/yr, which is double the record production of slightly less than four billion gallons (15 billion liters) of ethanol in 2005. Because of the huge potential of corn stover for biofuel production, the majority of research and development efforts on lignocellulosic biomass conversion have been focused on utilization of this untapped agricultural resource.

---

[†]Theoretical ethanol yield is 0.51 kg ethanol/kg sugar. One kilogram polymer of C-6 sugars produces 1.11 kg monomeric sugars and 1 kg polymer of C-5 sugars produces 1.14 kg monomeric sugars.

Projections are made for future increase in agricultural crop yields and thus increase in agricultural residues available for biofuel production. Under the most favorable conditions it is estimated that by the mid-twenty-first century, 345 million dry MT/yr agricultural biomass residues will be available in the United States. Corn stover still dominates with 232 million dry MT/yr.

The quantities of agricultural residues of the main crops available for biofuel production in various regions of the world and their theoretical ethanol production are summarized in Tables 4.6 and 4.7, respectively. It was reported that in Africa for several crops—namely corn, barley, oat, and sorghum—because of the low yield, all of the residues have to be left on the field for soil protection and hence are not available for biofuel production.[14]

### 4.3.3 Agricultural Processing By-Products

In postharvest processing of crops large quantities of by-products are generated. The carbohydrates contained in these by-products can be converted to ethanol. Table 4.8 summarizes the carbohydrate composition of several crop processing by-products and their theoretical ethanol yields.

Corn fiber is a by-product of the wet-milling process whereas DDGS is a by-product of the dry-grind process. Both are currently sold for use in animal feed. However, with the current rate of increase in ethanol production in the United States, the feed market may soon be saturated and conversion of the excess corn fiber and DDGS to additional ethanol will be of great interest. Another advantage is that they are already available on site. Rice hull has great potential as a feedstock for ethanol production because it is available in very large quantities. Typically rice hull accounts for about 20 percent of the grain weight.[2] In 2005 rice production in the world was 153 million MT, of which Asia contributed 137 million MT.[5] The rice hulls available in Asia have a theoretical ethanol production of 49.7 billion liters (13.1 billion gallons). Since in Asia rice hulls are only used for animal feed and nonfood purposes, their use for ethanol production will not raise a moral issue in this region of the world where poverty and hunger are still widespread.

### 4.3.4 Dedicated Energy Crops

In addition to forest and agricultural biomass residues, dedicated energy crops are another lignocellulosic biomass source for biofuel production. Among a rather large number of herbaceous and woody crops that can be used for biofuel production, switchgrass (*Panicum virgatum*), poplar (*Populus* spp.), and willow (*Salix* spp.) have been the focus of most research efforts. The carbohydrate composition of these energy crops and their theoretical ethanol production yields are summarized in Table 4.9.

| Residue (million MT) | Africa | Asia | Europe | North America | Central America | South America | Oceania |
|---|---|---|---|---|---|---|---|
| Corn stover | Nil | 33.90 | 28.61 | 133.66 | Nil | 7.20 | 0.24 |
| Barley straw | Nil | 1.97 | 44.24 | 9.85 | 0.16 | 0.29 | 1.93 |
| Oat straw | Nil | 0.27 | 6.83 | 2.80 | 0.03 | 0.21 | 0.47 |
| Rice straw | 20.93 | 667.59 | 3.92 | 10.95 | 2.77 | 23.51 | 1.68 |
| Wheat straw | 5.34 | 145.20 | 132.59 | 50.05 | 2.79 | 9.80 | 8.57 |
| Sorghum straw | Nil | Nil | 0.35 | 6.97 | 1.16 | 1.52 | 0.32 |
| Bagasse | 11.73 | 74.88 | 0.01 | 4.62 | 19.23 | 63.77 | 6.49 |

Source: Adapted from Kim and Dale, 2003.

TABLE 4.6 Worldwide Quantities of Agricultural Residues Available for Biofuel Production

| Residue | Africa | Asia | Europe | North America | Central America | South America | Oceania |
|---|---|---|---|---|---|---|---|
| Corn stover | Nil | 14.44 | 12.19 | 56.94 | Nil | 3.07 | 0.10 |
| Barley straw | Nil | 0.85 | 19.16 | 4.27 | 0.07 | 0.13 | 0.84 |
| Oat straw | Nil | 0.15 | 3.85 | 1.58 | 0.02 | 0.12 | 0.26 |
| Rice straw | 8.83 | 281.72 | 1.65 | 4.62 | 1.17 | 9.92 | 0.71 |
| Wheat straw | 2.56 | 69.55 | 63.51 | 23.97 | 1.34 | 4.69 | 4.11 |
| Sorghum straw | Nil | Nil | 0.14 | 2.73 | 0.45 | 0.60 | 0.13 |
| Bagasse | 5.24 | 33.47 | 0.16 | 2.07 | 8.60 | 28.51 | 2.90 |

*Theoretical ethanol yields calculated using quantities listed in Table 4.6 and composition listed in Table 4.5.
*Source:* Adapted from Theoretical Ethanol Yield Calculator, U.S. Department of Energy.

**TABLE 4.7** Worldwide Theoretical Ethanol Production from Agricultural Residues of the Main Crops (Billion Liters)*

| Crop processing by-products | Cellulose (%) | Hemicellulose (%) | Starch (%) | Theoretical ethanol yield* | |
|---|---|---|---|---|---|
| | | | | (L/kg dry) | (gal/MT dry) |
| Corn fiber | 16 | 40 | 18 | 0.54 | 129 |
| DDGS | 24 | 26 | | 0.36 | 87 |
| Rice hull | 30 | 20 | | 0.36 | 87 |
| Soybean hull | 46 | 18 | | 0.46 | 111 |
| Oat hulls | 30 | 34 | | 0.47 | 112 |
| Cottonseed hull | 59 | | | 0.42 | 102 |
| Barley hull† | 34 | 37 | | 0.52 | 124 |

*It is assumed that the hemicellulose fractions are all polymers of xylose.
†Kim. 2007. Agricultural Research Service, USDA, personal communication.
*Source*: Adapted from Abbas et al., 2006.

**TABLE 4.8** Carbohydrate Composition of Crop Processing By-Products and Their Theoretical Ethanol Yields

Currently, none of the aforementioned energy crops is grown for bioenergy production in the United States, although hybrid poplar is produced commercially on about 200,000 acres of land for use as a fiber source by the pulp and paper industry, and switchgrass is grown on some Conservation Reserve Program (CRP) lands as a forage crop. Among the three potential energy crops, switchgrass has the lowest production cost. Poplar and willow have higher production cost because of the higher establishment costs and the longer period of time before harvest is possible.[35] Switchgrass has great potential for biofuel production. Alamo switchgrass, which is most suited for the Southeastern part of the United States, can be produced at 5.4 to 7.2 dry MT per acre. This is quite favorable compared to hay production at an

| Energy crops | Cellulose (%) | Hemicellulose (%) | Theoretical ethanol yield* | |
|---|---|---|---|---|
| | | | (L/kg dry) | (gal/MT dry) |
| Hybrid poplar | 39.8 | 18.4 | 0.42 | 101 |
| Willow | 43.0 | 21.3 | 0.47 | 112 |
| Switchgrass | 32.2 | 24.4 | 0.41 | 99 |

*It is assumed that the hemicellulose fractions are all polymers of xylose.
*Source*: Adapted from Sassner et al., 2006 and Esteghlalian et al., 1997.

**TABLE 4.9** Carbohydrate Composition of Dedicated Energy Crops and Their Theoretical Ethanol Yields

annual average of 2.3 dry MT per acre.[36] In the projected production of switchgrass in the year 2008 two scenarios are considered: the wildlife management scenario where minimal fertilizer application is used to maximize wildlife diversity, and the production management scenario where the objective is to achieve maximum production (App. 4.4).[23] The total projected productions under the two scenarios are 55 and 171 million dry MT/yr, respectively. Therefore, under the production management scenario, the theoretical yield of ethanol from switchgrass is 19.7 billion gallons (74.6 billion liters)/yr. As in the case of traditional agricultural crops, projections are made for future increase in yields for energy crops. It is estimated that by the mid-twenty-first century, 342 million dry MT of energy crops will be available for biofuel production.[23] Since energy crops have never been produced commercially on sufficiently large scale for biofuel production, the projected figures will need to be revised when more field data becomes available. Another problem with using these crops for biofuel production is, unlike wood, which is available all year round, dedicated energy crops are seasonal feedstocks and as such require storage to ensure continuous operation. Storage of large quantities of energy crops will add significantly to the final cost of biofuel production.

## 4.4 Plant Oils and Animal Fats

Plant oils and animal fats can be converted to biodiesel. There are several processes for biodiesel production. The current process of choice uses transesterification of vegetable oils. Soybean oil is the predominant feedstock in the United States and Canada, whereas in the European Union (EU) rapeseed oil is used most widely for biodiesel production. Typically one MT of biodiesel can be produced from 2.672 MT of rapeseed[37] or 5.98 MT ton of soybean.[38]* Rapeseed gives a higher biodiesel yield than soybean simply because it produces more oil per unit mass. Rapeseed can produce 37 kg oil/kg seed compared to 14 kg oil/kg seed for soybean.[39] Production of biodiesel in the EU is much higher than that in the United States. The production of biodiesel in Germany and France, which are the two largest biodiesel producers in the EU, is compared with the production of biodiesel in the United States in Fig. 4.9.

Production of soybeans and soybean oil in the United States are shown in Fig. 4.10. Figure 4.11 shows production of rapeseed in Germany and France. With the large quantities of soybeans produced in the United States, shortage of feedstock for biodiesel production should not be a concern. In Germany, however, supply of feedstock for biodiesel production may become insufficient to meet the demand in the near future.

---

*The biodiesel yield for soybean given in Stroup, 2004 is converted to MT soybean required per MT biodiesel, assuming a weight of 60 lb/bu and a density of 0.86 kg/L for biodiesel.

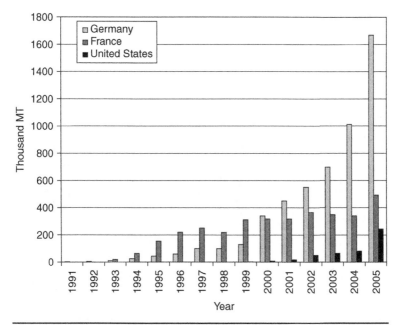

**Figure 4.9** Production of biodiesel in Germany, France, and the United States. (*Adapted from USDA Foreign Agricultural Service, European Biodiesel Board Statistics, and National Biodiesel Board Fact Sheets.*)

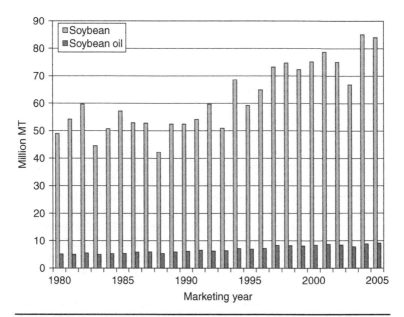

**Figure 4.10** Production of soybeans and soybean oil in the United States. (*Adapted from Ash and Dohlman, 2006.*)

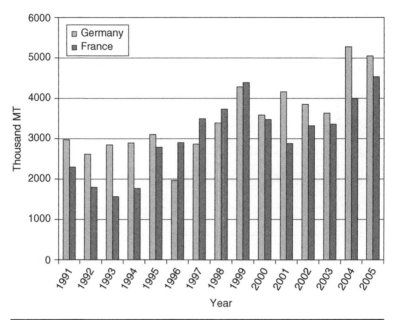

**Figure 4.11** Production of rapeseed in Germany and France. (*Adapted from USDA Foreign Agricultural Service and European Biodiesel Board Statistics.*)

Other vegetable oils besides soybean and rapeseed oils and animal fats such as beef tallow also can be used as feedstocks for biodiesel production. A feedstock suitable for biodiesel production must meet several criteria. First of all, the feedstock must be dry. Water contents as low as 1 percent can increase soap production and cause incompletion of the transesterification process. Crude soybean oil normally contains about 0.3 percent water but the water can easily be removed by flash evaporation. Free fatty acids (FFAs) of the feedstock must be less than 0.5 percent. The ASTM specification for biodiesel limits phosphorus content to 10 ppm. Thus, high phosphorus feedstock such as crude soybean oil (600 to 900 ppm in the form of phospholipids) must be pretreated to remove the phosphorus. The U.S. Environmental Protection Agency (EPA) required sulfur level in biodiesel to drop to 15 ppm in 2006. Most biodiesels already meet this requirement except some that are made from animal greases and fats. The so-called iodine value is a measurement of quality of a biodiesel feedstock. Iodine value is a crude measurement of degree of saturation. Saturation of a feedstock affects the quality of the biodiesel made from it. For example, biodiesel made from a highly saturated animal fat tends to gel more readily at low temperatures than one made from a vegetable feedstock like soybean oil.[43] On the other hand, high contents of polyunsaturated fatty acids (PFA) in some algal oils may increase polymerization of the biodiesel derived from them during combustion. The most preferred feedstocks are those that have high contents of saturated and monosaturated fatty acids and low contents of PFA.[44]

Fatty acid composition of common vegetable oils and beef tallow are shown in Table 4.10. Microalgal oils are another potential feedstock for biodiesel production. Several microalgae can accumulate oils, which are suitable for biodiesel production, up to 60 percent (dry basis) of their biomass. The major types of fatty acids produced by microalgal species collected at the National Renewable Energy Laboratory (NREL) are listed in Table 4.11.

Using microalgal oil for biodiesel production has many technically interesting features. These include (1) The carbon dioxide exhaust from coal-burning power plants can be used as carbon source for the microalgae at conversion efficiency up to 90 percent,[38] thus alleviating an environmental problem of great concern. (2) The residual biomass after oil extraction can be used for ethanol production, which is another biofuel. (3) The $CO_2$ produced in ethanol fermentation can be used as an additional source of feedstock for microalgal production. The early study performed at NREL concluded that due to the high costs of large-scale algae production, the process of producing biodiesel from cultivated microalgae in open ponds was not economical. Recently a closed-loop photobioreactor system for growing microalgae using carbon dioxide from smokestacks was developed by GreenFuel Technologies Corporation in the United States. This new development resurrected the interests in the use of microalgal oils as a feedstock for biodiesel production with ethanol, methane, and proteins as potential coproducts. Attempts to commercialize the technology are under way.

## 4.5 Miscellaneous Feedstocks

### 4.5.1 Animal Wastes

Animal wastes are a potential source of biofuel production. These wastes contain both undigested grains (starch) and undigested straw (lignocellulose). Animal wastes normally are used for fertilizers on farms. The remaining can be used for methane production by anaerobic digestion. The gas produced typically contains 60 percent methane and 40 percent carbon dioxide. Potential gas yield normally is expressed as $M^3$ per kg of volatile solids (VS) destroyed by the bacteria in the anaerobic digestion process. Volatile solids are determined as the total weight loss upon heating a sample until constant weight at 550°C. The potential biogas production from various animal wastes in the United States is shown in Table 4.12. It should be noted that the potential gas production for the same type of animal waste will be different for different regions of the world because of the different diets fed to the animals. Because of potential high transportation costs, biogas produced from animal wastes will be most economically used for on-farm energy requirements.

| Oils and fats | Fatty acid composition (% by weight) | | | | | | | | |
|---|---|---|---|---|---|---|---|---|---|
| | 14:0 | 16:0 | 18:0 | 20:0 | 22:0 | 24:0 | 18:1 | 22:1 | 18:2 | 18:3 |
| Corn | | 11.67 | 1.85 | 0.24 | 0.00 | 0.00 | 25.16 | 0.00 | 60.60 | 0.48 |
| Cotton seed | | 28.33 | 0.89 | 0.00 | 0.00 | 0.00 | 13.27 | 0.00 | 57.51 | 0.00 |
| Crambe | | 2.07 | 0.70 | 2.09 | 0.80 | 1.12 | 18.86 | 58.51 | 9.00 | 6.85 |
| Peanut | | 11.38 | 2.39 | 1.32 | 2.52 | 1.23 | 48.28 | 0.00 | 31.95 | 0.93 |
| Rapeseed | | 3.49 | 0.85 | 0.00 | 0.00 | 0.00 | 64.40 | 0.00 | 22.30 | 8.23 |
| Soybean | | 11.75 | 3.15 | 0.00 | 0.00 | 0.00 | 23.26 | 0.00 | 55.53 | 6.31 |
| Sunflower | | 6.08 | 3.26 | 0.00 | 0.00 | 0.00 | 16.93 | 0.00 | 73.73 | 0.00 |
| Beef tallow | 2–8 | 24–37 | 14–29 | | | | 40–50 | | 1–5 | |

*Source:* Adapted from Ma and Hanna, 1999.

**TABLE 4.10** Fatty Acid Composition of Common Vegetable Oils and Beef Tallow

| Strain | Under nitrogen-sufficient conditions | Under nitrogen-deficient conditions |
|---|---|---|
| Ankistrodesmus | 16:0, 16:4, **18:1**, **18:3** | 16:0, **18:1**, 18:3 |
| Botryococcus braunii | **16:0**, **18:1**, 18:2, **18:3** | 16:0, **18:1**, **18:3**, 20:5 |
| Dunaliella bardawil | Not determined | 12:0, 14:0/14:1, **16:0**, **18:1**, 18:2, 18:3 |
| Dunaliella salina | 14:0/14:1, **16:0**, 16:3, 16:4, 18:2, **18:3** | **16:0**, 16:3, 18:1, 18:2, **18:3** |
| Isochrysis sp. | 14:0/14:1, 16:0, 16:1, **18:1**, 18:3, **18:4**, 22:6 | **14:0/14:1**, **18:1**, 18:2, 18:3, 18:4, 22:6 |
| Nannochloris sp. | 14:0/14:1, 16:0, 16:1, 16:2, 16:3, 20:5 | Not determined |
| Nitzschia sp. | 14:0/14:1, 16:0, 16:1, 16:2, 16:3, 20:6 | Not determined |

*Note*: 1. Fatty acids in bold accumulate at 15 percent or higher. 2. Under nitrogen-deficient conditions, the total fatty acid accumulation in *Ankistrodemus* increases from 24.5 to 40.3 percent, *Isochrysis* from 7.1 to 26.0 percent, and *Nannochloris* from 20.8 to 35.5 percent, whereas that in *Dunaliella* decreases from 25.3 to 9.2 percent. Thus, nitrogen deficiency conditions are favorable for *Ankistrodemus* and *Isochrysis* since not just more fatty acids are produced but the quality of the oils also is improved (lower PFA accumulation).
*Source*: Adapted from Sheehan et al., 1998.

**TABLE 4.11** Major Fatty Acids of Various Microalgal Oils

| | Swine (68 kg) | Dairy (545 kg) | Poultry (1.82 kg) | Beef (454 kg) |
|---|---|---|---|---|
| Gas yield ($M^3$/kg VS destroyed) | 0.75 | 0.48 | 0.54 | 0.94 |
| VS voided (kg/day) | 0.32 | 4.3 | 0.020 | 2.27 |
| % reduction of VS | 49 | 31 | 56 | 41 |
| Potential gas production ($M^3$/animal unit-day) | 0.12 | 0.64 | 0.0059 | 0.88 |
| Energy production rate (kJ/animal-hr) | 109 | 599 | 5.54 | 818 |

*Source*: Adapted from Fulhage et al., 1993.

**TABLE 4.12** Potential Biogas Production from Animal Wastes in the United States

### 4.5.2 Municipal Solid Waste

Municipal solid waste (MSW) is another potential feedstock for biofuel production. In 2005, 223 million MT was generated in the United States. Typically MSW contains about 36 percent paper and paperboard products and 12 percent yard trimmings.[47] All of these materials, after separation from other components such as metals and plastics, can be converted to biofuels using similar processes used for conversion of lignocellulosic feedstocks.

## References

1. Dairy Research and Technology Center, University of Alberta. Barley grain for dairy cattle. Available from: http://www.afns.ualberta.ca/Hosted/DRTC/Articles/Barley_Dairy.asp.
2. Juliano, B. O. 1993. "Rice in human nutrition." United Nations Food and Agriculture Organization (FAO).
3. Coproducts for fuel ethanol production reference database. 2003. Agriculture and Agri-Food Canada. Available from: http://sci.agr.ca/publications/cfar/appen2-annex2_e.htm.
4. Wheals, A. E., et al. 1999. "Fuel ethanol after 25 years." *Trends Biotechnol.* 17: 482–487.
5. United Nations FAOSTAT.
6. Green Car Congress. China Halts Expansion of Corn Ethanol Industry; Focus on Biomass Feedstocks. December 20, 2006.
7. USDA Feed Outlook, 2007.
8. National Corn Growers Association (NCGA). World of Corn 2005.
9. (S&T)² Consultants Inc. 2003. The addition of ethanol from wheat to GHGenius.
10. Hicks, K. B. 2006. "Utilization of barley for fuel ethanol production." Presentation to the North Dakota Barley Growers Association, Bismarck, ND.
11. National Grain Sorghum Producers.
12. Corredor, D. Y., et al. 2006. *Cereal Chem.* 83(1):17–21.
13. Thomas, K. C., and W. M. Ingledew. 1990. "Fuel alcohol production: Effects of free amino nitrogen on fermentation of very-high-gravity wheat mashes." *Appl Environ Microbiol.* 56(7):2046–2050.
14. Kim, S., and B. E. Dale. 2003. "Global potential bioethanol production from wasted crops and crop residues." *Biomass Bioenerg.* 26:361–375.
15. Wu, X, et al. 2006. "Ethanol production from pearl millet using Saccharomyces cerevisiae." *Cereal Chem.* 83(2):127–131.
16. International Starch Institute. Available from: http://www.starch.dk/isi/starch/cassava.htm.
17. Latner, K., C. O'Kray, and J. Jiang. 2006. Peoples Republic of China–Bio-fuels—An alternative future for agriculture. USDA Foreign Agricultural Service (FAS) Global Agriculture Information Network (GAIN) report number CH6049.
18. Nguyen, T. L. T., S. H. Gheewala, and S. Garivait. 2007. "Full chain energy analysis of fuel ethanol from cassava in Thailand." *Environ Sci Technol.* 41: 4135–4142.
19. Diop, A. 1998. "Storage and processing of roots and tubers in the tropics." United Nations Food and Agriculture Organization (FAO).
20. Murray, D. 2005. "Ethanol's potential: Looking beyond corn." Earth Policy Institute. Available from: http://www.earth-policy.org/Updates/2005/Update49.htm.
21. Shapouri, H., and M. Salassi. 2006. "The economic feasibility of ethanol production from sugar in the United States." USDA Office of the Chief Economist (OCE).
22. Francis, M. K. 2006. "EU-25–Sugar–The economics of bioethanol production in the EU." USDA Foreign Agricultural Service (FAS) Global Agriculture Information Network (GAIN) report number E36081.
23. Perlack, R. D., et al. 2005. "Biomass Feedstock for a Bioenergy and Bioproducts Industry: The Technical Feasibility of a Billion-ton Annual Supply," a joint study sponsored by the U.S. Department of Energy and U.S. Department of Agriculture.

24. Smeets, E. M. W., and A. P. C. Faaij. 2006. "Bioenergy potentials from forestry in 2050–An assessment of the drivers that determine the potentials." *Clim Change.* In press.
25. Biomass Feedstock Composition and Property Database, U.S. Department of Energy. Available from: http://www1.eere.energy.gov/biomass/feedstock_databases.html.
26. Thiago, L. R. L. D. S., and R. C. Kellaway. 1982. "Botanical composition and extent of lignification affecting digestibility of wheat and oat straw and paspalum hay." *Anim Feed Sci Technol.* 7:71–81.
27. Lequart, C., et al.1999. "Hydrolysis of wheat bran and straw by an endoxylanase: Production and structural characterization of cinnamoyl-oligosaccharides." *Carbohydr Res.* 319:102–111.
28. Linde, M., M. Galbe, and G. Zacchi. 2006. "Steam pretreatment of acid-sprayed and acid-soaked barley straw for production of ethanol." *Appl Biochem Biotechnol.* 129–132, 546–562.
29. Herrera, A., et al. 2004. "Effect of the hydrochloric acid concentration on the hydrolysis of sorghum straw at atmospheric pressure." *J Food Eng.* 63: 103–109.
30. Zhu, S., et al.2005. "Comparison of three microwave/chemical pretreatment processes for enzymatic hydrolysis of rice straw." *Biosyst Eng.* 93:279–283.
31. Theoretical Ethanol Yield Calculator, U.S. Department of Energy. Available from:http://www1.eere.energy.gov/biomass/ethanol_yield_calculator.html.
32. Abbas, C., et al. 2006. "Process for the production of animal feed and ethanol and novel feed." Patent application WO 2006/113683 A2.
33. Sassner, P., M. Galbe, and G. Zacchi. 2006. "Bioethanol production based on simultaneous saccharification and fermentation of steam-pretreated *Salix* at high dry-matter content." *Enzyme Microb Technol.* 39:756–762.
34. Esteghlalian, A., et al. 1997. "Modeling and optimization of the dilute-sulfuric-acid pretreatment of corn stover, poplar and switchgrass." *Bioresour Technol.* 59:129–136.
35. De La Torre Ugarte, D. G., et al. 2003. "The Economic Impacts of Bioenergy Crop Production on U.S. Agriculture." U.S. Department of Agriculture Agricultural Economic report no. 816.
36. Bransby, D. "Switchgrass Profie." Available from: http://bioenergy.ornl.gov/papers/misc/switchgrass-profile.html.
37. "EU: Biodiesel industry expanding use of oilseeds." The USDA Foreign Agricultural Service (FAS). Updated September 5, 2003.
38. Stroup, R. L. 2004. "Feedstock considers for future U.S. producers." *Biodiesel Mag.* Jan/Feb 2004 issue.
39. Biodiesel. Wikipedia, the free encyclopedia.
40. European Biodiesel Board (EBB) Statistics. Available from: http://www.ebb-eu.org/stats.php.
41. The National Biodiesel Board Fact Sheets. Available from: http://www.biodiesel.org/resources/fuelfactsheets/default.shtm.
42. Ash, M., and E. Dohlman. 2006. "Oil crops situation and outlook yearbook 2006." The USDA Economic Research Service (ERS).
43. Van Gerpen, J., et al. 2004. "Biodiesel Production Technology." National Renewable Energy Laboratory report number NREL/SR-510-36244.
44. Sheehan, J., et al. 1998. "A look back at the U.S. Department of Energy's aquatic species program–biodiesel from algae." National Renewable Energy Laboratory, U.S. Department of Energy.
45. Ma, F., and M. A. Hanna. 1999. "Biodiesel production: A review." *Bioresour Technol.* 70:1–15.
46. Fulhage, C., D. Sievers, and J. R. Fischer. 1993. "Generating methane gas from manure." Department of Agricultural Engineering, University of Missouri. Available from: http://muextension.missouri.edu/explore/agguides/agengin/g01881.htm.
47. "Municipal Solid Waste in the United States: 2005 Facts and Figures." 2001. Office of Solid Waste and Emergency Response, U.S. Environmental Protection Agency (EPA) report number EPA530-R-06-011.

| State | Area harvested (1000 acres) | Yield (bushels/acre) | Production (1000 bushels) |
|---|---|---|---|
| Alabama | 200 | 119 | 23,800 |
| Arizona | 22 | 195 | 4,290 |
| Arkansas | 230 | 131 | 30,130 |
| California | 110 | 172 | 18,920 |
| Colorado | 950 | 148 | 140,600 |
| Delaware | 154 | 143 | 22,022 |
| Florida | 28 | 94 | 2,632 |
| Georgia | 230 | 129 | 29,670 |
| Idaho | 60 | 170 | 10,200 |
| Illinois | 11,950 | 143 | 1,708,850 |
| Indiana | 5,770 | 154 | 888,580 |
| Iowa | 12,500 | 173 | 2,162,500 |
| Kansas | 3,450 | 135 | 465,750 |
| Kentucky | 1,180 | 132 | 155,760 |
| Louisiana | 330 | 136 | 44,880 |
| Maryland | 400 | 135 | 54,000 |
| Michigan | 2,020 | 143 | 288,860 |
| Minnesota | 6,850 | 174 | 1,191,900 |
| Mississippi | 365 | 129 | 47,085 |
| Missouri | 2,970 | 111 | 329,670 |
| Montana | 17 | 148 | 2,516 |
| Nebraska | 8,250 | 154 | 1,270,500 |
| New Jersey | 62 | 122 | 7,564 |
| New Mexico | 55 | 175 | 9,625 |
| New York | 460 | 124 | 57,040 |
| North Carolina | 700 | 120 | 84,000 |
| North Dakota | 1,200 | 129 | 154,800 |
| Ohio | 3,250 | 143 | 464,750 |
| Oklahoma | 250 | 115 | 28,750 |

*Source*: USDA National Agricultural Statistics Service (NASS) Crop Production 2005 Summary.

**APPENDIX 4.1** Corn Production in the United States in 2005

| State | Area harvested (1000 acres) | Yield (bushels/acre) | Production (1000 bushels) |
|---|---|---|---|
| Oregon | 25 | 160 | 4,000 |
| Pennsylvania | 960 | 122 | 117,120 |
| South Carolina | 285 | 116 | 33,060 |
| South Dakota | 3,950 | 119 | 470,050 |
| Tennessee | 595 | 130 | 77,350 |
| Texas | 1,850 | 114 | 210,900 |
| Utah | 12 | 163 | 1,956 |
| Virginia | 360 | 118 | 42,480 |
| Washington | 80 | 205 | 16,400 |
| West Virginia | 28 | 109 | 3,052 |
| Wisconsin | 2,900 | 148 | 429,200 |
| Wyoming | 49 | 140 | 6,860 |
| United States | 75,107 | 148 | 11,112,072 |

**APPENDIX 4.1** (*Continued*)

| State | Forest residues (thousand dry tons) | Urban wood residues (thousand dry tons) |
|---|---|---|
| Alabama | 2,318 | 438 |
| Alaska | 670 | 59 |
| Arizona | 54 | 477 |
| Arkansas | 2,607 | 285 |
| California | 1,182 | 3,539 |
| Colorado | 64 | 409 |
| Connecticut | 71 | 341 |
| Delaware | 46 | 77 |
| District of Colombia | 0 | 51 |
| Florida | 1,613 | 1,522 |
| Georgia | 3,226 | 838 |
| Hawaii | 0 | 121 |
| Idaho | 792 | 117 |
| Illinois | 602 | 1,213 |
| Indiana | 783 | 649 |
| Iowa | 326 | 290 |
| Kansas | 122 | 301 |
| Kentucky | 1,864 | 412 |
| Louisiana | 3,070 | 430 |
| Maine | 2,622 | 121 |
| Maryland | 239 | 566 |
| Massachusetts | 81 | 623 |
| Michigan | 1,157 | 1,085 |
| Minnesota | 2,034 | 450 |
| Mississippi | 3,470 | 279 |
| Missouri | 1,669 | 556 |
| Montana | 639 | 96 |
| Nebraska | 65 | 171 |
| Nevada | 5 | 210 |

*Source*: Milbrandt, A. 2005. "A Geographic Perspective on the Current Biomass Resource Availability in the United States," Technical Report NREL/TP-560-39181, National Renewable Energy Laboratory.

**APPENDIX 4.2** Distribution of Forest and Urban Wood Residues in the United States

| State | Forest residues (thousand dry tons) | Urban wood residues (thousand dry tons) |
|---|---|---|
| New Hampshire | 894 | 114 |
| New Jersey | 26 | 811 |
| New Mexico | 64 | 173 |
| New York | 1,008 | 1,852 |
| North Carolina | 2,717 | 756 |
| North Dakota | 24 | 61 |
| Ohio | 722 | 1,153 |
| Oklahoma | 594 | 342 |
| Oregon | 944 | 347 |
| Pennsylvania | 1,523 | 1,123 |
| Rhode Island | 7 | 99 |
| South Carolina | 1,572 | 424 |
| South Dakota | 113 | 68 |
| Tennessee | 1,197 | 557 |
| Texas | 1,869 | 2,093 |
| Utah | 27 | 207 |
| Vermont | 450 | 59 |
| Virginia | 2,180 | 738 |
| Washington | 938 | 612 |
| West Virginia | 1,222 | 167 |
| Wisconsin | 1,824 | 497 |
| Wyoming | 53 | 54 |

**APPENDIX 4.2** (*Continued*)

| State | Total agricultural residues (thousand dry tons) |
|---|---|
| Alabama | 355 |
| Alaska | 0 |
| Arizona | 318 |
| Arkansas | 4,351 |
| California | 1,505 |
| Colorado | 1,406 |
| Connecticut | 0 |
| Delaware | 222 |
| District of Colombia | 0 |
| Florida | 2,960 |
| Georgia | 904 |
| Hawaii | 359 |
| Idaho | 1,622 |
| Illinois | 17,775 |
| Indiana | 8,143 |
| Iowa | 21,401 |
| Kansas | 6,907 |
| Kentucky | 1,562 |
| Louisiana | 3,933 |
| Maine | 0 |
| Maryland | 530 |
| Massachusetts | 0 |
| Michigan | 3,253 |
| Minnesota | 12,910 |
| Mississippi | 1,988 |
| Missouri | 5,450 |
| Montana | 1,415 |
| Nebraska | 9,917 |
| Nevada | 4 |

*Source*: Milbrandt, A. 2005. "A Geographic Perspective on the Current Biomass Resource Availability in the United States," Technical Report NREL/TP-560-39181, National Renewable Energy Laboratory.

**APPENDIX 4.3** Distribution of Total Crop Residues in the United States

| State | Total agricultural residues (thousand dry tons) |
|---|---|
| New Hampshire | 0 |
| New Jersey | 83 |
| New Mexico | 152 |
| New York | 460 |
| North Carolina | 1,355 |
| North Dakota | 5,989 |
| Ohio | 4,537 |
| Oklahoma | 1,489 |
| Oregon | 514 |
| Pennsylvania | 735 |
| Rhode Island | 0 |
| South Carolina | 300 |
| South Dakota | 4,663 |
| Tennessee | 1,362 |
| Texas | 5,524 |
| Utah | 80 |
| Vermont | 0 |
| Virginia | 455 |
| Washington | 1,584 |
| West Virginia | 29 |
| Wisconsin | 4,009 |
| Wyoming | 96 |

**APPENDIX 4.3** (*Continued*)

| State | Quantity under wildlife management scenario (million dry tons) | Quantity under production management scenario (million dry tons) |
|---|---|---|
| Alabama | 3.1 | 6.6 |
| Arkansas | 2.2 | 5.5 |
| Connecticut | 0.0 | 0.2 |
| Delaware | 0.0 | 0.03 |
| Florida | 0.0 | 1.3 |
| Georgia | 1.3 | 4 |
| Illinois | 0.8 | 7.7 |
| Indiana | 0.0 | 5 |
| Iowa | 0.0 | 8.3 |
| Kansas | 2.9 | 11.4 |
| Kentucky | 3.0 | 5.1 |
| Louisiana | 3.7 | 5.8 |
| Maryland | 0.0 | 0.3 |
| Massachusetts | 0.0 | 0.2 |
| Michigan | 1.2 | 4.2 |
| Minnesota | 0.4 | 5.8 |
| Mississippi | 4.3 | 9.3 |
| Missouri | 2.5 | 12.8 |
| Montana | 0.0 | 2.8 |
| Nebraska | 1.9 | 5.2 |
| New Hampshire | 0.0 | 0.2 |
| New Jersey | 0.0 | 0.1 |
| New York | 0.0 | 3.4 |
| North Carolina | 0.6 | 1.6 |
| North Dakota | 1.9 | 16.8 |
| Ohio | 3.8 | 9.7 |
| Oklahoma | 3.6 | 8.1 |
| Pennsylvania | 0.0 | 2.3 |

*Source*: Adapted from De La Torre Ugarte, et al., 2003.

**APPENDIX 4.4** Projected Production of Switchgrass for 2008 in the United States

| State | Quantity under wildlife management scenario (million dry tons) | Quantity under production management scenario (million dry tons) |
|---|---|---|
| Rhode Island | 0.0 | 0.005 |
| South Carolina | 1.3 | 2.4 |
| South Dakota | 5.6 | 12.8 |
| Tennessee | 6.5 | 9.4 |
| Texas | 4.5 | 9.1 |
| Vermont | 0.0 | 0.3 |
| Virginia | 1.3 | 2.6 |
| West Virginia | 0.3 | 1.2 |
| Wisconsin | 3.6 | 6.1 |
| Wyoming | 0.0 | 0.5 |

**APPENDIX 4.4** *(Continued)*

# CHAPTER 5
# Ethanol Production

## 5.1 Ethanol Production from Sugar and Starch Feedstocks

### 5.1.1 Microorganisms

#### 5.1.1.1 *Saccharomyces cerevisiae*

The yeast *Saccharomyces cerevisiae* is the universal organism for fuel ethanol production using starch and sugar feedstocks. The sugars that are metabolizable by this organism include glucose, fructose, mannose, galactose, sucrose, maltose, and maltotriose. Ethanol production by *S. cerevisiae* is carried out via the glycolytic pathway (also known as the Embden-Myerhof-Parnas or EMP pathway) (Fig. 5.1).

In the simplest form, production of ethanol from glucose can be expressed by the following equation:

$$C_6H_{12}O_6 + 2\,Pi + 2\,ADP \rightarrow 2\,C_2H_5OH + 2\,CO_2 + 2\,ATP + 2\,H_2O$$

Glucose → 2 ethanol + 2 carbon dioxide + energy

From the above equation, it can be calculated that the theoretical yield is 0.511 g ethanol produced per gram glucose consumed. This yield can never be realized in practice since not all of the glucose consumed is converted to ethanol but part of it is used for cell mass synthesis, cell maintenance, and production of by-products such as glycerol, acetic acid, lactic acid, and succinic acid. Under ideal conditions, however, 90 to 95 percent of the theoretical yield can be achieved.[1]

Sugar transport is the first and also most important step in ethanol production. It has been suggested to be the rate-limiting step of glycolysis in yeast.[2] *Saccharomyces cerevisiae* has a complex and highly effective system for sugar transport. There are 20 genes for hexose transporters on the genome. The rate of transport can exceed $10^7$ glucose molecules per cell per second in fast fermenting cells.[3] The hexose transporters in *S. cerevisiae* can be divided into three classes:

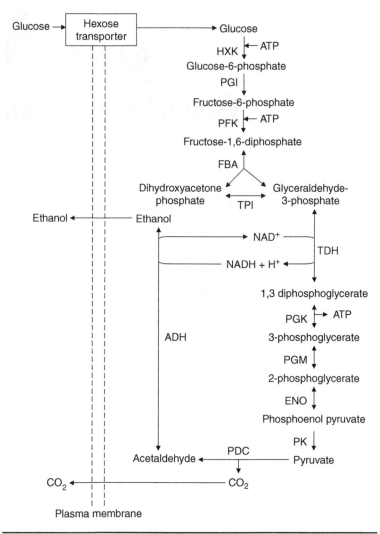

**FIGURE 5.1** The glycolytic pathway for ethanol production in *S. cerevisiae*. (HXK: hexokinase; PGI: phosphoglucose isomerase; PFK: phosphofructokinase; FBA: fructose bisphosphate aldolase; TDI: triose phosphate isomerase; TDH: triose phosphate dehydrogenase; PGK: 3-phosphoglycerate kinase; PGM: phosphoglycerate mutase; ENO: enolase; PK: pyruvate kinase; PDC: pyruvate decarboxylase; ADH: alcohol dehydrogenase.)

low-affinity transporters ($K_{m\ (glucose)}$ = 50–100 mM), moderate-affinity transporters ($K_{m\ (glucose)}$ about 10 mM), and high-affinity transporters ($K_{m\ (glucose)}$ = 1–2 mM).[4] Despite the complexity of the system, all the hexose transporters in *S. cerevisiae* use facilitated diffusion for glucose transport.[5] This mode of transport requires only a concentration gradient across the plasma membrane. Sucrose, the major sugar in sugarcane and sugar beet, is hydrolyzed to glucose and fructose by the enzyme invertase, which is located between the cell membrane and the cell wall. The two monosaccharide products then are taken up by the cell. Fructose enters the glycolytic pathway via its conversion to fructose-6-phosphate by hexokinase. The two sugars obtained in starch hydrolysis, maltose and maltotriose, pass intact through the cell membrane and are hydrolyzed intracellularly to glucose by the enzyme α-glucosidase. Maltotetraose and higher polysaccharides (dextrins) are not metabolized by S. cerevisiae.[6]

Glucose and sucrose are the two most preferred substrates among the sugars metabolized by S. cerevisiae. In industrial applications, where many sugars are present, glucose and sucrose are always consumed first. In fact, the presence of these two sugars causes repression in uptake and metabolism of other sugars.[7] This catabolite repression is the result of both competition for sugar transporters and down-regulation of the genes involved in the uptake and metabolism of the other sugars. For example, the uptake of fructose is slowed down by glucose since both sugars are transported into the cell by the same carriers, which have higher affinity for glucose. In addition, glucose also can repress the expression of specific fructose transporters.[7] Glucose represses the expression of the maltose utilization genes even when maltose is present.[8,9] Glucose also causes repression of the galactose utilization genes even in the presence of that sugar.[10] Catabolite repression is not immediately relieved after glucose and sucrose are depleted. In fact the effect may last for several hours.[11] The repression of utilization of other sugars by glucose and/or sucrose may have a negative effect on fermentation rates in industrial ethanol production.

Saccharomyces cerevisiae requires certain minerals (e.g., Ca, Mg, Mn, Co, Fe, Cu, K, Na, Zn) for growth and ethanol fermentation.[12] A number of organic compounds also have been observed to improve ethanol production.[13] Most of the required nutrients normally are already available in industrial feedstocks for ethanol production.[14]

Saccharomyces cerevisiae is inhibited by its own product, that is, ethanol. Ethanol increases fluidity and permeability of cell membrane and causes leakages of ions and small metabolites.[15,16] The kinetics of growth and fermentation inhibition by ethanol is represented by the following equations, respectively.[17]

$$\mu_{Sg} = \mu_{Sog}[1-(S/S_{max})]^n$$

$$f_{Sf} = f_{Sof}[1-(S/S^*_{max})]^{n^*}$$

where $\mu_{Sog}$ and $\mu_{Sg}$ are the specific growth rates in the absence and presence of a concentration $S$ of ethanol, respectively, $f_{Sof}$ and $f_{Sf}$ are the specific fermentation rates in the absence and presence of a concentration $S$ of ethanol, respectively, $S_{max}$ and $S^*_{max}$ are the maximum concentrations of ethanol that allow growth and fermentation, respectively, and $n$ and $n^*$ are the ethanol toxic power calculated for growth and fermentation inhibition, respectively.

The values of n and n* can be determined from experimental data. In practice, the values of $S_{max}$ and $S^*_{max}$ are about 10 percent (v/v) and 15 percent (v/v), respectively. Ethanol inhibition in *S. cerevisiae* can be reversed by addition of calcium.[17] However, addition of calcium to the fermentor must be controlled tightly since high Ca:Mg ratios may cause antagonism of essential biochemical functions of magnesium and negatively affect growth and ethanol production.[6]

Acetic and lactic acid can also inhibit *S. cerevisiae*. In ethanol fermentation these acids are minor by-products and accumulate only to relatively low concentrations, which are well below the inhibitory levels. However, in industrial ethanol production completely sterile conditions are never maintained and contamination by lactic acid and acetic acid bacteria can raise the concentrations of these acids to harmful levels. The inhibitory effects of lactic acid and acetic acid on growth and ethanol production are synergistic[18] and pH dependent.[19]

Oxygen plays an important role in *S. cerevisiae* metabolism.[6] Ethanol inhibition of growth is reduced under micro-aerobic conditions compared to anaerobic conditions.[20] Aeration conditions also impact the synthesis of by-products. At a respiratory quotient (RQ) controlled at 17 or below, glycerol synthesis was almost eliminated.[21] However, excess oxygen will result in significantly higher cell mass production and lower ethanol yield.[6] Thus, in practice optimal aeration conditions must be carefully determined and maintained to achieve optimal ethanol production.

### 5.1.1.2 *Zymomonas mobilis*

Although it is not currently used commercially, *Zymomonas mobilis* is considered the most effective organism for production of ethanol. It can produce ethanol at much faster rates than *S. cerevisiae*.[21] The interesting aspect about this organism is it is an obligately fermentative organism but metabolizes sucrose, glucose, and fructose to pyruvate via the Entner-Doudoroff (ED) pathway, which is used mostly by strictly aerobic organisms such as *Pseudomonas*.[23] The ED pathway is shown in Fig. 5.2.

The most striking difference between the pathway used by *Z. mobilis* and the one used by *S. cerevisiae* is the absence of the step catalyzed by the enzyme PFK in the ED pathway. PFK is a key enzyme of glycolysis. In yeast this enzyme is tightly regulated.[24] The absence of the enzyme PFK in *Z. mobilis* allows ethanol production in this organism to be decoupled from energy generation.[23] In other words, cell growth is not required for ethanol production and high cell concentrations are not needed for high ethanol yield. Thus, ethanol yield closer to the theoretical value is possible. Indeed ethanol yield

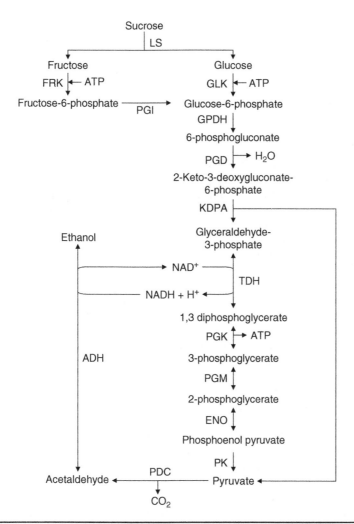

FIGURE 5.2  The pathway for ethanol production in *Z. mobilis*. (LS: levansucrase; GLK: glucokinase; FRK: fructokinase; PGI: phosphoglucose isomerase; GPDH: glucose-6-phosphate dehydrogenase; PGD: 6-phosphogluconate dehydratase; KDPA: 2-keto-3-deoxy-6-phosphogluconate aldolase; TDI: triose phosphate isomerase; TDH: triose phosphate dehydrogenase; PGK: 3-phosphoglycerate kinase; PGM: phosphoglycerate mutase; ENO: enolase; PK: pyruvate kinase; PDC: pyruvate decarboxylase; ADH: alcohol dehydrogenase.)

independence of cell concentrations and conversion efficiencies approaching 97 percent of the theoretical have been demonstrated.[25] Other advantages of *Z. mobilis* include high sugar uptake and ethanol production rates, high ethanol tolerance, and no requirements for aeration during the fermentation process for optimal ethanol production.[25,26]

Although it has the capability of producing ethanol from glucose at high yield and rates, Z. mobilis also has a number of disadvantages that have prevented it from being used for industrial ethanol production. The biochemistry of sucrose metabolism has hampered efforts to use Z. mobilis for ethanol production from cane and beet molasses. Sucrose has to be converted to glucose and fructose by the enzyme levansucrase and possibly by other enzymes[23] before the two monomeric sugars can be used for ethanol production. Levansucrase also catalyzes the polymerization of fructose to form levan, a (2-6)-$\beta$-D-fructofuranosyl linear polysaccharide with a number of (2-1)-$\beta$-D-fructofuranosyl branch points. Other fructose polymer products may also be formed by that enzyme. Levan synthesis requires the presence of sucrose or another suitable sugar, for example, raffinose, to initiate the formation of the chain. Allowing sucrose to stay in the fermentor over extended periods, for example, by allowing the rate of sucrose hydrolysis to slow down or by using high sucrose concentrations in the fermentation medium, will lead to increased levan formation.[23] Another by-product, sorbitol, is also produced when sucrose is used for ethanol fermentation. The formation of these two by-products results in lower ethanol yields from sucrose. The presence of levan and other fructose polymers may also cause problems for downstream processing such as fouling of distillation columns and other equipment.

Both S. cerevisiae and Z. mobilis produce organic acid by-products but in different proportions. The ratios of acetic acid:lactic acid normally are 16:1 for Z. mobilis and 8:1 for S. cerevisiae. Thus, the pH of the fermentation medium used for S. cerevisiae can drop to about 3 whereas in the case of Z. mobilis it tends to stabilize at 4.5. Since growth of most potential contaminating bacteria is almost completely inhibited at pH 3 but not quite at pH 4.5, the fermentation media for Z. mobilis will need to be sterilized to ensure high ethanol yields.[27] On the industrial scale this is undesirable because sterilization of large volumes of media will add significantly to the operating costs, especially since ethanol production has a very small profit margin.

Another factor that favors the use of S. cerevisiae is the convenience in handling of this organism. Starting yeast cultures can be purchased in many forms, and reconstitution of the cultures in many cases only requires a rehydration step in a mixing tank with little control. On the other hand, if Z. mobilis is used for commercial ethanol production, a stock culture of desired traits must be maintained in the plant and several seed fermentors with tight control may be required to raise the inoculum to a suitable level for use in the production fermentors. Traditionally ethanol production is a very simple process, and ethanol producers, the majority of whom are farm co-ops, want to have a process that is simplified as much as possible.

Although Z. mobilis currently is not used for commercial production of ethanol, interest in this organism remains high. Research efforts to develop improved ethanol production processes using this organism have not been lacking. For example, advanced bioreactor systems using immobilized cells to take advantage of the low cell yield requirement of Z. mobilis have been developed and tested at various scales.[28] Recombinant Z. mobilis strains also have been developed for ethanol production from lignocellulosic biomass. Recently a consortium of academic and industrial Korean researchers published the entire 2.06 Mb gene sequence of Z. mobilis ZM4.[29] This information will allow improvement of ethanol production and also exploitation of this organism for development of other important industrial products.

### 5.1.2 Process Technology

#### 5.1.2.1 Sugar Feedstocks

Both raw juice and molasses from sugarcane and sugar beets can be used for ethanol production. The compositions of selected sugarcane and sugar beet juice are shown in Table 5.1.

The characteristics of some cane and beet molasses are shown in Table 5.2.

Brazil leads the world in production of ethanol using sugar feedstocks. In Brazil, both sugarcane juice and molasses are used for ethanol production. Normally an ethanol plant is built next to a sugar mill or is part of the mill. A block diagram describing the operation of such facilities is shown in Fig. 5.3.

The juice is extracted from sugarcane by either squeezing (roll mills) or diffusion (diffuser). Part of the juice is used for sugar manufacture and the remaining is used for ethanol production. Molasses, which is a low-value by-product, also is used for ethanol production. The solid residue from the extraction step, which is referred to as bagasse, is burned to generate energy for use in the plant. Sugar manufacturing is not within the scope of this book and therefore is not discussed in this chapter. The readers are referred to the web sites www.sucrose.com and www.kpatents.com/sugar_app.php for details on the sugar manufacturing process.[34,35]

In Brazil, ethanol normally is obtained by fermentation of cane juice or a mixture of cane molasses and juice. Before going to the fermentors the sugar solution must go through purification and pasteurization. Purification normally involves treatment with lime, heating, and later decantation, similar to the treatment use in sugar manufacture. Pasteurization involves heating and immediate cooling. The cooling typically includes two stages. In the first stage the hot sugar solution is passed through a heat exchanger in counter-current flow to the cold solution. At the end of this stage the hot solution is cooled to about 60°C. In the second stage the sugar solution is cooled further to 30°C using water as the cooling fluid. The sugar concentration normally is adjusted to approximately 19° Brix.[36]

|  | Concentration (g/100 g) | |
| --- | --- | --- |
| Components | Sugarcane juice* | Sugar beet juice[†] |
| Solid | 13.7 | 17.3 |
| Sucrose | 12 | 16.5 |
| Raffinose |  | 0.07 |
| Monosaccharides | 0.63 | 0.15 |
| Polysaccharides | 0.028 | 0.019 |
| Lactate | 0.016 |  |
| Acetate | 0.035 |  |
| Sulfate | 0.039 | 0.02 |
| Phosphate | 0.033 | 0.047 |
| Nitrate |  | 0.015 |
| Nitrite |  | 0.005 |
| Aconitate | 0.09 |  |
| K | 0.11 | 0.125 |
| Na | 0.005 | 0.015 |
| Cl |  | 0.003 |
| Ca | 0.04 |  |
| Mg | 0.028 |  |
| Total-N |  | 0.105 |
| Betaine-N |  | 0.046 |
| Amino acid-N |  | 0.026 |
| Ammonia-N |  | 0.006 |
| Amide-N |  | 0.011 |

*Data were taken from Enterprise mill in Louisiana in 2004. The juice was extracted by milling. Data are also available for juice obtained with a diffuser. The compositions in both cases are very similar.
[†]Data were taken from Nippon Beet Sugar Manufacturing, Sapporo, Japan.
Source: Adapted from Polanco, Rein, and White, 2006, and Ogbonna, Mashima, and Tanaka, 2001.

**TABLE 5.1** Compositions of Selected Sugarcane and Sugar Beet Juice

***NOTE:*** *Brix is a crude measure of sugar contents of a molasses product. By definition it is a measure of what the sugar content of a liquid would be if all the dissolved and suspended solids in the liquid were sugar. It was used widely before the advent of advanced instrumentation such as high*

| Item | Cane molasses | Beet molasses |
|---|---|---|
| Brix | 79.5 | 79.5 |
| Total solids (%) | 75.0 | 77.0 |
| Specific gravity | 1.41 | 1.41 |
| Total sugars (%) | 46.0 | 48.0 |
| Crude protein (%) | 3.0 | 6.0 |
| Nitrogen free extract (%) | 63.0 | 62.0 |
| Ash (%) | 8.1 | 8.7 |
| Ca (%) | 0.8 | 0.2 |
| P (%) | 0.08 | 0.03 |
| K (%) | 2.4 | 4.7 |
| Na (%) | 0.2 | 1.0 |
| Cl (%) | 1.4 | 0.9 |
| S (%) | 0.5 | 0.5 |
| *Trace minerals* | | |
| Cu (mg/kg) | 36 | 13 |
| Fe (mg/kg) | 249 | 117 |
| Mn (mg/kg) | 35 | 10 |
| Zn (mg/kg) | 13 | 40 |
| *Vitamins* | | |
| Biotin (mg/kg) | 0.36 | 0.46 |
| Choline (mg/kg) | 745 | 716 |
| Pantothenic acid (mg/kg) | 21.0 | 7.0 |
| Riboflavin (mg/kg) | 1.8 | 1.4 |
| Thiamine (mg/kg) | 0.9 | — |

*Source*: Adapted from Curtin, 1983.

**TABLE 5.2** Characteristics of Cane and Beet Molasses

*pressure liquid chromatography (HPLC) and still is used in many places today. As shown in Table 5.2, typically cane molasses has a Brix of approximately 80° and contains about 46 percent sugar by weight.*

The fermentation can be either batch/fed-batch or continuous. In Brazil, approximately 70 percent of the ethanol plants use a batch process.[37] The Melle-Boinot process, which was developed in the 1930s, is most commonly used in Brazil. Its main characteristic is total yeast recycle, normally by centrifugation.[36] The continuous process with cell

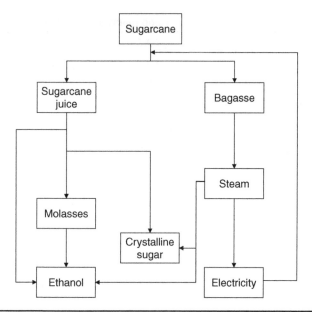

**FIGURE 5.3** Block diagram of an integrated sugar mill-ethanol fermentation plant. (*Adapted from Lipinsky, 1978.*)

recycle was developed in the 1980s to replace the batch process in a number of ethanol plants.[38] Andrietta and Maugeri reported on the replacement of a batch process with a continuous process.[38] The old fed-batch plant used 24 fermentors having working volume of 200 $M^3$ each to produce 400 $M^3$ of 96 percent (v/v) ethanol per day. The new continuous plant had four continuous flow stirred tank reactor (CSTR) fermentors having different individual volumes in series which gave a total volume of 2500 $M^3$, producing about 440 $M^3$ of 96 percent (v/v) ethanol per day. Thus, the new plant had productivity about twice the productivity of the old plant. Typical parameters of a batch fermentation process with cell recycle are summarized in Table 5.3. Normally before returning to the fermentor the recycled cells are diluted with water,

| Process parameters | Values |
| --- | --- |
| Cell density | 8–17% (w/v)[37] |
| Temperature | 33–35°C[37]; 32°C[36] |
| Ethanol concentration | 8–11% (v/v)[37]; 7–10% (v/v)[36] |
| Fermentation time | 6–10 h[37]; 4–12 h[36] |
| Ethanol yield | 90–92% theoretical[37] |

**TABLE 5.3** Typical Parameters of Batch Fermentation Process for Ethanol Production from Sugarcane Juice and Molasses

and sulfuric acid is added to pH 2.5 or lower (pH 2) if there is bacterial contamination.[36] The very high cell densities allow reduced growth, high ethanol yield, and very short fermentation time. The yeast can be recycled up to three times per day for up to 200 days.[37] Often, a nitrogen source is required in the fermentation process. The most preferred nitrogen source is urea. In cane molasses fermentation, ammonium sulfate, which is used quite often as a nitrogen source in industrial fermentations, should be avoided, since it may add to the scaling problem by forming calcium sulfate. Another common nitrogen source in industrial fermentation, liquid ammonia, also should not be used since it will raise the pH, which subsequently will encourage bacterial contamination.

In Brazil, about 55 percent of sugarcane harvested is used for ethanol and the rest for sugar production. One hectare of land produces about 81–82 MT of sugarcane and about 7000 L of ethanol.[39] The total cost of ethanol production in Brazil during the 2005 crop year was $1.10/gal ($0.291/L), with variable costs of $0.89/gal ($0.235/L) and fixed costs of $0.21/gal ($0.055/L).[40]

The use of sugar beets for ethanol production has not been practiced on large scale as widely as sugarcane. Details of commercial processes therefore are not readily available. Interests in ethanol production from sugar beets recently started to grow, especially in the European Union (EU) and Japan.[31,41] Because of the similarities in chemical compositions of sugar solutions extracted from sugarcane and sugar beets, it is expected that those obtained from sugar beets also can be fermented to ethanol by industrial yeast at high efficiencies. Indeed it has been demonstrated that 9.5 percent (v/v) ethanol could be produced by a flocculating yeast strain, *S. cerevisiae* IR2, immobilized on loofa sponge in about 15 h using a raw sugar beet juice containing sucrose at 16.5 percent by weight.[31] In these demonstrations the sugar beet juice did not require any pH adjustment or nutrient addition. It has been estimated that in the EU, ethanol could be produced from sugar beets at 86 L/MT of feedstock or approximately 5000 L/ha of land.[41] Production costs of ethanol in Germany were estimated for sugar beet feedstock and compared to those for wheat.[41] These results are shown in Table 5.4.

A process model for combined sugar and ethanol production from sugar beets has recently been developed.[42] A simplified process flow diagram (PFD) for that process is shown in Fig. 5.4 and a typical mass balance is shown in Table 5.5.

### 5.1.2.2 Starch Feedstocks

Among potential starch feedstocks for ethanol production, corn is used most widely, especially in the United States. Other grains, such as wheat and barley, also are used as feedstocks for ethanol production. The process technology developed for corn can easily be adapted to other grains. Corn ethanol plants need only minor modifications to

| Plant capacity | 50 million liters | | 200 million liters | |
| --- | --- | --- | --- | --- |
| Raw material | Wheat | Sugar beets | Wheat | Sugar beets |
| Feedstock cost | 0.28 | 0.35 | 0.28 | 0.35 |
| By-product credit | 0.07 | 0.07 | 0.07 | 0.07 |
| Net feedstock cost | 0.21 | 0.28 | 0.21 | 0.28 |
| Labor cost | 0.04 | 0.04 | 0.01 | 0.01 |
| Other operating and energy cost | 0.20 | 0.18 | 0.20 | 0.17 |
| Net investment cost | 0.10 | 0.10 | 0.06 | 0.06 |
| Total | 0.55 | 0.60 | 0.48 | 0.52 |
| Total gasoline-per-equivalent liter* | 0.81 | 0.88 | 0.71 | 0.77 |

*The concept of gallon equivalents using a gallon of gasoline compared to another fuel is one that allows for a comparison of energy content based on British thermal units (Btu). The energy content of 1 gal of reformulated gasoline with 10 percent MBTE (RFG) is 112,000 Btu and that of 1 gal of ethanol is 76,100 Btu (Source: *National Association of Fleet Administrators*—http://www.nafa.org). Thus, 1 gal of RFG is equivalent to 1.47 gal of ethanol.

*Source*: Adapted from USDA Foreign Agricultural Service, GAIN report number E36081.

**TABLE 5.4** Production Costs (in $/L) of Ethanol from Sugar Beets and Wheat in Germany

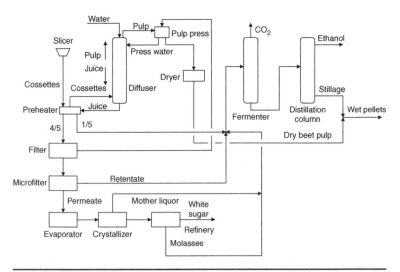

**FIGURE 5.4** Simplified process flow diagram of combined sugar and ethanol production from sugar beets. (*Redrawn from process diagram and description in www.kpatents.com/sugar_app.php and Henke et al., 2006.*)

|  | Mass (MT/h) | Dry matter (%) | Sucrose (%) | Purity (%) |
|---|---|---|---|---|
| *Extraction* | | | | |
| Sugar beet | 417 | 24.0 | 17.0 | |
| Raw juice | 474 | 16.4 | 14.8 | 90.2 |
| Desugarized pulp | 459 | 7.0 | 1.1 | |
| Pulp press water | 391 | 1.1 | | |
| Water for extractio | 126 | | | |
| Pressed pulp | 69 | 32.4 | 0.9 | |
| *Juice purification and sugar production* | | | | |
| Raw juice to fermenter | 95 | 16.4 | 14.8 | 90.2 |
| Raw juice to membrane | 379 | 16.4 | 14.8 | 90.2 |
| Permeate | 300 | 15.4 | 15.0 | 97.4 |
| Retentate (to fermenter) | 79 | 20.2 | 14.2 | 70.3 |
| Thick juice (from permeate) | 71 | 65.0 | 63.4 | 97.5 |
| White sugar | 22 | | | 95.1 |
| Mother liquor (to fermentor) | 31 | 76.1 | 72.4 | |
| *Ethanol and feed pellets production* | | | | |
| Ethanol | 22 | | | |
| Carbon dioxide | 24 | | | |
| Stillage | 160 | 7.7 | 0.6 | |
| Wet pellets | 205 | 16.9 | 0.8 | |
| Dry pellets | 38 | 85.0 | 3.8 | |

*Source*: Adapted from Henke et al., 2006.

**TABLE 5.5** Example of Mass Balance in a Process for Combined Sugar and Ethanol Production from Sugar Beets

handle other grain feedstocks. Wheat and barley have been used for ethanol production in Canada and Europe for many years.[43]

Ethanol is produced from corn by either the wet milling or dry milling process. The dry milling also is referred to as the dry-grind process. The key steps of these processes are shown in Fig. 5.5.

The main difference between the two processes is that in the dry milling process the whole corn is ground and fed to the fermentor for ethanol fermentation, whereas in the wet milling process the corn components are fractionated first and then only the starch fraction is used in ethanol fermentation. As a result the wet milling process

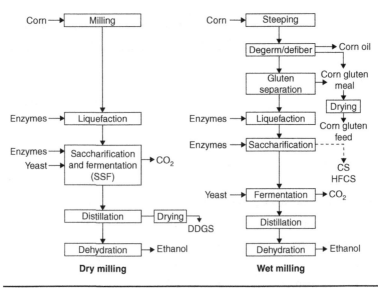

**FIGURE 5.5** Comparison of key steps of the wet milling and dry milling process. (*Adapted from Bothast and Schlicher, 2005.*)

requires much higher capital investment, and ethanol plants using this process are much larger than those using the dry milling process. The large size of the wet milling ethanol plants is needed to justify the rate of return on capital investment (see the section on corn ethanol process economics). Both processes generate a number of ethanol-related coproducts. These coproducts include distillers dried grains with solubles and carbon dioxide in the dry milling process, and corn oil, corn gluten meal, corn gluten feed, and carbon dioxide in the wet milling process. All of these coproducts are of relatively moderate values. However, the wet milling plants can easily be modified to produce other products such as corn syrups and high fructose corn syrups, which can be produced independent of ethanol production. These independent coproducts are not just economically beneficial but also can be strategically important, especially during the times of reduced ethanol market demand.

**Wet Milling Process** Wet milling was developed more than 150 years ago for corn starch processing.[45] The first step in this process involves soaking of corn grains, which have been cleaned to remove foreign matters such as dirt and chaff, in water containing 0.1 to 0.2 percent $SO_2$ at 52°C (125°F) for 24 to 40 h. Steeping softens the kernel and breaks the disulfide bonds in the protein matrix of the endosperm to release the starch granules. Soluble nutrients also are released into the steep water, which normally is referred to as light steep water (LSW). The softened grains then are ground gently to break up the kernels. The less dense germs are recovered in a hydroclone system.

The recovered germs receive further processing to remove loose starch and gluten and excess water, and then are dried and cooled for storage. Corn oil can be extracted from the germs on-site. Alternatively, the germs can be sold to crushers for corn oil extraction. The spent germs are processed into a feed product called corn germ meal (CGM). This product contains at least 20 percent protein and 1 percent fat, and up to 12 percent fiber. Due to its high fiber contents, the CGM usually is added to ruminant diets.

After germ separation, the slurry goes through an intense grinding, which further loosens the starch and gluten from the remaining fiber. The slurry is screened to remove the fiber, which then is washed and pressed to about 60 percent moisture. The LSW from the steeping step is concentrated in evaporators to produce heavy steep water (HSW). This nutrient-rich concentrated product is typically dried together with the fiber to be sold as corn gluten feed (CGF) to the livestock industry. The CGF contains about 21 percent protein.

Following fiber removal, the gluten, which is lighter, is separated from the starch in a centrifuge. The separated gluten then is concentrated from 2–3 oz/gal (light gluten) to 16–20 oz/gal (heavy gluten) in a second centrifuge. The heavy gluten is cooled to 35°C and filtered on a rotary vacuum filter to produce a gluten cake of 60 percent moisture. The cake is dried to 10 percent moisture to produce corn gluten meal (CGM). This feed product contains at least 60 percent protein and 1 percent fat, and up to 3 percent fiber. It is used in poultry feed because of its high protein and xanthophyll and low fiber contents.

The starch is washed and processed through a series of up to 14 hydroclones to remove impurities. The final product, which is 99.5 percent pure starch, then is used for ethanol fermentation in dedicated plants or processed further to produce modified starch, corn syrups (CS), and high fructose corn syrups (HFCS) in integrated plants.

All fermentation processes used in wet milling plants today are continuous.

Figure 5.6 is a simplified process flow diagram of a cascaded continuous fermentation process with yeast recycle. Like other continuous fermentation processes, this process is subject to contamination. To minimize the contamination problem, low pH, normally around 3.5, must be maintained. The low pH requires fermentors constructed of stainless steel, which results in high capital costs.

**Dry Milling Process** Because wet milling plants are complex and capital-intensive, most new and smaller ethanol plants use the dry milling process. The process begins by adding process water to the milled corn grains, adjusting the pH to about 6, and adding a thermostable $\alpha$-amylase. The next step is starch liquefaction. This step has been reviewed in detail by Lewis.[47]

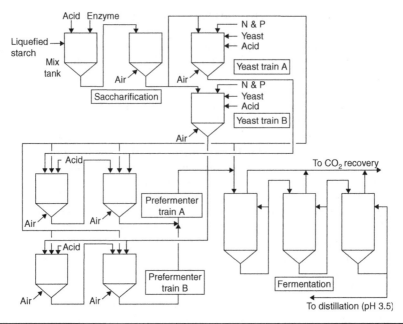

**FIGURE 5.6** Simplified process flow diagram of a wet milling cascaded continuous fermentation process. (*Redrawn from Madson and Monceaux, 1999.*)

The slurry first is heated to either 60 to 70°C (warm cook) or 80 to 90°C (hot cook). As the temperature rises, the starch granules swell and become gelatinized. The swelling and hydration of the starch granules cause dramatic increase of the slurry viscosity and loss of crystallinity of the granule structures. This preliquefaction step lasts about 30 to 45 min. In the following liquefaction step, three processes normally are practiced.

In the first process, the entire required enzyme dosage is added initially and the slurry is heated to 85 to 95°C, and then held at that temperature until a DE of 15 is achieved or a negative iodine test result is observed.

In the second process, the slurry is forced through a jet cooker maintained at 110 to 120°C at a residence time of 5 to 7 min. In some ethanol plants the temperature of the jet cooker as high as 140 to 150°C is preferred. After leaving the jet cooker, the slurry is discharged into a flash tank at atmospheric pressure and held at about 90°C for up to 3 h. In this process, some of the required enzyme is added initially and the rest is added to the flash tank.

In the third process approximately half of the enzyme is added initially and the slurry is heated to 150°C by direct steam injection and held for a short period to gelatinize the starch. The slurry then is cooled to 85 to 95°C by flashing and the rest of the enzyme required is added. At the end of the liquefaction, the starch is hydrolyzed to short-chain dextrins.

After liquefaction, the slurry is cooled to 32°C and its pH is adjusted to 4.5. Glucoamylase then is added and the slurry is transferred into the fermentor. Once inside the fermentor it is normally referred to as mash. Then yeast is added. Often urea also is added as a nitrogen source. The fermentation process is referred to as simultaneous saccharification and fermentation (SSF). In the fermentor dextrins are hydrolyzed by glucoamylase to glucose, which immediately is taken up by the yeast and fermented to ethanol and carbon dioxide. Since there is no accumulation of glucose, feedback inhibition of glucoamylase will not occur and complete conversion of starch will result. Some ethanol plants hold the mash at 55°C, which is the optimum temperature of commonly used commercial glucoamylases, to achieve partial hydrolysis of dextrins before cooling it down and adding the yeast to start the SSF process. The SSF is a batch process in which the mash normally is allowed to ferment for about 50 to 60 h. At the end of this period, an ethanol concentration of about 15 percent by volume is obtained. Figure 5.7 is a simplified process flow diagram of a typical dry milling ethanol fermentation process, which includes yeast propagation and simultaneous saccharification and fermentation (SSF).

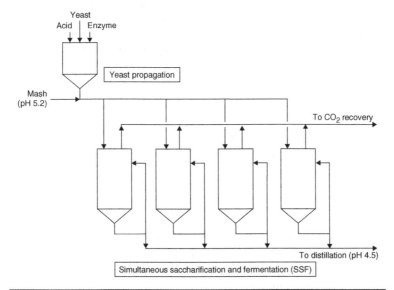

**FIGURE 5.7** Simplified process flow diagram of a dry milling ethanol fermentation process with yeast propagation and simultaneous saccharification and fermentation. (*Redrawn from Madson and Monceaux, 1999.*)

**NOTE:** *The DE or dextrose equivalent is a measure of the extent of starch hydrolysis. It represents the percentage of the glycosidic linkages hydrolyzed with respect to the total number of glycosidic linkages present initially. Thus, glucose has a DE of 100, maltose has a DE of 50, and starch has a DE of practically zero.*

In the dry-milling process described above, the saccharification step, which is the major source of contamination, is not carried out externally as in the wet-milling process. With the risk of contamination being significantly reduced, the process does not require low pH. Normally the pH can be maintained at 4.5. Thus, the fermentors can be constructed of carbon steel instead of stainless steel, which helps drive down the capital costs.[46]

The dry milling process also generates two coproducts, carbon dioxide and distillers dried grains with solubles (DDGS). Large quantities of carbon dioxide are produced in the fermentation process. At some ethanol plants, the carbon dioxide produced in the fermentor is recovered and sold to carbonated beverage bottling companies or other food processing industries. The stillage obtained at the bottom of the distillation columns contains oil, nonfermentable solids (fiber and proteins), and yeast. Centrifuges or presses normally are used to separate this stream into a liquid called *thin stillage* and a slurry called *wet distillers grains* (WDG). Between 15 and 30 percent of the thin stillage is recycled to the slurry tank as make-up water to reduce water usage. The recycled thin stillage normally is referred to as backset. The remaining is concentrated in multiple-effect evaporators into a syrup containing 20 to 25 percent solids. The syrup is mixed with the WDG to produce a high-nutrient cattle feed called *wet distillers grains with solubles* (WDGS). To increase shelf-life and reduce transportation costs, WGDS is dried to 10 to 12 percent moisture to produce DDGS. Drying of WDGS to produce DDGS is very energy-intensive. The process may consume up to one-third of the total energy consumption of the entire dry milling plant.[44]

**Theoretical Ethanol Yield**  The hydrolysis of starch to produce glucose can be expressed by the following equation:

$$(C_6H_{10}O_5)_n + nH_2O \rightarrow nC_6H_{12}O_6 \tag{5.1}$$

where, n is the number of glucose residues in the starch molecule.

The amount of glucose produced from 1 kg of starch is $180n/(162n + 18)$. When n is 2, as in maltose, the conversion factor is 1.053. When n becomes very large, this factor approaches 1.111.

Ethanol fermentation using glucose as substrate can be expressed by the following equation:

$$C_6H_{12}O_6 \rightarrow 2\,C_2H_5OH + 2\,CO_2 \tag{5.2}$$

Thus, 1 kg glucose will produce 0.511 kg ethanol.

**Example 5.1**  Calculate theoretical ethanol yield from 1 bu of corn which has 15 percent moisture and contains 70 percent starch on a dry basis.

**Solution**  One kg of corn contains: 1 kg − 0.15 kg = 0.85 kg dry solids.

Starch content of 1 kg of corn: $0.70 \times 0.85$ kg = 0.595 kg
Theoretical ethanol yield from 1 kg of corn is

$$0.595 \text{ kg} \times 1.111 \times 0.511 = 0.338 \text{ kg ethanol/kg corn}$$

The specific gravity of ethanol at 20°C is 0.79 kg/L. Therefore, the volume of ethanol produced from 1 kg of corn is

$$[(0.338 \text{ kg}) \div 0.79 \text{ (kg/L)}] \div 3.785 \text{ (L/gal)} = 0.113 \text{ gal}$$

The weight of 1 bu of corn is 56 lb.
Therefore, the theoretical yield of ethanol from 1 bu of corn is

$$0.113 \text{ (gal/kg)} \times 0.454 \text{ kg/lb} \times 56 \text{ lb/bu} = 2.87 \text{ gal/bu}$$

In practice, the actual amount of ethanol produced from 1 bu of corn is known and the overall process efficiency can easily be calculated.

**Example 5.2** Determine ethanol fermentation efficiency for the corn described in Example 5.1 (15 percent moisture; 70 percent starch on dry basis) in laboratory.

**Solution** Let's assume that in this experiment we prepared a mash having 30 percent total solids on dry basis and after 72 h of fermentation, analysis of the final liquid sample by high pressure liquid chromatography (HPLC) showed an ethanol concentration of 13.1 g/L.

Basis: 1 kg of mash.
One kg of mash contains 300 g total solids and 700 g water.
Starch content: 300 g × 0.70 = 210 g
Glucose production by starch hydrolysis: 210 g × 1.111 = 233.3 g
Water consumption in starch hydrolysis: 210 g × 0.111 = 23.3 g
Theoretical ethanol production: 233.3 g × 0.511 = 119.0 g
Volume of the ethanol produced: 119.0 g ÷ 0.79 (g/mL) = 150.8 mL
Total liquid volume: 700 mL − 23.3 mL + 150.8 mL = 827.5 mL
Ethanol concentration expected: 119.0 g ÷ 827.5 mL = 0.144 g/mL or 14.4 g/L.
Therefore, the fermentation efficiency is: (13.1 ÷ 14.4) × 100% = 91.0%.

---

**NOTE:** *In this example we assume that when a given volume of ethanol is dissolved in a given volume of water, the total liquid volume is equal to the sum of the individual volumes. In reality, because of the hydrogen bonds formed between the ethanol and water molecules the actual volume will be smaller than the volume obtained by addition of the two individual volumes. However, using data on the densities of aqueous ethanol solutions[48] it can be calculated that our assumption resulted in an error of only 1 percent. We also have determined in our laboratory that for a 15 g/L ethanol solution at 30°C over a time course of 72 h, the loss of ethanol to the vapor phase was less than 1 percent.*

An indirect laboratory method to determine total ethanol production is to measure total carbon dioxide production and calculate the corresponding ethanol production using the stoichiometric equation [Eq. (5.2)]. It should be noted that in addition to ethanol, the yeast *S. cerevisiae* also produces other products whose syntheses involve carbon dioxide.

For example, synthesis of acetic acid produces carbon dioxide whereas synthesis of succinic acid results in its net consumption. However, these by-products normally are produced at very low concentrations (less than 1 g/L), and the error they cause in calculation of total ethanol production based on total carbon dioxide production is small, not more than 2 percent, as demonstrated in several experiments where the total production of ethanol and that of carbon dioxide were carefully measured.[49] The synthesis of glycerol, which is the only coproduct of ethanol fermentation by yeast that normally is produced in significant amounts, does not involve carbon dioxide.

**Recovery of Ethanol**  Ethanol from the fermentation broth is recovered by distillation followed by a dehydration step. Ethanol and water forms an azeotropic mixture of 95 percent ethanol and 5 percent water by volume. In the past, ternary azeotropic distillation using an agent such as benzene, cyclohexane, diethyl ether, and *n*-pentane was employed to produce anhydrous ethanol.[46] Molecular sieves currently are used for dehydration of 95 percent ethanol.

Molecular sieves are synthetic zeolite adsorbents having cylindrical or spherical shapes manufactured from materials such as potassium aluminosilicates.[46] These adsorbent beads have precise pore size, which allows them to adsorb and remove one molecular size from a bulk mixture containing molecules with a larger size. A synthetic zeolite having pore size of 3 angstroms (Å) is used for ethanol dehydration. Since water molecules are 2.8 Å and ethanol molecules are 4.4 Å, water molecules are strongly adsorbed and ethanol molecules are excluded.[50] A process normally referred to as *pressure swing adsorption* (PSA) is used for molecular sieve dehydration of 95 percent ethanol. Each unit of this process consists of two packed columns. Figure 5.8 illustrates the operation of a typical unit. The process begins with column 1 on line (drying the 95 percent feed vapor) and column 2 in regeneration mode. Ninety five percent ethanol enters the top of the first column under moderate pressures, passes through the bed, and leaves the bottom as anhydrous ethanol. About 60 to 85 percent of this stream leaves the system as the final product. The remaining is fed to column 2 as a regeneration purge stream. Column 2 is maintained under vacuum to shift the equilibrium to allow the adsorbed water to desorb. The wet ethanol stream leaving the top of column 2 is recycled and fed to the distillation column at an appropriate position. One column typically stays on line to dehydrate the 95 percent ethanol feed stream for 3 to 10 min before being

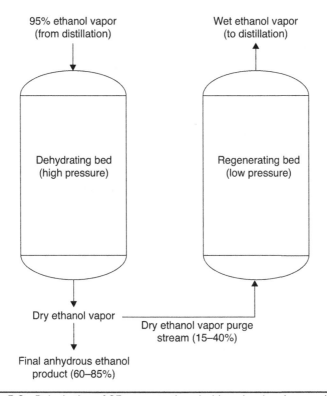

**FIGURE 5.8** Dehydration of 95 percent ethanol with molecular sieves using the PSA process. (*Redrawn from Bibb Swain, 1999.*)

regenerated. The PSA process avoids using high temperatures for regeneration and allows the useful life of the molecular sieves to be extended to several years.[46]

**Process Economics** In 2003, the U.S. Department of Agriculture surveyed 21 dry mill ethanol plants to estimate their 2002 production costs, including both variable (feedstock and plant operation) and capital expenses.[51] The wet mill plants did not participate in this survey. The results of the survey are summarized in Table 5.6.

The net feedstock cost is 57 percent of the total production cost. Thus, any increase in corn cost due to increased corn demand for ethanol production will significantly reduce net profits of ethanol plants. Among the operating costs, fuel cost is the highest cost component and accounts for 33 percent of the total operating cost. Thus, any increase in fuel cost, especially for natural gas, which is the major fuel used in ethanol plants, will also have similar effects.

The effects of plant size on economics of both wet mill and dry mill processes are investigated and presented in a report.[52] These results are

| Cost component | U.S. Average ($/gal) |
|---|---|
| Feedstock costs | 0.8030 |
| *By-product credits* | |
| DDGS | 0.2520 |
| Carbon dioxide | 0.0060 |
| Net feedstock costs | 0.5450 |
| *Cash operating expenses* | |
| Electicity | 0.0374 |
| Fuels | 0.1355 |
| Waste management | 0.0059 |
| Water | 0.0030 |
| Enzymes | 0.0366 |
| Yeast | 0.0043 |
| Chemicals | 0.0229 |
| Denaturant | 0.0348 |
| Maintenance | 0.0396 |
| Labor | 0.0544 |
| Administrative costs | 0.0341 |
| Others | 0.0039 |
| Total | 0.4124 |
| Total cash & net feedstock costs | 0.9574 |

**TABLE 5.6** Average 2002 Cost of Ethanol Production in 21 Dry Mill Plants

summarized in Tables 5.7 to 5.11. The results clearly show for equal plant size much higher capital investments are required for wet mill plants and as a result of this high capital requirement, a wet mill plant should have capacity of at least 40 million gallons per year to be profitable.

The locations and capacities of ethanol plants using corn and other minor feedstocks in the United States as of April 2008 are given in the appendix (Table 5A.1 and Fig. 5A.1).

**Recent Developments** Research has been continued in earnest to improve the corn ethanol technology. These new developments are reviewed briefly in this section.

1. New corn feedstocks

    New corn feedstocks for ethanol production have recently been developed by several seed companies. Pioneer

# Ethanol Production

| Cost component | 15 million gal/yr | 30 million gal/yr | 40 million gal/yr |
|---|---|---|---|
| Denaturant | $0.03 | $0.03 | $0.03 |
| Enzymes | $0.06 | $0.05 | $0.05 |
| Chemicals | $0.03 | $0.03 | $0.03 |
| Fuel | $0.12 | $0.11 | $0.11 |
| Electricity | $0.06 | $0.06 | $0.06 |
| Labor | $0.07 | $0.05 | $0.04 |
| Maintenance | $0.04 | $0.03 | $0.03 |
| Other expenses | $0.03 | $0.02 | $0.02 |
| SGA* | $0.04 | $0.03 | $0.03 |
| Total per gallon | $0.46 | $0.41 | $0.39 |

*Sales/general/administrative costs.

**TABLE 5.7** Effect of Plant Size on Operating Costs of Dry Mill Ethanol Plants

| Cost component | 30 million gal/yr | 40 million gal/yr | 100 million gal/yr |
|---|---|---|---|
| Denaturant | $0.03 | $0.03 | $0.03 |
| Enzymes | $0.07 | $0.06 | $0.05 |
| Chemicals | $0.05 | $0.04 | $0.04 |
| Fuel | $0.08 | $0.08 | $0.08 |
| Electricity | $0.07 | $0.07 | $0.07 |
| Labor | $0.07 | $0.06 | $0.05 |
| Maintenance | $0.06 | $0.06 | $0.05 |
| Water Treatment | $0.03 | $0.03 | $0.02 |
| Other expenses | $0.04 | $0.04 | $0.03 |
| SGA* | $0.04 | $0.03 | $0.03 |
| Total per gallon | $0.53 | $0.46 | $0.44 |

*Sales/general/administrative costs.

**TABLE 5.8** Effect of Plant Size on Operating Costs of Wet Mill Ethanol Plants

| Cost component | 15 million gal/yr | $/bu | 30 million gal/yr | $/bu | 40 million gal/yr | $/bu |
|---|---|---|---|---|---|---|
| Capital cost | $27,000,000 | $4.77 | $38,000,000 | $3.25 | $45,000,000 | $2.98 |
| SGA* | $524,040 | $0.09 | $930,000 | $0.08 | $1,120,420 | $0.07 |
| Depreciation (15.7 yr)† | $1,719,740 | $0.30 | $2,420,380 | $0.21 | $2,866,240 | $0.19 |
| Loan payment‡ | $1,296,000 | $0.23 | $1,824,000 | $0.16 | $2,160,000 | $0.14 |
| Annual feedstock throughput @ 2.65 gal/bu | 5.66 million bushels | | 11.32 million bushels | | 15.10 million bushels | |

*Sales/general/administrative costs.
†12 years on equipment and 20 years on building.
‡60% financing at 6% interest for 20 years.

**TABLE 5.9** Dry-Mill Plant Size vs. Financing Costs

| Cost component | 30 million gal/yr | $/bu | 40 million gal/yr | $/bu | 100 million gal/yr | $/bu |
|---|---|---|---|---|---|---|
| Capital cost | $81,100,000 | $6.89 | $96,400,000 | $6.15 | $167,000,000 | $4.26 |
| SGA* | $1,059,000 | $0.09 | $1,255,000 | $0.08 | $2,745,000 | $0.07 |
| Depreciation (15.7 yr)[†] | $5,165,000 | $0.44 | $6,140,100 | $0.39 | $10,637,000 | $0.27 |
| Loan payment[‡] | $3,893,000 | $0.33 | $1,824,000 | $0.29 | $8,016,000 | $0.20 |
| Annual feedstock throughput @ 2.65 gal/bu | 11.77 million bushels | | 15.69 million bushels | | 39.22 million bushels | |

*Sales/general/administrative costs.
[†]12 years on equipment and 20 years on building.
[‡]60% financing at 6% interest for 20 years.

TABLE 5.10  Wet-Mill Plant Size vs. Financing Costs

|  | 15 million gal | | | 30 million gal | | | 40 million gal | | | 100 million gal | | |
|---|---|---|---|---|---|---|---|---|---|---|---|---|
| Process | $/G | $/b | ROI | $/G | $/b | ROI | $/G | $/b | ROI | $/G | $/b | ROI |
| Dry mill | 0.059 | 0.160 | 8.2% | 0.171 | 0.460 | 33.8% | 0.205 | 0.540 | 45.6% | | | |
| Wet mill | | | | 0 | 0 | 0.0% | 0.104 | 0.270 | 10.8% | 0.203 | 0.520 | 30.2% |

**TABLE 5.11** ROI Comparisons by Plant Size and Process Technologies

developed corn hybrids that are best suited for dry mill ethanol production. These hybrids are characterized by *high total fermentables* (HTF). Research performed at Pioneer indicated that HTF was a more accurate measure of potential ethanol production in dry mill plants than total starch or extractable starch. Selected hybrids were shown to increase ethanol yield by as much as 4 percent at commercial scale compared to "commodity" corn. In a 40 million gallon per year facility, this may represent up to 1 to 2 million dollars in additional profit. The company also developed an analytical tool based on near infrared (NIR) technology to measure HTF and potential ethanol yields of different corn grains and made it available to the ethanol industry on a royalty-free basis.[53]

Monsanto developed several corn hybrids called *processor preferred high fermentable corn* (HFC), which were shown to give 2.7 percent increase in ethanol yield at commercial scale in tests performed in both the United States and Canada.[54]

Syngenta used transgenic technology to develop an amylase corn that produces and stores $\alpha$-amylase in the kernel.[55] The enzyme is activated in the presence of water at temperatures above 70°C. It was demonstrated that when this amylase corn was added to normal corn at 1 to 10 percent by weight, equal ethanol yield was observed compared to the control using all normal corn taken through a typical dry mill process with exogenous addition of $\alpha$-amlyase.[56]

Renessen, a 50-50 joint venture between Cargill and Monsanto, developed Mavera, a corn hybrid with higher oil and lysine contents.[57] The corn hybrid is used in a process developed jointly by Cargill and Renessen, where the corn is fractionated into a highly fermentable fraction (HFF) and a high oil fraction (HOF). The HFF is fermented to produce ethanol and a high protein low oil DDGS coproduct. Corn oil is extracted from the HOF and can be used for either food or biodiesel production. The residues of the oil extraction process are used for swine and poultry feed. The technology has been tested in pilot plants and is expected to be brought up to commercial scale in 2007.

2. New enzymes for starch hydrolysis

Genencor developed Stargen products, which contain an $\alpha$-amylase and a glucoamylase. The synergistic effects of the two enzymes allow starch to be hydrolyzed at temperatures of typical yeast fermentation, thus allowing elimination of the cooking step, which results in significant energy savings. Calcium requirement for $\alpha$-amylase stability at high temperatures also is eliminated, which will greatly reduce the scaling problems. Elimination of the cooking step also replaces the requirement for double adjustments of pH with a single pH adjustment, which will simplify process control and plant operation.[58]

Novozymes also developed low temperature $\alpha$-amylase/glucoamylase products, which also help eliminate the cooking step. These enzyme products are used exclusively in a patented process developed by Poet (formerly Broin).[59] Technical details of the Novozymes $\alpha$-amylase/glucoamylase products are not readily available.

Innovase, a 50-50 joint venture between Diversa and Dow, developed Ultra-Thin, which is a hybrid $\alpha$-amylase capable of operating stably and effectively at pH 4.5. Thus, double pH adjustments are no longer required and only a single pH adjustment is necessary. In the development of this enzyme, the directed evolution process was used to screen DNA from a variety of unusual environments. Three enzymes were selected and then a gene reassembly process was used to isolate fragments from each of these and combine them to create the hybrid enzyme.[60]

3. New production processes

Poet developed the BFRAC process for corn fractionation. Satake, a grain processing company in Japan, collaborated with Poet to develop the equipment for large-scale implementation.[61] The fractionated raw starch then is processed in the BPX process in which the cooking step is eliminated[62] by using the low-temperature starch hydrolysis enzymes developed by Novozymes (see above). Commercialization of the technologies was begun in 2003. Since then the technologies have been implemented at 11 Poet ethanol plants with combined capacity of 525 million gallons per year. The reported average increase in final ethanol concentration is 20 percent.[63]

The USDA's Agricultural Research Service (ARS) at the Eastern Regional Research Center (ERRC) has supported studies for improvement of fermentation processes for corn ethanol production. A continuous dry-grind fermentation and stripping process was developed.[64] The process has been demonstrated in a pilot plant. A process simulation model was developed for a 15 million gallon per year plant. The results obtained with the model indicated a savings of $0.03/gal of ethanol compared to a state-of-the-art plant using a conventional dry-grind process.

The group at the ERRC also has developed the enzymatic milling (E-milling) process.[65,66] This process begins with soaking of corn in water for brief periods (less than 6 h), followed by a coarse grind and addition of a protease to release the starch granules from the endosperm. The E-milling process does not require sulfur dioxide to obtain starch yield equivalent to a conventional wet-milling process. However, complete removal of sulfur dioxide from the process will result in loss of its important antimicrobial effects. Thus, low levels of sulfur dioxide may still be needed. The E-milling process has been successfully tested at pilot scale.

## 5.2 Ethanol Production from Lignocellulosic Feedstocks

### 5.2.1 Basic Concept

Lignocellulosic feedstocks consist of three main components, cellulose, hemicellulose, and lignin. Technologies for conversion of these feedstocks to ethanol have been developed on two platforms, which can be referred to as the sugar platform and the synthesis gas (or syngas) platform. The basic steps of these platforms are shown in Fig. 5.9.

In the sugar platform, cellulose and hemicellulose are first converted to fermentable sugars, which then are fermented to produce ethanol. The fermentable sugars include glucose, xylose, arabinose, galactose, and mannose. Hydrolysis of cellulose and hemicellulose to generate these sugars can be carried out by using either acids or enzymes. Pretreatments of the biomass are needed prior to hydrolysis.

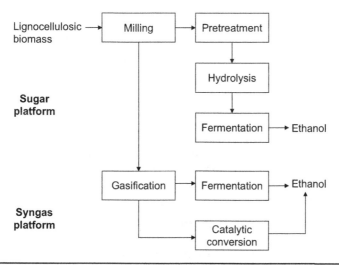

FIGURE 5.9 Basic concept of ethanol production from lignocellulosic feedstocks.

The main objectives of the pretreatment process are to speed up the rates of hydrolysis and increase the yields of fermentable sugars. In all pretreatment processes, these goals are accomplished by modifying the structure of the polymer matrix in the biomass, thus making the carbohydrate fractions more susceptible to acid attack or more accessible to enzyme action. Lignin normally is considered a waste and is burned to supply thermal energy.

In the syngas platform, the biomass is taken through a process called *gasification*. In this process, the biomass is heated with no oxygen or only about one-third the oxygen normally required for complete combustion. The biomass subsequently is converted to a gaseous product, which contains mostly carbon monoxide and hydrogen.[67] The gas, which is called synthesis gas or syngas, then can be fermented by specific microorganisms or converted catalytically to ethanol.

In the sugar platform, only the carbohydrate fractions are utilized for ethanol production, whereas in the syngas platform, all three components of the biomass are converted to ethanol feedstock.

### 5.2.2 The Sugar Platform

#### 5.2.2.1 Biomass Pretreatment

An ideal biomass pretreatment process should meet the following requirements:

- High rates of hydrolysis and high yields of fermentable sugars
- Minimal degradation of the carbohydrate fractions
- No production of compounds that are inhibitory to microorganisms used in the subsequent fermentation step
- Inexpensive materials of construction
- Mild process conditions to reduce capital costs
- Recycle of chemicals to reduce operating costs
- Minimal wastes

There currently is no single pretreatment process that meets all of the above requirements. In selection of a pretreatment process, all of the above requirements should be considered. The biomass to be processed should also be considered since a pretreatment process may be a good choice for one type of biomass but may not be suitable for another. In this section, the pretreatment processes that have been developed are discussed.

**Concentrated Sulfuric Acid Hydrolysis** The use of concentrated acid, especially sulfuric acid, for cellulose hydrolysis has been known for a long time. The process developed by Arkenol by far has the best chance of reaching commercialization. The Arkenol's process[68] is shown in Fig. 5.10. In this process, decrystallization of cellulose and

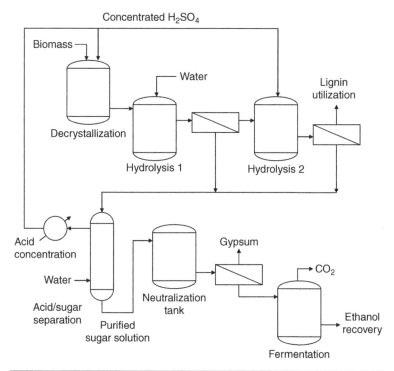

**FIGURE 5.10** The Arkenol's concentrated acid process for biomass hydrolysis. (*Adapted from Farone and Cuzens, 1996.*)

hemicellulose is carried out by adding 70 to 77 percent sulfuric acid to a biomass that has been dried to 10 percent moisture. The acid is added to achieve a ratio of acid to total cellulose plus hemicellulose of 1.25:1 and the temperature is maintained at about 50°C. The concentrated acid disrupts the hydrogen bonding between cellulose chains and converts it to an amorphous state, which is extremely susceptible to hydrolysis. Dilution of the acid to 20 to 30 percent will cause hydrolysis of both cellulose and hemicellulose to monomeric sugars. The hydrolysate is separated from the residual biomass by pressing. The partially hydrolyzed biomass then undergoes a second hydrolysis, which uses conditions similar to the first one. This second hydrolysis releases the rest of the sugars. The key unit operation that separates the Arkenol's process from other concentrated sulfuric acid processes is the acid/sugar separation, which allows recycle and reuse of sulfuric acid. The sugars are separated from the acid in a moving-bed chromatography column containing a cross-linked polystyrene cation exchange resin. The product is a liquid containing at least 15 percent sugar and less than 3 percent acid. The sugar solution then is neutralized and the acid from the ion-exchange column is reconcentrated in a triple-effect evaporator and recycled.

One drawback is large quantities of gypsum, which is generated in the neutralization of the sugar stream, will need to be removed. Lignin is burned to generate thermal energy for use in the plant. Arkenol claims that sugar recovery in the acid/sugar separation column is at least 98 percent and acid lost in the sugar stream is less than 3 percent. Experiments performed with rice straw hydrolysate made by Arkenol's process using immobilized cells of recombinant *Zymomonas mobilis* 31821 (pZB5) showed no inhibition of ethanol production. In a continuous fluidized-bed reactor, 92 percent conversion of xylose and complete conversion of glucose were achieved.[69]

It is reported that BlueFire Ethanol plans to use Arkenol's process in their first biomass ethanol plant, which will be built in California. The intended feedstocks include municipal solid waste (MSW), rice and wheat straws, wood waste, and other agricultural residues. Other partners of this project include the Japanese engineering firm JGC and MECS (formerly Monsanto EnviroChem Systems).[70]

**Dilute Sulfuric Acid Hydrolysis** Dilute acids also can be used for pretreatment of lignocellulosic biomass. However, dilute acids can only partially hydrolyze biomass to monomeric sugars. Following dilute acid treatment, the enzyme cellulase is needed for hydrolysis of the remaining carbohydrates in the treated biomass. Sulfuric acid is the acid that is used most widely. Dilute acid pretreatment can be a simple single-stage process in which biomass is treated with dilute sulfuric acid at suitable acid concentrations and temperatures for a period of time. Examples of pretreatment conditions for softwoods are 0.4 percent sulfuric acid, 200 to 230°C, and 1 to 5 min.[71] When Douglas fir chips were treated under these conditions, 90 to 95 percent of the hemicellulose and 20 percent of the cellulose was solubilized, and 90 percent of the remaining cellulose was hydrolyzed to glucose by cellulase. To reduce enzyme requirements, a two-stage process was developed at the National Renewable Energy Laboratory (NREL) in Golden, Colorado.[72] A schematic diagram of this process is shown in Fig. 5.11.

The biomass solid content in the first-stage acid impregnation is 30 to 35 percent by weight. After drying, the biomass arrives at the first-stage hydrolysis at 40 to 60 percent solids by weight. The biomass is treated at 130 to 220°C in the first-stage hydrolysis. It is then discharged into a flash tank at 120 to 140°C. The treated biomass is washed with water in a counter-current extractor. At least 95 percent of the soluble sugars are recovered from the extractor. The pH of the first hydrolysate is adjusted to a suitable value and taken to the fermentation section. The washed solids are fed to the second-stage acid impregnation at 40 to 60 percent solids by weight. The solids are treated under more severe conditions (190 to 240°C) to hydrolyze part of the remaining cellulose to glucose. The solids then are discharged into a flash tank

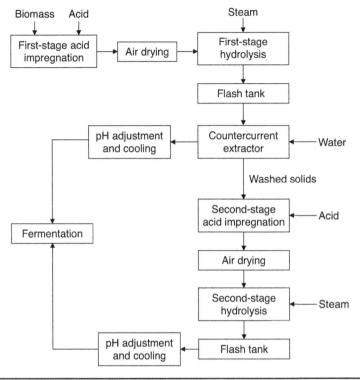

**Figure 5.11** The NREL's two-stage dilute sulfuric acid pretreatment process.

at 120 to 140°C. The soluble sugars are extracted and the hydrolysate is pH adjusted for fermentation in a similar manner to the one described for the first stage. The residual solids are subject to enzyme hydrolysis to generate more fermentable sugars for fermentation. Compared to the single-stage process, the overall sugar yield is approximately 10 percent higher and the enzyme requirement is reduced by about 50 percent in the two-stage process.[73]

The dilute acid hydrolysis process is very effective in releasing fermentable sugars from several types of biomass. However, it has a major drawback. In dilute acid treatment of biomass, a number of degradation compounds are formed. These compounds, which include furan derivatives such as furfural and 5-hydroxy-methyl-furfural (5-HMF); organic acids such as acetic, formic, and levulinic acid; and phenolic compounds[74] are inhibitory to ethanol-producing organisms intended for conversion of biomass-derived sugars. Several methods have been developed for detoxification of dilute acid hydrolysates. The two most promising detoxification methods among these are treatment with calcium hydroxide at pH 9 to 10, which normally is referred to as overliming, and anion exchange.

Anion exchange effectively removes all three groups of inhibitory compounds, but also some of the fermentable sugars. In addition, this is an expensive process. Overliming causes degradation of the inhibitory compounds furfural and 5-HMF, but also results in loss of sugars.[75] This detoxification process is expensive also since it requires pH adjustment to values as high as 9 or 10, followed by readjustment back to about 5 for fermentation. In addition, it generates large quantities of gypsum, which adds disposal cost to operating expenses. If dilute acid is used for biomass pretreatment, the best option seems to be development of microorganisms that can tolerate the levels of inhibitory compounds in the hydrolysates.

**Steam Explosion** In the steam-explosion pretreatment process, biomass is exposed to superheated steam in a reactor. The high-pressure steam penetrates the biomass and initiates an autohydrolysis reaction. The organic acids, which are formed initially from the acetyl groups in the biomass, catalyze hydrolysis of most of the hemicellulose fraction to soluble sugars.[76] After a specific reaction time, a ball valve is rapidly opened to discharge the biomass explosively into a collection tank at a much lower pressure. Upon hydrolysis by cellulases, the biomass treated by the steam-explosion process yields much higher fermentable sugars than the untreated material. For example, enzymatic hydrolysis of poplar chips treated by this process achieves 90 percent of theoretical yield compared to 15 percent obtained with untreated material.[77] It is believed that improvement of enzyme hydrolysis is caused by removal of most of the hemicellulose and some of the lignin rather than disruption of the biopolymer structure by the explosion.[78]

The important process variables in steam explosion are reaction time, temperature, particle size, and moisture content.[79] Good results can be obtained at either high temperature and short reaction time (270°C, 1 min) or low temperature and long reaction time (190°C, 10 min). The general rule is, when larger particle size is used, reaction time is increased and temperature is decreased to avoid overcooking of the biomass outer surface and formation of inhibitory compounds similar to those formed in dilute acid hydrolysis. Excess water also should be avoided. If the pores contain water, heat will be transferred by conduction and rapid steam penetration will not occur. This will delay initiation of autohydrolysis.[80]

Sulfuric acid can be added to improve enzymatic hydrolysis.[81] The amount of acid used normally is not more than 1 percent. The acid can be added either by spraying the biomass with a dilute acid solution or by soaking the biomass in it.

Although it is not very effective for treatment of softwoods, steam explosion is considered as one of the most cost-effective pretreatment processes for hardwoods and agricultural residues.[78] It is reported that the Canadian company SunOpta will supply steam-explosion equipment and process technology to ethanol producer Abengoa for use in their first commercial plant for production of ethanol from cereal straw. It also is reported that the Canadian biomass ethanol company Iogen is using a modified steam-explosion pretreatment process.[82]

**Ammonia Treatment** Bases such as sodium hydroxide, potassium hydroxide, and ammonia can be used for biomass pretreatment. Base solutions cause swelling of biomass, which subsequently leads to decrease in the degree of polymerization, decrease in crystallinity, disruption of the lignin structure, and separation of structural linkages between lignin and carbohydrates.[83] Among the bases investigated, ammonia has the highest potential for use in commercial processes since it can be recovered and recycled due to its high volatility. Thus, it reduces chemical cost and waste treatment cost.

A flow-through process called Ammonia Recycle Percolation (ARP) was developed for pretreatment of corn stover.[84] In this process, 15 percent (by weight) ammonia is pumped through a bed of biomass maintained at 170°C and 2.3 MPa. Although up to 85 percent delignification and almost theoretical yield of glucose in enzyme hydrolysis can be achieved, significant amounts of xylan also are lost. For fractionation of biomass the process can be modified by adding a hot-water wash step to extract up to 86 percent of the hemicellulose prior to ammonia delignification.[85] However, the concentrations of solubilized xylan in the wash water are only about 20 g/L, which is too low for economical ethanol fermentation. In a batch process called *soaking in aqueous ammonia* (SAA), corn stover is simply soaked in 29.5 wt% ammonia for extended periods. Up to 74 percent lignin can be removed whereas 85 percent xylan and almost all glucan are retained.[86] Thus, conversion of glucan and xylan to fermentable sugars can be performed in a single step. Enzyme hydrolysis of the treated material gives 85 percent theoretical yield for xylose and 95 percent theoretical yield for glucose. The effectiveness of the SAA process is strongly dependent on the pretreatment temperature. For example, at 25°C 10 days are needed whereas at 60°C the reaction time can be reduced to only 12 h.[87]

The ammonia fiber/freeze explosion/expansion (AFEX) process uses anhydrous ammonia instead of aqueous ammonia.[88] The concept of AFEX is similar to steam explosion. Biomass in the AFEX process is brought into contact with anhydrous ammonia at temperatures ranging from 60°C to 120°C. After a specific reaction time, the pressure is rapidly released to explode the treated biomass into a collection chamber. The important process parameters include moisture content of the biomass, ammonia dosage, temperature, and reaction time. The AFEX process has many advantages. Similar to the ARP and SAA process, the ammonia used in the AFEX process can be recovered and recycled due to its high volatility. The AFEX process does not generate a waste. After treatment, the only exit stream is a gas mix containing ammonia and water vapor. All biomass components remain with the treated solids. Thus, there is no loss of any carbohydrate fraction. Since all of the ammonia will quickly evaporate, there is no need for pH adjustment of the treated material over a wide range before it can be used in subsequent enzyme hydrolysis and ethanol fermentation. Enzyme hydrolysis of AFEX-treated biomass can produce glucose with greater than 90 percent theoretical yield and xylose with up to 80 percent theoretical yield.[89] There is no formation of inhibitory compounds. Hydrolysates obtained by enzyme hydrolysis of AFEX-treated corn stover did not show any sign of inhibition when used in ethanol fermentation.[90] The AFEX process, however, also has some drawbacks. Compared to the SAA process, the AFEX process is more complicated to implement at commercial scale. High pressures of anhydrous ammonia in the pretreatment reactor may require expensive pressure vessels. Finally, since there is no lignin removal, solid loadings in the ethanol fermentor may be more restricted than in the case of biomass treated by other processes with significant delignification.

**Lime Treatment** Biomass can also be pretreated with lime to improve subsequent enzyme hydrolysis to fermentable sugars. Typical lime loading is 0.1 g $Ca(OH)_2$ per gram biomass. A minimum amount of 5 g water per gram biomass is needed.[91] Lime treatment can be performed at temperatures below 100°C to avoid the use of expensive pressure vessels, but at such low temperatures the required treatment times normally are very long. For example, corn stover treated with excess lime at 0.5 g $Ca(OH)_2$ per gram biomass at 55°C with aeration needed 4 weeks.[92] At the end of the treatment, 87.5 percent of the lignin was removed and some of the carbohydrate fractions were also solubilized. The total yields of glucose and xylose after enzyme hydrolysis were 93.2 percent and 79.5 percent, respectively. Lime treatment requires long reaction time but is a simple and inexpensive pretreatment process.

**Alkaline Peroxide Treatment** The use of alkaline solutions of hydrogen peroxide was one of the early attempts to develop pretreatment processes for biomass. In this treatment, large fractions of the hemicellulose and lignin are solubilized whereas most of the cellulose remains intact. The cellulose in the residual solid can be hydrolyzed with enzymes at very high rates and near theoretical yields. The optimum pH is 11.5, which is the pKa for the dissociation of $H_2O_2$. When this pretreatment was applied to corn stover, most of the hemicellulose and as much as 50 percent of the lignin were solubilized. The residual solid fraction, which still contains most of the original cellulose, was hydrolyzed with cellulase to above 90 percent of theoretical glucose yield.[93] The solubilized hemicellulose in the liquid fraction can be recovered, for example, by adding excess ethanol.[94] The recovered hemicellulose, which contains as much as 50 percent xylose, can be further purified and converted into valuable products. Alternatively, it can be hydrolyzed with enzymes or acids to produce xylose, which then can be fermented to ethanol by xylose-metabolizing microorganisms. Lignin also can be recovered for production of high-value products. The high cost of hydrogen peroxide may be inhibitory to a process with ethanol being the sole product. However, the other option, that is, conversion of the recovered hemicellulosic fraction to higher-value products, may be a viable process for use in a biorefinery.

**Wet Oxidation**   In the wet oxidation process, biomass is treated with water and air or oxygen at elevated temperatures and pressures.[95] Similar to the alkaline peroxide treatment process, large fractions of hemicellulose and lignin are solubilized during wet oxidation, leaving a solid residue high in cellulose. The cellulose in the residual solid can be hydrolyzed with enzymes at high rates and yields.[96] The main advantage of the wet oxidation over the alkaline peroxide process is replacement of hydrogen peroxide by air, which helps reduce the chemical costs significantly. However, if the solubilized hemicellulose is not recovered, the effect on the overall process economics caused by loss of potential fermentable sugars should not be overlooked.

**Organosolv Fractionation**   The organosolv process uses hot organic solvents such as ethanol at acidic pH to fractionate biomass components. It was first considered for paper making, but recently it has also been considered for pretreatment of lignocellulosic feedstocks for ethanol production.[97] In this process, lignin and hemicellulose are solubilized, leaving a cellulose-rich residue, which can be hydrolyzed with enzymes at high rates and to almost theoretical glucose yield and used for ethanol production. Hemicellulose and lignin can be recovered for production of high-value coproducts. When this process was applied to hybrid poplar under the conditions of 180°C, 60 min reaction time, 1.25 percent sulfuric acid, and 60 percent aqueous ethanol, it yielded a solid residue containing 88 percent of the original cellulose. The total recoveries of hemicellulose and lignin were 72 percent and 74 percent, respectively.[97]

**Concentrated Phosphoric Acid Fractionation**   This process was developed just recently based on the well-known technique for generating amorphous cellulose with concentrated phosphoric acid. A proposed process flow diagram[98] is shown in Fig. 5.12. The claimed advantages of this process include:

- Solubilization of cellulose and hemicellulose in biomass can be carried out at a moderate temperature (50°C) using concentrated phosphoric acid (>82 percent).
- The solubilized cellulose and hemicellulose can be separated based on the insolubility of cellulose in water. The cellulose obtained from several biomass materials (corn stover,

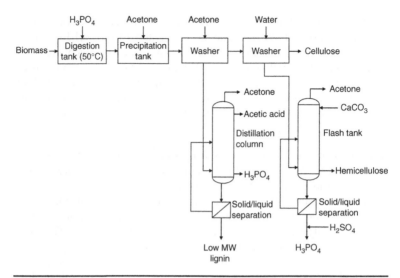

**FIGURE 5.12** Fractionation of biomass using concentrated phosphoric acid. (*Adapted from Zhang et al., 2007.*)

switchgrass, hybrid poplar) can be hydrolyzed with enzymes at very high rates and to almost theoretical yield. The hemicellulose can be converted to high-value products.

- Phosphoric acid and acetone can be recovered by simple processes and recycled.
- Lignin can be recovered and potentially can be used to generate high-value products.

The process concept is interesting but it has only been tested at a very modest scale (1 g of biomass) in the laboratory. It needs to be validated at a larger scale where meaningful process data can be generated for economic evaluation.

**Ionic Liquid Fractionation** Ionic liquids are salts that are nonvolatile liquids at or near room temperature and are stable up to 300°C. Crystalline cellulose can be solubilized in ionic liquids and then precipitated in amorphous form by adding an anti-solvent such as water or ethanol.[99] The treated cellulose can be hydrolyzed with enzymes at initial rates 90 times those observed for the untreated cellulose and to greater than 90 percent theoretical glucose yield. The ionic liquids can be regenerated simply by evaporating off the anti-solvent. Cellulose in corn stover can be dissolved in ionic liquids.[99] Ionic liquids also can be used to selectively solubilize lignin in biomass materials.[100] Up to 86 percent delignification of bagasse has been achieved with ionic liquids.[101] The recovered lignin can be used to make high-value products, whereas the residual solids can be subjected to enzyme hydrolysis for fermentable sugar production. Currently the biggest problem of using ionic liquids in industrial applications is their extremely high costs.[102] If these costs can be brought down substantially, for example, by making ionic liquids from inexpensive and renewable feedstocks such as glucose,[103] fractionation of biomass with ionic liquids may become a viable pretreatment process.

### 5.2.2.2 Enzyme Hydrolysis

Cellulose can be hydrolyzed to glucose by the enzyme complex called *cellulase*, which is excreted by organisms capable of degrading cellulose. This enzyme complex consists of an endo-$\beta$-(1→4)-glucanase (Cx-cellulase) and an exo-$\beta$-(1→4)-glucanase (cellobiohydrolase). The Cx-cellulase breaks the bonds in the amorphous regions of the cellulose molecules, and the cellobiohydrolase removes cellobiose from the nonreducing ends. The combined action of the two enzymes causes the degradation of cellulose. Another enzyme, called *cellobiase* [$\beta$-(1→4)-glucosidase] is required to hydrolyze cellobiose to glucose. This enzyme is excreted by fungi and a number of aerobic bacteria growing on cellulose.[104] Industrial cellulases already contain some cellobiase activity but normally at relatively low levels. Therefore, in enzymatic hydrolysis of biomass, a cellobiase needs to be added to ensure high yields of glucose.

Enzyme hydrolysis of hemicellulose is much more complex. Complete breakdown of this heterogeneous biopolymer requires the action of several hydrolytic enzymes. These enzymes collectively are called *hemicellulases* and consist of endo-enzymes, which cleave internal glycosidic bonds, exo-enzymes, which remove sugar residues from the nonreducing ends, and esterases, which attack nonglycosidic ester linkages. The hemicellulose hydrolytic enzymes include endo-$\beta$-1,4-xylanase, exo-$\beta$-D-xylosidase, $\alpha$-L-arabinofuranosidase, endo-1,5-$\alpha$-L-arabinanase, $\alpha$-glucuronidase, acetyl esterases, which consist of acetylxylan esterase and acetyl esterase, and phenolic acid esterases, which consist of feruloyl esterase and p-coumaroyl esterase. Figure 5.13 shows the activity of the enzymes required for complete hydrolysis of wheat straw xylan and their sites of attack.[105]

Commercial cellulases normally also contain hemicellulase activities. In fact, the hemicellulose in pretreated corn stover has been routinely hydrolyzed to 70 to 80 percent theoretical yield with commercial cellulases.[84–92] However, because of their complex nature, not all biomass hemicelluloses can be effectively hydrolyzed with commercial enzymes as in the case of corn stover. The best strategy for complete hydrolysis of corn fiber to fermentable sugars is to use dilute sulfuric acid under relatively mild conditions, for example 0.25 percent acid by weight for 1 h at 121°C, to hydrolyze the hemicellulose fraction, then using commercial cellulases to hydrolyze the cellulose in the residual solids.[106]

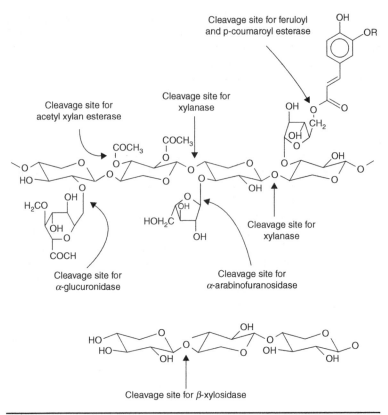

**FIGURE 5.13** The enzymes required for hydrolysis of wheat straw xylan and their sites of attack. (*Adapted from Schmidt et al., 1998.*)

#### 5.2.2.3 Fermentation

Hydrolysis of lignocellulosic biomass generates a mixture of sugars, which includes glucose, xylose, arabinose, galactose, and mannose, with glucose, xylose, and arabinose being the most predominant ones. Unfortunately, the yeast *Saccharomyces cerevisiae*, which is used for industrial ethanol production from sugar and starch feedstocks, can only metabolize glucose to ethanol under anaerobic conditions. Therefore, many attempts have been made to develop microorganisms that can metabolize other sugars in addition to glucose and convert them to ethanol at high yields and rates. The strategy generally includes the following steps:

- Select a suitable host which is either a good ethanol producer but has narrow sugar range or a poor ethanol producer but is capable of utilizing a wide range of sugars
- Transform the host with plasmids harboring the genes required for broadening the range of sugar utilization or improvement of ethanol production
- Perform proof-of-concept experiments

- Integrate the required genes into the chromosome of the host to improve genetic stability
- Inactivate the biosynthetic pathways leading to unwanted by-products to improve ethanol yield

All of the recombinant strains created using the above approach are mesophilic organisms. Several thermophilic organisms having wide ranges of sugar utilization and capability of producing ethanol also have been isolated and mutants lacking pathways for synthesis of by-products then were obtained from the wild-type strains.

Another approach is to develop organisms that can metabolize cellulose and hemicellulose directly, thus eliminating the need of enzymes for biomass hydrolysis. This is the newest approach taken to develop ethanol-producing organisms for lignocellulosic feedstocks and therefore, has not advanced as far as the others. Nevertheless, several strains with interesting characteristics have been developed.

In addition to the capability of converting a wide range of biomass-derived sugars to ethanol, the organism of choice should also possess the following traits:

- Tolerance of inhibitory compounds formed during the pretreatment
- Tolerance of high ethanol levels
- No requirements for expensive nutrients
- High growth rates and high ethanol yields under conditions unfavorable for contaminants, that is, low pH and high temperatures

**Mesophilic Sugar-Utilizing Strains**  Some of the mesophilic sugar-utilizing strains are as follows:

*Zymomonas mobilis*  *Zymomonas mobilis* attracted attention as a potential host because of its effectiveness in ethanol production from glucose and high ethanol tolerance. The first recombinant *Z. mobilis* strain capable of utilizing xylose for ethanol production was developed at the National Renewable Energy Laboratory (NREL). Xylose metabolism was made possible by cloning and expression of four *E. coli* genes: xylose isomerase (*xylA*), xylulose kinase (*xylB*), transketolase (*tktA*), and transaldolase (*talB*) in a *Z. mobilis* host.[107] The genes were expressed on a plasmid using strong constitutive promoters from the host. The proposed pathway for xylose metabolism is shown in Fig. 5.14.

The stoichiometry of ethanol production can be expressed by the equation

$$3 \text{ xylose} + 3 \text{ ADP} + 3 \text{ Pi} \rightarrow 5 \text{ ethanol} + 5 \text{ CO}_2 + 3 \text{ ATP} + 3 \text{ H}_2\text{O}$$

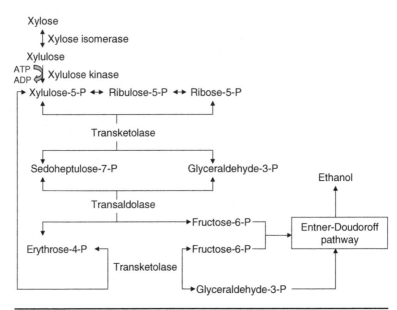

**FIGURE 5.14** The pathway for xylose metabolism in recombinant *Z. mobilis*. (*Adapted from Zhang et al., 1995.*)

The theoretical yield is 0.511 g ethanol per gram xylose or 1.67 mol ethanol per mole xylose.

The same strategy used to develop xylose-metabolizing strains was used to develop strains capable of utilizing arabinose as the sole carbon source.[108] Five *E. coli* genes L-arabinose isomerase (*araA*), L-ribulose kinase (*araB*), L-ribulose-5-phosphate-4-epimerase (*araD*), transketolase (*tktA*), and transaldolase (*talB*) were inserted into the host using a plasmid. The first three enzymes convert arabinose to xylulose-5-phosphate, which then is converted to intermediates of the ED pathway by the transketolase and transaldolase. Strain ATCC39676 (pZB206) was able to convert 25 g/L arabinose to ethanol at 98 percent theoretical yield, but the rates of arabinose utilization were rather slow. The arabinose conversion pathway is shown in Fig. 5.15.

One of the key issues of plasmid-bearing strains is their antibiotic requirement for plasmid maintenance. In the case of the recombinant *Z. mobilis* strains the required antibiotic is tetracycline. Addition of

$$\text{L-arabinose} \xrightarrow{araA} \text{L-ribulose} \xrightarrow[araB]{ATP\ ADP} \text{L-ribulose-5-P} \xrightarrow{araD} \text{D-xylulose-5-P}$$

**FIGURE 5.15**[109] The pathway for conversion of arabinose to the metabolizable intermediate, D-xylulose-5-P, in recombinant *Z. mobilis*.

| Kinetic parameter | pH 5.0 | pH 5.5 |
|---|---|---|
| $\mu_{max}$ (h$^{-1}$) | 0.34 | 0.34 |
| $q_s$ [g sugar/(g cell·h)] | 0.72 | 0.71 |
| $q_p$ [g ethanol/(g cell·h)] | 0.32 | 0.32 |
| $Q_p$ [g ethanol/(L·h)] | 0.59 | 0.61 |
| Process yield (%) | 82.6 | 84.1 |

*Source*: Adapted from Mohagheghi et al., 2002.

**TABLE 5.12**  Kinetic Parameters of Batch Fermentation of *Z. mobilis* Integrated Strain AX101 in Synthetic Medium Containing Glucose, Xylose, and Arabinose at 40, 40, and 20 g/L at 30°C

antibiotics to the fermentor does not just add to the chemical costs but may also cause problem in wastewater treatment. To alleviate this problem several chromosomally integrated *Z. mobilis* strains were developed.[109] These strains do not require tetracycline to maintain genetic stability since all the genes required for metabolism of xylose and arabinose were integrated into the host chromosome instead of being inserted into a plasmid. Among the integrated *Z. mobilis* strains, AX101 was found to be the most genetically stable one. Strain AX101 was able to maintain capability of metabolizing xylose and arabinose after 160 generations. Strain AX101 has been tested in synthetic media containing glucose, xylose, and arabinose at 40, 40, and 20 g/L at pH 5.0 and 5.5 and 30°C in batch fermentation. Although all three sugars were consumed, there was a preferential order of consumption: glucose first, xylose second, and arabinose last. Conversion of xylose and arabinose also were not complete. Some of the kinetic parameters obtained with the synthetic medium[110] are summarized in Table 5.12.

*Saccharomyces cerevisiae*  The choice of *Saccharomyces cerevisiae* as a host for biomass ethanol production is obvious since this yeast has been used universally for commercial production of ethanol from starch and sugar feedstocks for years. It has been shown that *S. cerevisiae* was unable to ferment xylose but it can ferment xylulose.[111] Thus, the first approach on creating a recombinant *S. cerevisiae* capable of utilizing xylose to produce ethanol was to introduce the gene for xylose isomerase, which converts xylose to xylulose, from a xylose-metabolizing organism into the yeast. However, attempts using bacterial xylose isomerase genes failed in producing either a functionally expressed enzyme or an enzyme with sufficient activity.[112]

Two other approaches then were taken. In the approach used by the group at Purdue University, three xylose-metabolizing genes, a xylose reductase (XR) gene, a xylitol dehydrogenase (XDH) gene, both from the xylose-metabolizing yeast *Pichia stipitis*, and a xylulose kinase (XK) gene from *S. cerevisiae*, were placed on highcopynumber plasmids and introduced into selected *Saccharomyces* hosts.[113] Although xylulose kinase already exists in the hosts, the levels of this enzyme need to be increased in the recombinant strain to alleviate the possible redox imbalance between XR and XDH, which causes accumulation of xylitol and lowers ethanol yield. The pathways for xylose conversion to metabolizable intermediates in recombinant *S. cerevisiae* are shown in Fig. 5.16. One of the strains created, *Saccharomyces* 1400(pLNH33), was able to produce almost 50 g/L ethanol from a synthetic mixture of glucose and xylose at 50 g/L each.[114] The Purdue group then successfully integrated the genes required for xylose metabolism into the host chromosome to create strain *Saccharomyces* 424A(LNH-ST). This strain was able to coferment glucose and xylose in hydrolysates made from various types of biomass and produced ethanol at yields ranging from 75 to 97 percent of theoretical value based on consumed sugars. The percent theoretical yields based on total initial sugars ranged from 36 to 86 percent.[115]

The group at Delft University in collaboration with the ethanol producer Nedalco developed a xylose-metabolizing yeast[116,117] by tranforming *S. cerevisiae* with plasmids harboring the xylose isomerase (XI) gene from the anaerobic fungus *Piromyces* sp. E2. The strain thus developed, *S. cerevisiae* RWB202-AFX, was able to ferment xylose to ethanol at 0.42 g/g sugar consumed (82 percent of theoretical yield). To improve xylose utilization, the group overexpressed all the enzymes required for conversion of xylulose to the intermediates of the glycolytic pathway, fructose-6-phosphate and glyceraldehydes-3-phosphate. In addition, the gene responsible for xylitol synthesis was deleted.[112] The new strain, *S. cerevisiae* RWB217, showed some improvement on growth on xylose but no improvement on ethanol production. Xylitol production was reduced but not completely eliminated.

The approach taken by the group at NREL to develop arabinose-metabolizing *Z. mobilis* strains was also used by the group at Delft to

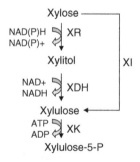

**Figure 5.16**[114]    The pathways for xylose conversion to metabolizable intermediates in recombinant *S. cerevisiae*.

develop arabinose-metabolizing *S. cerevisiae* strains.[118] Thus, the genes for arabinose metabolism from *Lactobacillus plantarum* and those for the nonoxidative pentose phosphate pathway from *S. cerevisiae* were overexpressed in a selected host. The resultant strain *S. cerevisiae* IMS0002 was able to ferment arabinose at 0.70 g/g dry cell weight-hour and produced ethanol at 0.29 g/g dry cell weight-hour and at 84 percent theoretical yield.

*Escherichia coli*    *Escherichia coli* is a good candidate for biomass ethanol production since it can metabolize a wide range of sugars and already is an ethanol producer. The only problem is ethanol yield by this organism is rather low, only 49.8 mol/mol glucose (0.13 g/g glucose).[119] The group at the University of Florida transformed *E. coli* with a plasmid harboring the *Z. mobilis* genes for pyruvate decarboxylase (*pdc*) and alcohol dehydrogenase (*adhII*). The thus created recombinant strain produced ethanol almost exclusively.[120] The two *Z. mobilis* genes were coexpressed under the regulation of the native *lac* promoter, and the construct was named the PET (*p*roduction of *et*hanol) operon. The most promising plasmid-based strain was *E. coli* ATCC11303 (pLOI297). The *Z. mobilis* genes then were integrated into the chromosome of strain *E. coli* ATCC11303.[121] The PET operon was inserted into the pyruvate formate lyase (*pfl*) gene to eliminate competition for pyruvate. The pathway for ethanol production in *E. coli* KO11 is shown in Fig. 5.17. Since the integrated DNA contained a chloramphenicol (Cm) resistance marker,

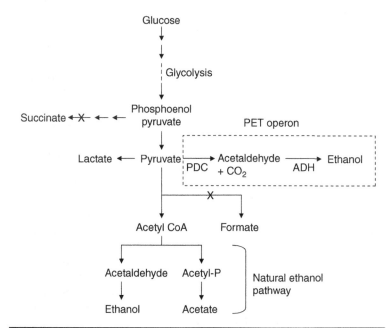

FIGURE 5.17[121]    The pathway for ethanol production in *E. coli* KO11.

mutants were selected based on increased Cm tolerance. The rationale was hyper Cm-tolerant strains might also possess high levels of expressed enzymes from the integrated PET operon. This strategy was successful and after strains with high ethanol production were obtained, the gene for succinate synthesis (*frd*—fumarate reductase) was disrupted to create the final strain *E. coli* KO11. This strain has been tested extensively both in synthetic sugar solutions and in actual biomass hydrolysates. The strain was found to be very stable in batch cultures and able to use an inexpensive nutrient such as corn steep liquor to metabolize glucose, xylose, arabinose, and galactose in hydrolysates and produce ethanol up to 50 g/L at near theoretical yield.[122–125] Thus, *E. coli* KO11 is a good organism for production of ethanol from biomass. However, the near neutral and narrow range of tolerable pH (ca pH 6 to pH 8) is a disadvantage, especially when simultaneous saccharification and fermentation (SSF) operation is considered (see the section on Process Integration).

*Klebsiella oxytoca* The PET operon also was introduced into *Klebsiella oxytoca*. *K. oxytoca* is of particular interest since, in addition to a wide variety of hexoses and pentoses, it also can effectively metabolize cellobiose and cellotriose.[126] This capability may reduce or eliminate the requirement of the expensive enzyme $\beta$-glucosidase that normally is supplemented to commercial cellulases in cellulose hydrolysis. Another advantage of *K. oxytoca* is it can grow well at pH ~ 5, which is the optimum pH range of commercial cellulases.[127] Thus it is well suited for use in the SSF process. Several plasmid-based strains were developed[128] using *K. oxytoca* strain M5A1. The PET operon also was inserted into the chromosome of this strain to create integrated strains.[129] The same strategy used for selection of hyper ethanol-producing recombinant *E. coli* was used for selection of hyper ethanol-producing recombinant *K. oxytoca*. Thus, mutants that demonstrated tolerance for high levels of chloramphenicol were isolated. Among these, strain P2 was the best one. *K. oxytoca* P2 was able to consume 100 g/L cellobiose to produce 45 g/L ethanol in 48 h at 88 percent theoretical yield without $\beta$-glucosidase.[129] It also has been successfully tested with solutions of mix sugars (glucose, xylose, and arabinose)[130] and actual hydrolysates, for example, from dilute acid and AFEX-pretreated sugarcane bagasse.[127]

More recently, attempts were made to give *K. oxytoca* P2 the capability of utilizing cellulose. Genes encoding endoglucanase activities from *Erwinia chrysanthemi* were inserted into the chromosome of *K. oxytoca* P2 using a surrogate promoter from *Z. mobilis* for expression.[131] In experiments using cellulose as substrate with addition of low levels of cellulases, the modified strains were able to produce 22 percent more ethanol than the original strain. These results are encouraging despite the fact that exogenous cellulase is still needed for efficient cellulose conversion.

**Thermophilic Sugar-Utilizing Strains** In addition to the common trait of wide range of sugar utilization, thermophilic ethanol-producing organisms offer a number of advantages, which include:

- Elimination of contamination problem due to high fermentation temperatures
- Lower ethanol recovery costs due to high volatility of ethanol at the fermentation temperatures
- Reduction of ethanol inhibition by continuous product removal under mild vacuum

*Bacillus stearothermophilus* *Bacillus stearothermophilus* is a facultative thermophilic organism that grows well at 65 to 70°C and produces ethanol as one of the main products.[132] Wild-type strains of *B. stearothermophilus* produce significant amounts of lactate via the NAD-linked L-lactate dehydrogenase (LDH) and smaller amounts of formate, acetate, and ethanol via the pyruvate formate lyase (PFL) pathway. A mutant strain, LLD-15, which is lacking LDH, is able to metabolize glucose, xylose, and cellobiose, and produce ethanol as one of the major products at 70°C.[133] Under anaerobic conditions, this strain metabolizes pyruvate primarily through the PFL pathway to yield (in moles per mole sucrose): formate, 4; acetate, 2; and ethanol, 2. However, under certain process conditions, for example, acidic pH and high sucrose, a significant amount of pyruvate is metabolized through the pyruvate dehydrogenase (PDH) pathway and the yield of ethanol is increased to 3.5 mol per mole sucrose. Strain LLD-15 has been proposed for ethanol production in a two-stage continuous process with an aerobic growth stage and an anaerobic production stage,[134] and a single-stage continuous process with partial cell recycle.[133] The main disadvantage of strain LLD-15 and other LDH⁻ mutant strains such as strain LLD-16 and T13 is their reversion to the LDH⁺ parent strain LLD-R if the process conditions are not properly maintained. To enhance the genetic stability and also improve ethanol yield of the LDH⁻ mutant strains, the gene encoding the enzyme pyruvate decarboxylase (PDC, *pdc*) from *Z. mobilis* was integrated into the chromosome to inactivate the *ldh* gene. Stability of the PDC enzyme was improved by integrating the gene encoding a more thermostable PDC from *S. cerevisiae* (*pdc5*) into the chromosome. Ethanol production close to the theoretical yield (0.50 g/g glucose consumed) was claimed.[135]

*Thermoanaerobacter mathranii* *Thermoanaerobacter* species have been considered for ethanol production for quite some time. Several ethanol-producing strains of *Thermoanaerobacter mathranii* recently were isolated from a hot spring in Iceland.[136] The best one, strain A3, was identified during screening for ethanol production on synthetic medium and hemicellulose hydrolysates.[137] This strain was able to metabolize glucose, xylose, arabinose, and xylan. Xylose is metabolized according to the equation

$$\text{Xylose} + H_2O \rightarrow 1.1 \text{ ethanol} + 0.4 \text{ acetate} + 0.06 \text{ lactate} + 1.81\ CO_2 + 0.9\ H_2$$

The theoretical ethanol yield from xylose, therefore, is 0.34 g/g xylose consumed, which is lower than the theoretical yields of recombinant *Z. mobilis* and *S. cerevisiae*. Several strains with improved ethanol production have been developed. Strain BG1L1, which was LDH deficient, was able to coferment glucose and xylose to produce ethanol at 0.45 g/g total sugars consumed. This strain also was able to grow and produce ethanol from mixed sugars in hydrolysates obtained with wheat straw and corn stover.[138]

**Cellulolytic and Xylanolytic Strains** Cellulolytic *S. cerevise* was created by coexpressing *Trichoderma reesei* endoglucanase II and cellobiohydrolase II and *Aspergillus aculeatus* β-glucosidase I on the cell surface of the yeast. The modified yeast strain was able to metabolize amorphous cellulose and produced 3 g/L ethanol from 10 g/L cellulose after 40 h at 0.45 g/g cellulose consumed or 88.5 percent of theoretical value.[139] Similarly, xylanase II from *T. reesei* QM9414 and β-xylosidase from *Aspergillus oryzae* NiaD300 were coexpressed on the surface of *S. cerevisiae* cells to create a xylanolytic yeast. The modified strain was able to hydrolyze xylan to xylose. The xylan-utilizing yeast strain then was transformed with plasmids harboring genes for xylose reductase and xylitol dehydrogenase from *Pichia stipitis* and xylulokinase from *S. cerevisiae*. The transformed strain was able to produce 7 g/L ethanol after 62 h from birchwood xylan at 0.30 g/g xylan consumed.[140]

Recently, a strain of *Clostridium phytofermentans* was isolated from forest soil. The isolated strain, $ISDg^T$, is an obligately anaerobic, mesophilic, cellulolytic bacterium. This strain is able to grow on cellulose, xylan, glucose, xylose, arabinose, galactose, and mannose. The optimum temperature for growth is 35 to 37°C and the major products include ethanol, acetate, $CO_2$, and $H_2$. Formate and lactate are the minor products.[141]

The microorganisms developed for production of ethanol from lignocellulosic biomass and the companies that are planning to use them in their process are summarized in Table 5.13.

### 5.2.2.4 Process Integration

Unlike ethanol production from sugar and starch feedstocks where the substrates, sucrose and glucose, are metabolized by the same pathway, the substrates for biomass ethanol fermentation include several sugars, which are metabolized by different pathways. Design of a process to maximize the conversion of these sugars to ethanol in the most

| Microorganism | Company |
|---|---|
| Z. mobilis | DuPont, Poet (formerly Broin) |
| S. cerevisiae (XR, XDH, XK) | Iogen |
| S. cerevisiae (XI) | Nedalco, Mascoma, SunOpta |
| E. coli | Verenium Corp. (formerly Celunol, BC International), BioEthanol Japan |
| K. oxytoca | Verenium Corp. |
| B. stearothermophilus | Colusa Biomass |
| T. mathranii | Biogasol |
| C. phytofermentans | SunEthanol |

*Note*: The information in this table was obtained by searching the name of the organism or the company on the web site http://www.greencarcongress.com.

**TABLE 5.13** Microorganisms for Production of Ethanol from Lignocellulosic Biomass and Companies That Have Plan for Using Them in Their Process

economical way therefore is not a simple task. In addition, there are many different types of feedstocks available for biomass ethanol production. Preparation of these feedstocks for conversion to ethanol, that is, pretreatment, requires different processes.

The key factors that must be considered in development of a process for production of ethanol from lignocellulosic biomass include:

- Feedstock availability and cost, including transportation cost
- Suitability of pretreatment process for the selected feedstock
- Technical efficiency and economic feasibility of bioconversion process for the pretreated feedstock
- Wastewater treatment and process water recycle

All of the above factors must be considered together before they are integrated into the final process. There currently is not even a single commercial plant for ethanol production from lignocellulosic biomass. However, several companies are near the final stage of starting construction of full-scale plants.

To illustrate process design and integration for ethanol production from lignocellulosic biomass, examples of processes developed by two biomass ethanol companies are presented and discussed in this section.

Figure 5.18 shows a simultaneous saccharification and fermentation (SSF) process proposed by Iogen[142,143] in 1999. Since fermentation of both C5 and C6 sugars are carried out in the same fermentor, this process is also known as *simultaneous saccharification and cofermentation* (SSCF). The pretreatment selected by Iogen is steam explosion with

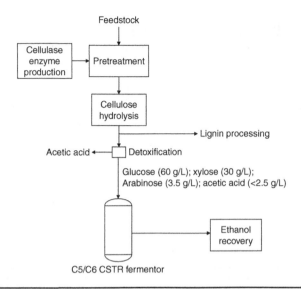

**FIGURE 5.18** SSF biomass to ethanol process as proposed by Iogen in 1999. (*Adapted from Tolan, 1999 and Lawford and Rousseau, 2003.*)

sulfuric acid catalyst. After a proprietary conditioning step the concentration of acetic acid in the slurry is below 2.5 g/L. In the SSF (or SSCF) process, enzyme hydrolysis of the pretreated biomass and subsequent fermentation of the released sugars are carried out together in the same vessel. The optimum conditions of most commercial cellulases are 50 to 55°C and pH 4.5 to pH 5. The thermophilic organisms discussed in the previous section cannot be used in an SSF process since at 70°C the enzymes will be quickly inactivated. The other organisms capable of utilizing C5 sugars are all mesophiles. Thus, the temperature in the reactor must be kept at the optimum temperature of the selected organism, which is around 30 to 35°C. At this temperature there still is sufficient enzyme activity for cellulose and hemicellulose hydrolysis. Among the organisms engineered for C5 sugar utilization, *Z. mobilis, S. cerevisiae,* and *K. oxytoca* have their optimum pH around 5 or lower, whereas *E. coli* has its optimum pH around 6. Thus, the recombinant *E. coli* is the least suitable organism for SSF.

The revised Iogen process using separate hydrolysis and fermentation (SHF)[143] is shown in Fig. 5.19. In this process, enzyme hydrolysis of the pretreated biomass and fermentation of the released sugars are carried out separately. Fermentation of the C5 and C6 sugars also are carried out in two separate fermentors. The separation of the three processing steps adds to the capital costs since instead of a single reactor three are needed. However, the big advantage is each of the three steps can be carried out at their corresponding optimum conditions. The separation of C5 and C6 fermentation allows the use of two organisms. For example, wild-type *S. cerevisiae*, which is the

# Ethanol Production

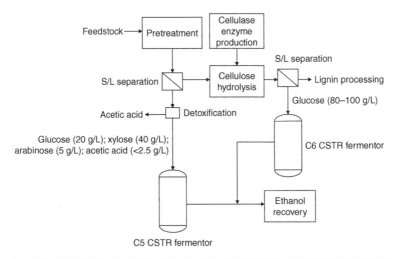

**FIGURE 5.19** The revised Iogen process using SHF and separation of C5 and C6 sugar fermentations. (*Adapted from Lawford and Rousseau, 2003.*)

most efficient organism for commercial ethanol production, can be used for glucose conversion, whereas a suitable organism can be used for conversion of the C5 sugars. The overall efficiency of this process may be higher than that of the SSF process and sufficient to justify the additional capital cost.

In addition to SSF and SHF, there is another process called *consolidated bioprocessing* (CBP).[144]. In this process, cellulase production, biomass hydrolysis, and ethanol fermentation are carried out together in a single reactor. A microorganism that can efficiently ferment cellulose directly to ethanol, such as *Clostridium phytofermentans*, will be most suitable for this process.

Figure 5.20 shows the integrated process proposed by BioGasol.[145,146] The proposed feedstock is wheat straw and the pretreatment process is wet oxidation with oxygen at 180°C and 20 bar. For glucose fermentation the BioGasol process uses the SSF configuration and wild-type *S. cerevisiae*. Xylose fermentation is carried out separately using immobilized *Thermoanaerobacter* BG1L1 in a continuous fermentor operated at 70°C and pH 7.0. The *Thermoanaerobacter* used in this process is a hydrolysate-adapted strain and can tolerate 8.3 percent (v/v) ethanol.[146] Hydrogen also is produced in the xylose fermentor and can be recovered for energy generation. The effluent from the distillation column is treated in a thermophilic anaerobic digestion process at 55°C where methane is produced and recovered for use as an energy source. The clean water is recycled and reused in the pretreatment process. The BioGasol process has been tested in a pilot plant producing 45 L ethanol per day for 138 days.[145] It was shown that there was no inhibition of either

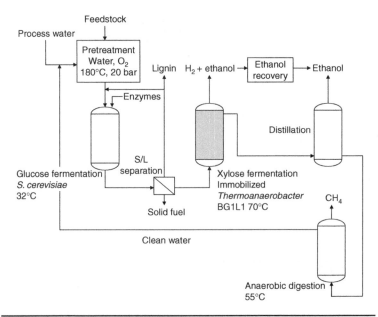

**FIGURE 5.20** The BioGasol process for production of ethanol from wheat straw. (*Adapted from Langvad, 2007 and Mikkelsen and Ahring, 2007.*)

*S. cerevisiae* or *Thermoanaerobacter* BG1L1 by degradation compounds in the hydrolysate and also no reversion of the *Thermoanaerobacter* strain. The pilot-plant test results showed ethanol yield of 0.39 to 0.42 g/g total sugars consumed and conversion of 90 to 96 percent for glucose and 72 to 80 percent for xylose. The mass balance of the overall process based on chemical oxygen demand (COD) is shown in Table 5.14.

#### 5.2.2.5 Process Economics

One characteristic of biomass ethanol production is it is very capital intensive. The capital cost of a biomass ethanol plant can easily be three times that of a corn dry-mill plant of the same capacity.[147] The

| Process component | Percent of total COD |
|---|---|
| Ethanol from glucose | 35% |
| Ethanol from xylose | 19% |
| Lignin for combustion | 20% |
| Biogas | 19% |
| Organic materials in process water | 3% |
| Oxidized in pretreatment | 5% |

*Source*: Adapted from Mikkelsen and Ahring, 2007.

**TABLE 5.14** Mass Balance in the BioGasol Process

main reason for high capital cost of biomass ethanol production facility is the expensive equipment associated with the pretreatment of feedstocks. In pretreatment processes that require detoxification of the sugar solution, this unit operation adds significantly to both capital and chemical costs. The cost of ethanol production from biomass also is strongly dependent on feedstock cost. The plant gate cost of ethanol produced from crop residues such as corn stover in a 50-million-gallon-per-year plant can jump from $1.40/gal at a feedstock cost of $40/ton to $1.65/gal if the feedstock cost increases to $60/ton.[146]

The pretreatment process using dilute acid hydrolysis normally is used as the reference in economic cost models. This process is selected for economic evaluation of biomass ethanol technology because it is the most extensively studied pretreatment process and considered by many, including the U.S. Department of Energy, as the process of choice for near-term commercialization.

The estimated capital costs for a 58-million-gallon-per-year (220,000 M$^3$/yr) ethanol plant using the dilute sulfuric acid pretreatment process are shown in Table 5.15.

| Cost category | Million $ (2006) |
| --- | --- |
| Feedstock handling (wood or switchgrass) | 12.7 |
| Pretreatment | 41.9 |
| Xylose fermentation | 10.9 |
| Cellulase production | 5.0 |
| Simultaneous saccharification and fermentation | 37.0 |
| Ethanol recovery | 7.1 |
| Off-site tankage | 7.2 |
| Environmental systems | 7.0 |
| Utilities (steam, electricity, water) | 90.0 |
| Miscellaneous | 8.5 |
| Fixed capital investment | 227.3 |
| Start-up costs | 11.4 |
| Working capital | 11.3 |
| Total capital investment | 250.0 |

*Note*: The ethanol yield is assumed to be 95 gal/dry ton (396 L/dry MT).
*Source*: Adapted from Solomon et al., 2007.

**TABLE 5.15** Estimated Capital Costs for a 58-million-gallon-per-Year (220,000 M$^3$/yr) Ethanol Plant

| Cost category | Million $/yr | Cents/L (Cents/gal) |
|---|---|---|
| Feedstock (wood or switchgrass) | 39.85 | 18.2 (69.0) |
| Enzymes | 11.50 | 5.3 (20.0) |
| Other raw materials (sulfuric acid, lime, glucose, nutrients) | 5.67 | 2.6 (9.8) |
| Gypsum disposal | 0.59 | 0.26 (1.0) |
| Electricity | −4.88 | −2.2 (−8.3) |
| Water | 0.21 | 0.11 (0.4) |
| Labor/supervision | 2.33 | 1.06 (4.0) |
| Maintenance | 7.70 | 3.49 (13.2) |
| Direct overhead | 1.42 | 0.63 (2.4) |
| General overhead | 7.04 | 3.17 (12.0) |
| Insurance and property taxes | 3.86 | 1.74 (6.6) |
| Total cash costs | 75.29 | 34.37 (130.1) |
| Annualized capital charge* | 50.00 | 22.59 (85.5) |
| Total production cost | 125.29 | 59.96 (215.6) |

*20% of total capital investment, and assuming a 10% after-tax rate of return on capital investment.
Source: Adapted from Solomon et al., 2007.

**TABLE 5.16** Estimated Ethanol Production Cost for a 58-Million-Gallon-per-Year (220,000 M$^3$/yr) Plant

The estimated cost of ethanol production is shown in Table 5.16.

## 5.2.3 The Syngas Platform

### 5.2.3.1 Biomass Gasification

Gasification of biomass can be considered as a two-stage process. It begins with pyrolysis, which then is followed by gasification. In pyrolysis, which occurs at around 500°C, biomass is decomposed to form gases and liquids; the remaining nonvolatile materials are referred to as *char*. The volatile materials and char then are converted to syngas in the second stage, gasification, which occurs at around 1000°C and higher[149] (Fig. 5.21).

The major chemical reactions of biomass gasification are summarized here.[149]

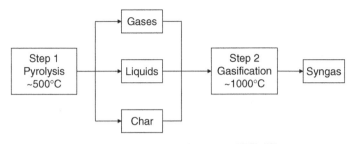

**FIGURE 5.21** The two stages of the biomass gasification process. (*Adapted from Ciferno and Marano, 2002.*)

Exothermic reactions:

| | | |
|---|---|---|
| Combustion | (Biomass volatiles/char) + $O_2 \rightarrow CO_2$ | (1) |
| Partial oxidation | (Biomass volatiles/char) + $O_2 \rightarrow CO$ | (2) |
| Methanation | (Biomass volatiles/char) + $H_2 \rightarrow CH_4$ | (3) |
| Water-gas shift | $CO + H_2O \rightarrow CO_2 + H_2$ | (4) |
| CO metanation | $CO + 3 H_2 \rightarrow CH_4 + H_2O$ | (5) |

Endothermic reactions:

Steam-carbon reaction (Biomass volatiles/char) + $H_2O \rightarrow CO + H_2$ (6)
Boudouard reaction (Biomass volatiles/char) + $CO_2 \rightarrow 2 CO$ (7)

Heat requirements by the endothermic reactions can be supplied directly or indirectly. In the directly heated gasification process both stages, that is, pyrolysis and gasification, are carried out in a single vessel. An oxidant, air or oxygen, is used for combustion of a portion of the biomass [reactions (1) and (2)]. The temperature of the reactor is controlled by the feed rate of the oxidant. If air is used, the syngas product will be diluted with nitrogen but the cost of oxygen is avoided. The indirectly heated gasification process uses a bed of hot solid particles, normally sand, which are fluidized using steam as the fluidizing fluid. The solids, that is, sand and char, are separated from the syngas by a cyclone and transferred to a second fluidized-bed reactor. The fluidizing fluid in the second reactor is air. The char is combusted to generate energy to heat the sand particles. The hot sand particles are separated from the flue gas and returned to the first reactor (the gasifier) to provide the heat required for pyrolysis. This approach generates a syngas that is practically nitrogen free.[149]

There are four types of gasifiers: fixed-bed updraft, fixed-bed downdraft, bubbling fluidized bed, and circulating fluidized bed. The operations of these gasifiers are summarized in Table 5.17 and their advantages and disadvantages are listed in Table 5.18.

| Gasifier type | Flow direction | Heat source | Biomass oxidant |
|---|---|---|---|
| Fixed-bed updraft | Down | Up | Combustion of char |
| Fixed-bed downdraft | Down | Down | Partial combustion of volatiles |
| Bubbling fluidized bed | Up | Up | Partial combustion of volatiles and char |
| Circulating fluidized bed | Up | Up | Partial combustion of volatiles and char |

Source: Adapted from Ciferno and Marano, 2002.

TABLE 5.17  Operations of the Four Main Types of Gasifiers

Before entering the gasifier the biomass has to go through size reduction and drying. The maximum allowed particle size and moisture content vary depending on the type of gasifier used for syngas production. The raw syngas from the gasifier needs to be cleaned before it can be used in the next processing step. Incompletely converted biomass and ash particles can be removed by cyclones, wet scrubbing, or high-temperature filters. Water scrubbing is the simplest method but it generates large quantities of contaminated water that needs to be treated. Tars, which are mostly polynuclear hydrocarbons, for example, pyrene and anthracene, can be removed by catalytic cracking. The reaction involved in catalytic tar cracking is

$$C_nH_{2m} + nH_2O \rightarrow nCO + (m+n)H_2$$

The composition of the syngas product varies depending on the source of biomass, gasifier type, and operating conditions of the gasification process. In a report prepared for the National Renewable Energy Laboratory, several gasifier manufacturers were surveyed.[149] The results on operating conditions and composition of the syngas products are summarized in Tables 5.19 and 5.20.

### 5.2.3.2 Syngas Fermentation

There are several microorganisms that can metabolize components of syngas and produce ethanol.[150] Among these, only *Clostridium ljungdahlii* has been developed into a commercial process (see the section on Process Integration). The two major products of this organism when growing on syngas are acetic acid and ethanol. The biosynthetic pathway for ethanol

# Ethanol Production

| Gasifier type | Advantages | Disadvantages |
|---|---|---|
| Fixed-bed updraft | • Simple, low cost process<br>• Able to handle biomass with high moisture and high inorganic content (e.g., municipal solid waste)<br>• Proven technology | • Syngas contains 10–20% tar by weight, requiring extensive clean-up |
| Fixed-bed downdraft | • Up to 99.9% of the tar formed is consumed, requiring minimal or no tar clean-up<br>• Minerals remain with the char/ash, reducing the need for a cyclone<br>• Proven, simple, and low-cost process | • Requires feed drying to low-moisture content (<20%)<br>• Syngas exiting at high temperatures, requiring a secondary heat recovery system<br>• 4–7% of the carbon remains unconverted |
| Bubbling fluidized bed | • Yields a uniform product gas<br>• Exhibits a nearly uniform temperature distribution throughout the reactor<br>• Able to accept a wide range of fuel particle sizes, including fines<br>• Provides high rates of heat transfer between inert material, fuel, and gas<br>• High conversion possible with low tar and unconverted carbon | • Large bubble size may result in gas bypass through the bed |
| Circulating fluidized bed | • Suitable for rapid reactions<br>• High heat transport rates possible due to high heat capacity of bed material<br>• High conversion rates possible with low tar and unconverted carbon | • Temperature gradients occur in direction of solid flow |

*Source*: Adapted from Ciferno and Marano, 2002.

TABLE 5.18 Advantages and Disadvantages of the Four Main Types of Biomass Gasifiers

| Gasifier type | BFB range | CFB range | Fixed-bed range | BCL/FERCO* | MTCI[†] | Shell[‡] |
|---|---|---|---|---|---|---|
| Feedstock | Various | Various | Various | Wood | Pulp | Coal |
| Throughput (MT/day) | 4.5–181 | 9–108 | 181–195 | 24 | 7 | 2155 |
| Pressure (bar) | 1–35 | 1–19 | 1 | 1 | 1 | 30 |
| Temperature (°C) | 650–950 | 800–1000 | 1300–1400 | 600–1000 | 790–815 | 1400 |
| Reactant 1 | $O_2$ or Air | Air | $O_2$ or Air | Air | | $O_2$ |
| Input (kg/kg feed) | 0.4–2.2 | 1.25–1.7 | | 0.08 | | 0.98 |
| Reactant 2 | Steam | | | Steam | Steam | Steam |
| Input (kg/kg feed) | 0.5–0.64 | | | 0.31 | 2.2 | ~0 |
| Gas Output (M³/h) | 335–8793 | 1181–12500 | 33,960 | 800 | | $1.48 \times 10^6$ |
| Exit temperature (°C) | 300–800 | 600–900 | | 820 | | 240 |

*Indirectly heated CFB with separate combustor.
[†]Indirectly heated BFB with separate combustor.
[‡]Fluid bed; entrained flow (no circulation).
*Source*: Adapted from Ciferno and Marano, 2002.

**TABLE 5.19**  Summary of Gasifier Operating Conditions

| Gasifier type | BFB range | CFB range | BCL/ FERCO* | MTCI[†] | Fixed-bed Purox | Shell[‡] |
|---|---|---|---|---|---|---|
| Feedstock | Various | Various | Wood | Pulp | MSW | Coal |
| $H_2$ | 5–26 | 7–20 | 14.9 | 43.3 | 23.4 | 24 |
| CO | 13–27 | 9–22 | 46.5 | 9.22 | 39.1 | 67 |
| $CO_2$ | 12–40 | 11–16 | 14.6 | 28.1 | 24.4 | 4 |
| $H_2O$ | <18 | 10–14 | Dry | 5.57 | Dry | 3 |
| $CH_4$ | 3–11 | <9 | 17.8 | 4.73 | 5.47 | 0.02 |
| $C_2^+$ | <3 | <4 | 6.2 | 9.03 | 4.93 | 0 |
| Tars | <0.11 | <1 | 0.31 | Scrubbed | | 0 |
| $H_2S$ | ~0 | ~0 | 800 | 0.08 | 0.05 | 1 |
| $O_2$ | <0.2 | 0 | 0 | 0 | 0 | 0 |
| $NH_3$ | 0 | 0 | 0 | 0 | | 0.04 |
| $N_2$ | 13–56 | 46–52 | 0 | 0 | | 1 |
| $H_2$/CO Ratio | 0.2–1.6 | 0.6–1.0 | 0.3 | 4.6 | 0.6 | 0.36 |

*Indirectly heated CFB with separate combustor.
[†]Indirectly heated BFB with separate combustor.
[‡]Fluid bed; entrained flow (no circulation).
Source: Adapted from ciferno and Marano, 2002.

**TABLE 5.20** Summary of Syngas Composition

and acetic acid production[151,152] is shown in Fig. 5.22. The important characteristics of the pathway include

- Formation of the methyl moiety of acetate on tetrahydrofuran
- Transfer of the methyl group to the carbon monoxide dehydrogenase (CODH) via a cobalt-containing corrinoid enzyme
- Condensation of the methyl group with a CO-derived carbonyl and coenzyme A to form acetyl-CoA
- Hydrolysis of acetyl-CoA to acetic acid
- Conversion of acetic acid to ethanol

The stoichiometric equations for formation of acetic acid from carbon monoxide can be written as

$$4CO + 2H_2O \rightarrow CH_3COOH + 2CO_2$$
$$4H_2 + 2CO_2 \rightarrow CH_3COOH + 2H_2O$$

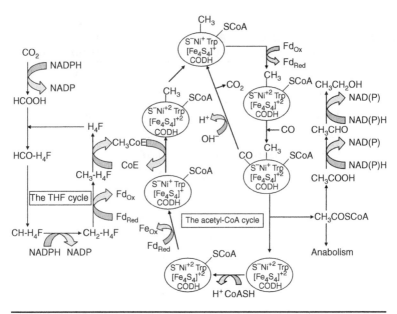

**Figure 5.22** The biosynthetic pathway for production of ethanol and acetic acid in *Clostridium ljungdahlii*. (Adapted from Hugenholtz and Ljungdahl, 1990 and Phillips et al., 1994.)

Similarly, the stoichiometric equations for ethanol formation can be written as

$$6CO + 3H_2O \rightarrow CH_3CH_2OH + 4CO_2$$

$$6H_2 + 2CO_2 \rightarrow CH_3CH_2OH + 3H_2O$$

The original *C. ljungdahlii* strain PETC (ATCC 49587) was isolated from chicken wastes at the University of Arkansas.[153] Other strains then were isolated[154,155] and used for development of a syngas fermentation process for ethanol production. The composition of the medium used for ethanol production is shown in Table 5.21.

The fermentation process[157] is a two-stage continuous process. In the first stage, the conditions are optimized for cell growth. In the second stage, cell recycle is used to maintain high cell concentration and the process parameters are adjusted to maximize the ethanol/acetate ratio. Two key strategies are used. The first strategy involves limiting pantothenic acid and the second one involves limiting cobalt. As shown in Fig. 5.22, excess NAD(P)H will lead to the production of ethanol from acetic acid. The acetyl-CoA cycle has an internal cycle, which is referred to as the CO cycle. If these two

| Component | Quantity per liter |
|---|---|
| 2 g/L FeCl$_2$.4H$_2$O | 10 mL |
| 85% H$_3$PO$_4$ | 0.05 mL |
| MPFN trace metals* | 20 mL |
| (NH$_4$)$_2$HPO$_4$ | 0.60 g |
| NH$_4$Cl | 2.00 g |
| NaCl | 0.20 g |
| KCl | 0.15 g |
| MgCl$_2$.6H$_2$O | 0.50 g |
| CaCl$_2$.2H$_2$O | 0.20 g |
| Cystein HCl.H$_2$O | 0.25 g |
| Vitamin solution† | Variable‡ |

*MPFN trace metal solution contains (per liter): 10 mL 85% H$_3$PO$_4$, 0.10 g ZnSO$_4$. 7H$_2$O, 0.03 g MnCl$_2$. 4H$_2$O, 0.3 g H$_3$BO$_3$, 0.20 g CoCl$_2$. 6H$_2$O, 0.02 g CuCl$_2$. H$_2$O, 0.04 g NiCl$_2$. 6H$_2$O, 0.03 g NaMoO$_4$. 2H$_2$O, 2.00 g FeCl$_2$. 4H$_2$O, 0.01 g Na$_2$SeO$_3$, and 0.10 g Na$_2$WO$_4$. 2H$_2$O.
†Vitamin solution contains 20.6 mg/L d-biotin, 50.6 mg/L thiamine HCl, and 50.5 mg/L d-pantothenic acid, calcium salt.
‡Varied from 0.3 to 0.5 mL at inoculation to 0.7 to 0.8 mL at high gas rates.

**TABLE 5.21** Composition of the Medium Used for Ethanol Production by *C. ljungdahlii*

cycles proceed at the same rates, ferredoxin is in a redox-equilibrium state. Since pantothenate is a component of the acetyl-CoA molecule, limiting pantothenic acid in the medium will cause reduction in the rate of the acetyl-CoA cycle with respect to the CO cycle. This will cause a build-up of reduced ferredoxin, which in turn will drive the reduction of NAD(P) to NAD(P)H, and subsequently will increase ethanol production from acetate. The same effect can be provided by limiting cobalt. Since cobalt is required by the corrinoid enzyme to transfer a methyl group from the THF cycle to the acetyl-CoA cycle, limiting cobalt will slow down the acetyl-CoA cycle, which in turn will cause a build-up of reduced ferredoxin and reduction of NAD(P) to NAD(P)H. Cobalt and pantothenate limitations can be used together with feeding excess carbon monoxide and excess hydrogen to fully utilize these two gases to improve ethanol production. However, the levels of carbon monoxide must be kept below the inhibitory levels. Ethanol concentration up to 33 g/L, ethanol/acetate ratio of 20/1, and productivity of 39 g ethanol/L/day have been demonstrated.[156]

### 5.2.3.3 Catalytic Conversion of Syngas

The two key reactions involved in the conversion of syngas to ethanol are summarized in Table 5.22.

Both reactions are highly exothermic and thermodynamically favorable. Thermodynamic analysis performed for a stoichiometric mixture of $H_2/CO = 2$ at a pressure of 30 bar suggested that the hydrogenation of CO should be carried out at 350°C or lower.[157]

In addition to the two reactions shown in Table 5.22, other reactions involved in syngas conversion are

$$CO + H_2O \leftrightarrow CO_2 + H_2 \quad \text{(water gas shift reaction)}$$

$$CO_2 + 4H_2 \rightarrow CH_4 + 2H_2O \quad \text{(methanation of CO)}$$

$$CO + 3H_2 \rightarrow CH_4 + H_2O \quad \text{(methanation of CO}_2\text{)}$$

If methanation is allowed to occur in parallel with the hydrogenation reactions, methane will be the most thermodynamically significant product.[157]

The four main groups of catalysts for ethanol production from CO and $CO_2$ are

- Rh-based catalysts
- Modified methanol synthesis catalysts (based on Cu)
- Modified Fischer-Tropsch type catalysts
- Modified Mo-based catalysts

Modified methanol synthesis catalysts give the highest ethanol formation from CO, but methanol is still the dominant alcohol product. Rh-based catalysts give highest ethanol selectivities, but the conversion of CO is low. Modified Fischer-Tropsch catalysts only give moderate ethanol selectivities.[157] Technical data on some of the catalysts for synthesis of mixed alcohols is summarized in Table 5.23.

| Reaction | $\Delta H_r°$ (kcal/mol) | $\Delta G_r°$ (kcal/mol) |
|---|---|---|
| Hydrogenation of CO to ethanol: | | |
| $2CO + 4H_2 \rightarrow C_2H_5OH + H_2O$ | −61.20 | −29.32 |
| Hydrogenation of $CO_2$ to ethanol: | | |
| $2CO_2 + 6H_2 \rightarrow C_2H_5OH + 3H_2O$ | −41.54 | −15.70 |

*Source*: Adapted from Spivey and Egbebi, 2007.

**TABLE 5.22** Reactions and Thermodynamics of Syngas Conversion to Ethanol

| Company | Country | Catalyst type | T (°C) | P (psi) | $H_2/CO$ ratio | Products | Conversion and selectivity |
|---|---|---|---|---|---|---|---|
| Lurgi | Germany | Modified MeOH | 250–420 | 725–1450 | 1–1.2 | 53.5 wt% MeOH 41.9 wt% $C_2$–$C_6$ | CO conversion = 20–60% |
| Union Carbide | United States | Rhodium | 300–350 | 1000–2500 | NA | NA | CO selectivity to EtOH = 60% |
| Sagami Research Center | Japan | Rhodium | 200–300 | 735 | 1.4 | Mainly MeOH, EtOH, $CH_4$ | CO conversion = 14% Selectivity to EtOH up to 61% Selectivity to alcohols = 70–75% |
| IFP (Substifuel) | France | Modified MeOH | 260–320 | 850–1450 | 1–2 | 30–50 wt% $C_2$–$C_4$ | CO conversion = 12–18% Selectivity to alcohols = 70–75% |
| Hoechst | Germany | Rhodium | 275 | 1455 | NA | NA | CO selectivity to EtOH = 74.5% |
| Snamprogetti | Italy | Modified MeOH | 260–420 | 2610–3822 | 0.5–3 | 20–40 wt% $C_2$–$C_4$ | CO conversion = 17% Selectivity to alcohols = 71% |

*Source:* Adapted from Spath and Dayton, 2003

**TABLE 5.23** Technical Data of Catalysts for Synthesis of Mixed Alcohols

| Company | Country | Catalyst type | T (°C) | P (psi) | H2/CO ratio | Products | Conversion and selectivity |
|---|---|---|---|---|---|---|---|
| Texaco (liquid phase system) | United States | Modified FT | 220–240 | 6615 | NA | 12–39 wt% non-alcohol oxygenates | Syngas conversion = 40%<br>Selectivity to products = 75% |
| Dow | United states | Modified FT | 200–310 | 1500–2000 | 1.1–1.2 | 30–70 wt% MeOH | CO conversion = 10–40%<br>Selectivity to alcohols = 85% |
| Ecalene | United states | NA | 290–360 | 145–1395 | NA | Main alcohol constituent is EtOH (45–75%) | NA |

TABLE 5.23 (*Continued*)

### 5.2.3.4 Process Integration

Similar to the production of ethanol from biomass through the sugar platform, there currently is no commercial plant for ethanol production from biomass-derived syngas. However, at least three companies are very near to starting construction of full-scale production plants using this feedstock. Gasification technology has been practiced widely and several commercial-scale gasifiers are available. These gasifiers can easily be integrated into the ethanol production plant. On the other hand, the size of the ethanol plant will be dictated by the size of the available gasifiers. The tendency is to design a modular unit and combine several units to meet the required plant capacity. Two examples of commercial processes are presented in this section to illustrate process integration for ethanol production from biomass-derived syngas.

The BRI (Bioengineering Resources, Inc.) process is shown in Fig. 5.23. This process uses a two-stage gasifier to generate syngas from biomass. The syngas then is fermented to ethanol by a strain of *C. ljungdahlii*.

The BRI's plant is modular. A single module has two gasifiers, each with a capacity of approximately 125 tons (113 MT) of feedstock per day. In addition to biomass, the gasifiers can also process other organic materials such as MSW, waste tires, and coal. Each module can process 85,000 tons (77,100 MT) biomass and produce 7 million gallons (26.5 million liters) ethanol and 5 MW of power annually. A midsize BRI plant requires 10 modules and approximately 30 acres.

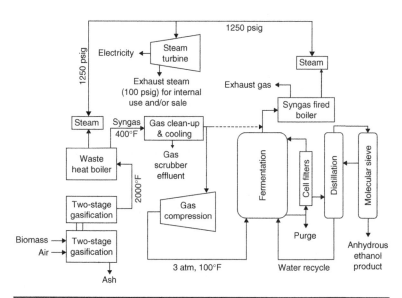

FIGURE 5.23 The BRI process for production of ethanol by fermentation of biomass-derived syngas. (*Adapted from http://www.greencarcongress.com/2006/05/bri_energy_seek.html.*)

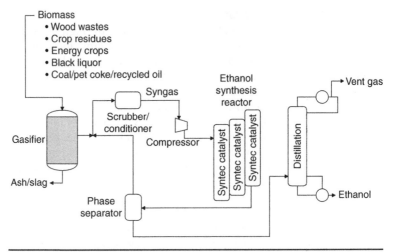

**FIGURE 5.24** The Syntec's process for production of ethanol by catalytic conversion of biomass-derived syngas. (*Adapted from www.syntecbiofuel.com/technology.html, 2007.*)

Such plant can process 1,000,000 tons (907,000 MT) of biomass annually to produce 80 million gallons (303 million liters) ethanol and 50 MW of power, of which 35 MW is excess to the operation of the plant.[160]

The Syntec's process (Syntec Biofuel Research Inc.) is shown in Fig. 5.24. This process uses a precious metal-based catalyst for syngas conversion. The products consist of ethanol, methanol, *n*-propanol, and *n*-butanol. The process is designed for biomass but with minor modification it can handle other feedstocks, which include solids such as coal and coke; liquids such as recycled oil; or gases such as biogas, natural gas, and coal bed methane. Catalysts based on nonprecious metals are currently under development. The expected ethanol yield is 114 gal/dry ton of biomass feedstock. The theoretical yield is given as 230 gal/dry ton. Thus, the efficiency of the Syntec's process is about 50 percent.[161,162]

### 5.2.3.5 Process Economics

Gasification is one of the key pieces in the production of ethanol from biomass-derived syngas. The capital costs of commercial gasifiers are given in Table 5.24.

There is not much information on the economics of the syngas fermentation process. A recent report describes an economic analysis for ethanol production from wood chips via syngas in Europe.[163] The conceptual system considered in this report is a modular unit processing 30,000 MT of biomass per year to produce 8812 MT (11.2 million liters or 2.95 million gallons) of 97 percent ethanol. This is a relatively small

| Technology type | Type | Size (MT/day) | Capital cost ($ 106) | Capital cost ($ 103/ MTPD) | Capital cost ($/GJ/h syngas) |
|---|---|---|---|---|---|
| BFB range | BFB | 170–960 | 2–36 | 13–45 | 21,600–54,900 |
| BFB average | BFB | 615 | 16.1 | 25 | 29,500 |
| BCL/FERCO* | CFB | 740–910 | 18.26 | 24.5–28.4 | 33,000–48,000 |
| MTCI† | BFB | 44 | 1.1 | 25.2 | |
| Shell‡ | | 2200 | 80.5 | 37.3 | 1400 |

*Notes*: See notes under Table 5.19.
*Source*: Adapted from Ciferno and Marano, 2002.

**TABLE 5.24** Capital Cost of Commercial Gasifiers

unit, about 40 percent the capacity of the BRI's module. The key process parameters are summarized here.

- Feedstock is wood chips with 20 percent moisture, which will be dried to 10 wt% prior to gasification.
- Oxygen is the fuel for gasification.
- Syngas is cleaned up by the OLGA process developed by the Energy Research Center of the Netherlands, which involves scrubbing of the gas with a special scrubbing oil.[164]
- Fermentation is carried out using *C. ljungdahlii* in either bubble columns or trickle-bed bioreactors at 39°C.
- Ethanol/acetate ratio in the fermentation product stream is 20/1.

The results of the economic analysis are summarized in Table 5.25.

The economics of ethanol production by catalytic conversion of biomass-derived syngas is examined in a report prepared by Delta T Corporation for the National Renewable Energy Laboratory.[165] The plant output is 15 million gallons per year. The results of the economic analysis are summarized in Table 5.26.

The total quantity of feedstock used in this analysis is 75,012 dry tons/yr. Thus, the ethanol conversion is 200 gal/ton, which is much higher than in the Syntec's process. If the conversion disclosed for the Syntec's process, that is, 114 gal/dry ton, is used, only 8.6 million gallons of ethanol are produced and the production cost will be $2.23/gal.

Production of ethanol from lignocellulosic feedstocks has not reached commercialization. However, construction of pilot/demonstration and

| Capital costs (€) | | |
|---|---|---|
| Equipment | | |
| Gasifier | 2,500,000 | |
| Gas cleaner | 1,000,000 | |
| Fermentors | 15,000,000 | |
| Distillation columns | 1,000,000 | |
| Heat exchangers | 600,000 | |
| Tanks | 150,000 | |
| Other direct investment | 1,000,000 | |
| Contingency (15% of total direct investment) | 3,187,500 | |
| Start-up (1% of total direct investment) | 212,500 | |
| Working capital | 40,000 | |
| Total fixed capital | 24,690,000 | |
| Operating costs (€) | | |
| Fixed costs | | |
| Personnel | 695,500 | |
| Support and maintenance | 1,062,500 | |
| Total fixed cost | 1,758,000 | |
| Variable costs | | |
| Wood chips | 1,200,000 | |
| Oxygen | 52,500 | |
| Water | 90,000 | |
| Electricity | 675,000 | |
| Waste disposal | 34,000 | |
| Total variable cost | 2,051,500 | |
| Returns (€) | | |
| Ethanol | 7,930,800 | |
| Ethanol selling price | | |
| At 0% ROI | 0.56 €/l (26.7 €/GJ) | ($2.93/gal) |
| At 14% ROI | 0.88 €/l (42 €/GJ) | ($4.60/gal) |

*Source*: Adapted from van Kasteren et al., 2005.

**TABLE 5.25** Summary of Conceptual Economic Analysis of Wood Chip-Derived Syngas Fermentation to Ethanol

# Ethanol Production

| Capital costs ($) | |
|---|---|
| Feed handling and processing | 1,511,670 |
| Oxygen plant | 3,000,000 |
| Syngas generation and clean-up | 3,706,690 |
| Ethanol production | 15,002,400 |
| Ethanol dehydration and processing | 1,829,550 |
| Ethanol storage and loading | 992,330 |
| Utilities | 1,078,620 |
| Wastewater treatment | 1,280,850 |
| Total installed equipment cost | 28,402,110 |
| Added cost to obtain total project investment | 19,798,630 |
| Total capital cost | 48,200,630 |
| Variable annual operating costs ($) | |
| Feedstock | 1,125,180 |
| Waste disposal | 101,400 |
| Electricity | 2,608,200 |
| Chemicals and catalyst | 2,606,483 |
| Make-up water | 100,110 |
| Total annual variable operating cost | 6,541,373 |
| Fixed annual operating cost | |
| Operating labor | 371,200 |
| Operating foreman | 85,065 |
| Supervision | 52,200 |
| Direct overhead (45% of L&M) | 228,809 |
| Maintenance (4% of fixed) | 1,928,025 |
| Plant overhead (6.5% of maintenance) | 125,322 |
| Insurance and taxes (1.5% of fixed) | 723,009 |
| Total annual fixed operating cost | 3,513,631 |
| Total annual operating cost | 10,055,004 |
| Total annualized capital cost | 9,034,920 |
| Total annual cost | 19,089,923 |
| Annual cost/Undenatured gal ethanol | 1.273 |

*Source*: Adapted from NREL report number NREL/SR-510-32381, 2002.

**TABLE 5.26** Summary of Conceptual Economic Analysis of Catalytic Conversion of Biomass-Derived Syngas to Ethanol

commercial plants has been started recently. Some pilot and demonstration plants currently are already in operation. The locations and capacities of the lignocellulosic biomass-based ethanol plants in the US and the types of feedstocks used in these plants are given in the appendix (Fig. 5A.2).

## Acknowledgments

One of the authors, Nghiêm Phú Nhuan, would like to express his sincere thanks to the following colleagues: Dr. Susanne Kleff (MBI International), Dr. Wilson T. Gautreaux (Mead-Westvaco Corporation), Dr. Duen-Gang Mou (Moubio Knowledge Works, Inc.), Dr. Elankovan Ponnampalam (Tate and Lyle), and Dr. Frank Taylor, Dr. Kevin Hicks, and Ms. Jhanel Wilson (Eastern Regional Research Center, USDA Agricultural Research Service), who spent their valuable time reviewing, and made numerous suggestions for improvement of the materials in, Chaps. 4 and 5.

## References

1. Thomas, K. C., S. H. Hynes, and W. M. Ingledew. 1996. "Practical and theoretical considerations in the production of high concentrations of alcohol by fermentation." *Process Biochem.* 31:321–331.
2. Becker, J-U., and A. Betz. 1972. "Membrane transport as controlling pacemaker of glycolysis in *Saccharomyces carlsbergensis*." *Biochimica Biophysica Acta.* 274:584–597.
3. Kruckeberg, A. L. 1996. "The hexose transporter family of *Saccharomyces cerevisae*." *Arch Microbiol.* 166:283–292.
4. Reifenberger, E., E. Boles, and M. Ciriacy. 1997. "Kinetic characterization of individual hexose transporters of Saccharomyces cerevisiae and their relation to the triggering mechanisms of glucose repression." *Eur J Biochem.* 245:324–333.
5. Van Maris, A J. A., et al. 2006. *Antonie Van Leeuwenhoek.* 90:391–418.
6. Russell, I. 2003. "Understanding yeast fundamentals." *The Alcohol Textbook*, 4th ed. K. A. Jacques, T. P. Lyons, and D. R. Kelsall (eds.). Nottingham University Press: Nottingham, UK.
7. Verstrepen, K. J., et al. 2004. "Glucose and sucrose: Hazardous fast-food for industrial yeast?" *Trends Biotechnol.* 22:531–537.
8. Hu, Z., et al. 1995. "MIG 1-dependent and MIG 1-independent glucose regulation of MAL gene expression in *Saccharomyces cerevisiae*." *Curr Genet.* 28:258–266.
9. Federoff, H. J., T. R. Eccleshall, and J. Marmur. 1983. "Carbon catabolite repression of maltase synthesis in *Saccharomyces carlsbergensis*." *J Bacteriol.* 156: 301–307.
10. Adams, B. 1972. "Induction of galactokinase in *Saccharomyces cerevisiae*: Kinetics of induction and glucose effects." *J Bacteriol.* 111:308–315.
11. Meneses, J. F. 2002. "A survey of industrial strains of *Saccharomyces cerevisiae* reveals numerous altered patterns of maltose and sucrose utilization." *J Inst Brew.* 108:310–321.
12. Jones, R. P., and P. F. Greenfield. 1984. "A review of yeast ionic nutrition: Growth and fermentation requirements." *Proc Biochem.* 19:48–60.
13. Alfenore, S., et al. 2002. "Improving ethanol production and viability of *Saccharomyces cerevisiae* by a vitamin feeding strategy during fed-batch process." *Appl Microbiol Biotechnol.* 60:67–72.

14. Kosaric, N., et al. 1983. "Ethanol fermentation." *Biotechnology*. Vol. 3. H. Dellweg (ed.). Verlag Chemie: Basel, FRG.
15. Salgueiro, S. P., I. Sa-Correia, and J. M. Novais. 1988. "Ethanol-induced leakage in *Saccharomyces cerevisiae*: Kinetics and relationship to yeast ethanol tolerance and alcohol fermentation productivity." *Appl Environ Microbiol.* 54:903–909.
16. Ingram, L. O., and T. M. Burke. 1984 "Effects of alcohols on microorganisms." *Advances Microbial Physiol.* 25:253–300.
17. Nabais, R. C., et al. 1988. "Influence of calcium ion on ethanol tolerance of Saccharomyces bayanus and alcoholic fermentation by yeasts." *Appl Environ Microbiol.* 54:2439–2446.
18. Narendranath, N. V., K. C. Thomas, and W. M. Ingledew. 2001. "Effects of acetic and lactic acid on the growth of *Saccharomyces cerevisiae* in a minimum medium." *J Ind Microbiol Biotechnol.* 26:171–177.
19. Graves, T., et al. 2006. *J Ind Microbiol Biotechnol.* 33:469–474.
20. Hoppe, G. K., and G. S. Hansford. 1984. "The effect of micro-aerobic conditions on continuous ethanol production by *Saccharomyces cerevisiae*." *Biotechnol Lett.* 6:681–686.
21. Franzen, C. J. 2003. "Metabolic flux analysis in RQ-controlled microaerobic ethanol production by *Saccharomyces cerevisiae*." *Yeast.* 20:117–132.
22. Panesar, P. S., S. S. Marwaha, and J. F. Kennedy. 2006. "*Zymomonas mobilis*: An alternative ethanol producer." *J Chem Technol Biotechnol.* 81:623–635.
23. Doelle, H. W., et al. 1993. "*Zymomonas mobilis*—science and industrial application." *Crit Rev Biotechnol.* 13:57–98.
24. Kopperschläger, G., and J. J. Heinisch. 1997. *Phosphofructokinase. Yeast Sugar Metabolism.* F. K. Zimmermann, and K. D. Entian (eds.). Technomic: Lancaster, PA.
25. Jeffres, T. W. 2005. "Ethanol fermentation on the move." *Nat Biotechnol.* 23:40–41.
26. Gunasekaran, P., and K. Chandra Raj. 1999. "Ethanol fermentation technology—*Zymomonas mobilis*. Available from: http://www.ias.ac.in/currsci/jul10/articles14.htm.
27. Karsch, T., U. Stahl, and K. Esser. 1983. "Ethanol production by *Zymomonas* and *Saccharomyces*, advantages and disadvantages." *Eur J Appl Microbiol Biotechnol.* 18:387–391.
28. Nghiem, N. P., and B. H. Davison. 2002. "Continuous ethanol production using immobilized-cell/enzyme biocatalysts in fluidized-bed reactor (FBR)." Oak Ridge National Laboratory report number ORNL/TL-2002/164.
29. Seo, J-S., et al. 2005. "The genome sequence of the ethanologenic bacterium *Zymomonas* ZM4." *Nat Biotechnol.* 23:63–69.
30. Polanco, L. S., P. W. Rein, and B. E. White. 2006. "A comparison of sugarcane juice quality from a mill and a diffuser." *Sugar J.* March issue, 12–20.
31. Ogbonna, J. C., H. Mashima, and H. Tanaka. 2001. "Scale up of fuel ethanol production from sugar beet juice using loofa sponge immobilized bioreactor." *Bioresour Technol.* 76:1–8.
32. Curtin, L. V. 1983. "Molasses—general considerations." *Molasses Anim Nutr.* National Feed Ingredients Association, West Des Moines, Iowa.
33. Lipinsky, E. S. 1978. "Fuels from biomass: Integration with food and materials systems." *Science.* 199:644–651.
34. Anonymous. Sugar Knowledge International (SKIL). Available from: http://www.sucrose.com/.
35. Anonymous. K-Patents Applications in Sugar Industry. Available from: http://www.kpatents.com/sugar_app.php/.
36. Anonymous. Copersucar. Available from: http://www.copersucar.com.br/institucional/ing/academia/alcool.asp/.
37. Wheals, A. E., et al. 1999. *Trends Biotechnol.* 17:482–487.
38. Andrietta, S. R., and F. Maugeri. 1994. "Optimum design of a continuous fermentation unit of an industrial plant for alcohol production." In: *Advances in Bioprocess Engineering.* E. Galindo, and O. T. Ramirez (eds.). Kluwer Academic Publishers: The Netherlands, pp. 47–52.

39. Pessoa, Jr., A., et al. 2005. "Perspectives on bioenergy and biotechnology in Brazil." *Appl Biochem Biotechnol.* 121–124, 59–70.
40. Martines-Filho, J., H. L. Burnquist, and C. E. F. Vian. 2006. "Bioenergy and the rise of sugarcane-based ethanol in Brazil." *Choices.* 21:91–96.
41. Anonymous. 2006. "EU-25 Sugar—the economics of bioethanol production in the EU." *Global Agriculture Information Network*, USDA Foreign Agricultural Service, GAIN report number E36081.
42. Henke, S., et al. 2006. "Model of a sugar factory with bioethanol production in program Sugars." *J Food Eng.* 77:416–420.
43. Williams, J. 2004. "Beyond corn." *Ethanol Producer Magazine.* July 2004.
44. Bothast, R. J., and M. A. Schlicher. 2005. "Biotechnological processes for conversion of corn into ethanol." *Appl Microbiol Biotechnol.* 67:19–25.
45. Butzen, S., and T. Hobbs. 2002. "Corn processing III: Wet milling." *Crop Insights.* Vol. 12, No. 15. Published by Pioneer Hi-Bred International, Inc.
46. Madson, P. W., and D. A. Monceaux. 1999. "Fuel ethanol production." In: *The Alcohol Textbook: A Reference for the Beverage, Fuel and Industrial Alcohol Industries.* K. A. Jacques, T. P. Lyons, and D. R. Kelsall (eds.). Nottingham University Press: Nottingham, UK, pp. 257–268.
47. Lewis, S. M. 1996. "Fermentation alcohol." In: *Industrial Enzymology.* T. Godfrey, and S. West (eds.). MacMillan Press: London, UK, pp. 12–48.
48. Liley, P. E., Y. S. Touloukian, and W. R. Gambill. 1963. "Physical and chemical data." In: *Chemical Engineers' Handbook*, 4th ed. R. H. Perry, C. H. Chilton, and S. D. Kirkpatrick (eds.). McGraw-Hill Book Company: New York, USA, pp. 3–83.
49. Van Halsema, E., W. De Laat, and H. Van Dijken. 2007. "Simple and convenient on-line monitoring of alcoholic fermentation in laboratory experiments." Presented at the 29th Symposium on Biotechnology for Fuels and Chemicals, Denver, CO, May 2007.
50. Bibb Swain, R. L. 1999. "Molecular sieve dehydrators. How they became the industry standard and how they work." In: *The Alcohol Textbook: A Reference for the Beverage, Fuel and Industrial Alcohol Industries.* K. A. Jacques, T. P. Lyons, and D. R. Kelsall (eds.). Nottingham University Press: Nottingham, UK, pp. 289–293.
51. Shapouri, H., and P. Gallagher. 2005. "USDA's 2002 ethanol cost-of-production survey." USDA's agricultural economic report number 841.
52. Whims, J. 2002. "Corn based ethanol costs and margins attachment 1." AgMRC Agricultural Marketing Resource Center, Department of Agricultural Economics, Kansas State University.
53. Butzen, S., and D. Haefele. 2006. "Dry-grind ethanol production from corn." *Crop Insights.* Vol. 16, No. 7. Published by Pioneer Hi-Bred International, Inc.
54. Fraley, R. 2006. "Monsanto whistle-stop summer investor field tour." Available from: http://www.monsanto.com/pdf/investors/2006/07-31-06.pdf
55. Lanahan, M. B., et al. 2006. "Self-processing plants and plant parts." Patent application U.S. 2006/0200877 A1.
56. Singh, V., et al. 2006. "Dry-grind processing of corn with endogenous liquefaction enzymes." *Cereal Chem.* 83:317–320.
57. Stern, M. 2007. "The Renessen corn processing system: Rebalancing the bioenergy/feed equation." *BIO Energy Panel Discussion: The Future of BIO Energy*, Chicago, Illinois, February 2007. Available from: http://ibio.org/images/renessen.pdf
58. Stargen product brochure published by Genencor International, a division of Danisco. STARGEN 001—Granular Starch Hydrolyzing Enzyme for Ethanol Production. Available from: http://www.beutilityfree.com/pdf_files/STARGEN_001.pdf.
59. Anonymous. 2004. "Biotech boost to ethanol production." Available from: http://www.greencarcongress.com/2004/11/biotech_boost_t.html.
60. Anonymous. 2005. "New enzyme for more efficient corn ethanol production." Available from: http://www.greencarcongress.com/2005/06/new_enzyme_for_.html.
61. Anonymous. 2005. "Broin Companies announces fractionation process for ethanol production." Posted July 05, 2005 on: http://www.grainnet.com/.

62. Lewis, S. M., and S. E. Van Hulzen. 2007. "Methods and systems for producing ethanol using raw starch and fractionation." U.S. patent application 2007/0037267 A1.
63. Anonymous. 2006. "Broin Companies reveals ethanol performance achievement of BPX process, makes technology available outside its group." Posted February 23, 2006 on: http://www.grainnet.com/.
64. Taylor, F., et al. 2000. "Dry-grind process for fuel ethanol by continuous fermentation and stripping." *Biotechnol Prog.* 16:541–547.
65. Johnston, D. B., and V. Singh. 2003. "Use of enzymes to reduce steep times and $SO_2$ requirements in a maize wet-milling process." U.S. patent 6,566,125.
66. Johnston, D. B., and V. Singh. 2005. "Enzymatic milling product yield comparison with reduced levels of bromelain and varying levels of sulfur dioxide." *Cereal Chem.* 82:523–527.
67. Anonymous. 2008. "Biomass gasification." Energy Efficiency and Renewable Energy, U.S. Department of Energy's web site. Available from: http://www1.eere.energy.gov/biomass/thermochemical_processes.html.
68. Farone, W. A., and J. E. Cuzens. 1996. "Method of producing sugars using strong acid hydrolysis of cellulosic and hemicellulosic materials." U.S. patent 5,562,777.
69. Nghiem, N. P., et al. 2000. "Ethanol production from rice straw hydrolysate by immobilized recombinant *Zymomonas mobilis* in a fluidized-bed reactor." Paper presented at the Pacific Chemical 2000, Honolulu, Hawaii.
70. Anonymous. 2006. "BlueFire Ethanol targeting 1.5B gallons per year of cellulosic ethanol by 2012." Available from: http://www.greencarcongress.com/2006/12/luefire_ethanol.html.
71. Nguyen, Q. A., et al. 1998. "Dilute acid pretreatment of softwoods." *Appl Biochem Biotechnol.* 70–72, 77–87.
72. Nguyen, Q. A., F. A. Keller, and M. P. Tucker. 2003. "Ethanol production with dilute acid hydrolysis using partially dried lignocellulosics." U.S. patent application 2003/0199049 A1.
73. Nguyen, Q. A., et al. 2000. "Two-stage dilute-acid pretreatment of softwoods." *Appl Biochem Biotechnol.* 84–86, 561–576.
74. Larsson, S., et al. 1999. "Comparison of different methods for the detoxification of lignocellulose hydrolyzates of spruce." *Appl Biochem Biotechnol.* 77–79, 91–103.
75. Purwadi, R., C. Niklasson, and M. J. Taherzadeh. 2004. "Kinetic study of detoxification of dilute-acid hydrolyzates by $Ca(OH)_2$." *J Biotechnol.* 114:187–198.
76. Lora, A. H., and M. Wayman. 1978. "Delignification of hardwoods by autohydrolysis and extraction." *Tappi J.* 61:47–50.
77. Grous, W. R., A. O. Converse, and H. E. Grethlein. 1986. "Effect of steam explosion pretreatment on pore size and enzymatic hydrolysis of poplar." *Enzyme Microb Technol.* 8:274–280.
78. Duff, S. J. B., and W. D. Murray. 1996. "Bioconversion of forest products industry waste cellulosics to fuel ethanol: A review." *Bioresour Technol.* 55:1–33.
79. Saddler, J. N., L. P. Ramos, and C. Breuil. 1993. "Steam pretreatment of lignocellulosic residues." In: *Bioconversion of Forest and Agricultural Residues*, Chap. 3. J. N. Saddler (ed.). CAB International: Oxford, UK.
80. Wright, J. D. 1988. "Ethanol from biomass by enzymatic hydrolysis." *Chem Eng Prog.* 8:62–74.
81. Linde, M., M. Galbe, and G. Zacchi. 2006. "Steam pretreatment of acid-sprayed and acid-soaked barley straw for production of ethanol." *Appl Biochem Biotechnol.* 129–132, 546–562.
82. Anonymous. 2005. "SunOpta providing steam explosion technology for world's first cereal-straw ethanol plant." Available from: http://www.greencarcongress.com/2005/08/sunopta_providi.html.
83. Fan, L. T., M. M. Gharpuray, and Y-H. Lee. 1987. In: *Cellulose Hydrolysis Biotechnology Monographs*, Springer: Berlin, p. 57.
84. Kim, T. H., J. S. Sunwoo, and Y. Y. Lee. 2003. "Pretreatment of corn stover by aqueous ammonia." *Bioresour Technol.* 90:39–47.

85. Kim, T. H., and Y. Y. Lee. 2006. "Fractionation of corn stover by hot water and aqueous ammonia treatment." *Bioresour Technol.* 97:224–232.
86. Kim, T. H., and Y. Y. Lee. 2005. "Pretreatment of corn stover by soaking in aqueous ammonia." *Appl Biochem Biotechnol.* 121–124, 1119–1132.
87. Kim, T. H. 2004. "Bioconversion of lignocellulosic material into ethanol." Ph.D. dissertation. Auburn University.
88. Dale, B. E. 1986. "Method for increasing the reactivity and digestibility of cellulose with ammonia." Patent U.S. 4,600,590.
89. Teymouri, F., et al. 2004. "Ammonia fiber explosion of corn stover." *Appl Biochem Biotechnol.* 113–116, 951–963.
90. Tiedje, T., et al. 2003. "Production of Succinic Acid and Ethanol from Ammonia Fiber Explosion (AFEX) Pretreated Biomass Hydrolysates." *25th Symposium on Biotechnology for Fuels and Chemicals*, Breckenridge, Colorado.
91. Wyman, C. E., et al. 2005. "Coordinated development of leading biomass pretreatment technologies." *Bioresour Technol.* 96:1959–1966.
92. Kim, S., and M. T. Holtzapple. 2005. "Lime pretreatment and enzymatic hydrolysis of corn stover." *Bioresour Technol.* 96:1994–2006.
93. Gould, L. M. 1984. "Alkaline peroxide delignification of agricultural residues to enhance enzymatic saccharification." *Biotechnol Bioeng.* 26:46–52.
94. Sun, R. C., and X. F. Sun. 2002. "Fractional and structural characterization of hemicelluloses isolated by alkaline and alkaline peroxide from barley straw." *Carbohydr Polym.* 49:415–423.
95. McGinnis, G. D., W. W. Wilson, and C. E. Mullen. 1983. "Biomass pretreatment with water and high-pressure oxygen. The wet-oxidation process." *Ind Eng Chem Prod Res Dev.* 22:352–357.
96. Bjerre, A. B., et al. 1995. "Pretreatment of wheat straw using combined wet oxidation and alkaline hydrolysis resulting in convertible cellulose and hemicellulose." *Biotechnol Bioeng.* 49:568–577.
97. Pan, X., et al. 2006. "Bioconversion of hybrid poplar to ethanol and co-products using an organosolv process: Optimization of process yields." *Biotechnol Bioeng.* 94:851–861.
98. Zhang, Y-H. P., et al. 2007. "Fractionating recalcitrant lignocellulose at modest reaction conditions." *Biotechnol Bioeng.* 97:214–223.
99. Dadi, A., C. A. Schall, and S. Varanasi. 2007. "Mitigation of cellulose recalcitrance to enzymatic hydrolysis by ionic liquid treatment." *Appl Biochem Biotechnol.* 136–140, 407–422.
100. Dadi, A. P., S. Varanasi, and C. A. Schall. 2007. "Fractionation of lignocellulosic biomass with ionic liquid pretreatment." *29th Symposium on Biotechnology for Fuels and Chemicals,* Denver, CO.
101. Upfal, J., D. R. Macfarlane, and A. S. Forsyth. 2005. "Solvents for use in the treatment of lignin-containing materials." Patent application WO 2005/017252 A1.
102. Anonymous. (2004). "Acceleration ionic liquid commercialization. Research needs to advance new technology." BCS Incorporated.
103. Poletti, L., et al. 2007. "Glucose-derived ionic liquids: Exploring low-cost sources for novel chiral solvents." *Green Chem.* 9:337–341.
104. Gottschalk, G. 1986. *Bacterial Metabolism.* 2 ed. Springer-Verlag: New York, NY, pp. 143–144.
105. Schmidt, A. S., et al. 1998. "Hydrolysis of solubilized hemicellulose derived from wet-oxidized wheat straw by a mixture of commercial enzyme preparations." Riso National Laboratory, Roskilde, Denmark.
106. Saha, B. C., and R. J. Bothast. 1999. "Pretreatment and enzymatic saccharification of corn fiber." *Appl Biochem Biotechnol.* 76:65–77.
107. Zhang, M., et al. 1995. "Metabolic engineering of a pentose metabolism pathway in ethanologenic *Zymomonas mobilis*." *Science.* 267:240–243.
108. Deanda, K., et al. 1996. "Development of an arabinose-fermenting *Zymomonas mobilis* strain by metabolic pathway engineering." *Appl Environ Microbiol.* 62:4465–4470.
109. Zhang, M., et al. 2007. "Zymomonas pentose-sugar fermenting strains and uses thereof." Patent U.S. 7,223,575.

110. Mohagheghi, A., et al. 2002. "Cofermentation of glucose, xylose, and arabinose by genomic DNA-integrated xylose/arabinose fermenting strain of Zymomonas mobilis AX101." *Appl Biochem Biotechnol.* 98–100, 885–898.
111. Gong, C. S., et al. 1981. "Production of ethanol from D-xylose by using D-xylose isomerase and yeasts." *Appl Environ Microbiol.* 41:430–436.
112. Kuyper, M., et al. 2005. "Metabolic engineering of a xylose-isomerase-expressing *Saccharomyces cerevisiae* strain for rapid anaerobic xylose fermentation." *FEMS Yeast Res.* 5:399–409.
113. Ho, N. W. Y., Z-D. Chen, and A. P. Brainard. 1998. "Genetically engineered *Saccharomyces* yeast capable of effective cofermentation of glucose and xylose." *Appl Environ Microbiol.* 64:1852–1859.
114. Krishnan, M. S., N. W. Y. Ho, and G. T. Tsao. 1999. "Fermentation kinetics of ethanol production from glucose and xylose by recombinant *Saccharomyces* 1400(pLNH33)." *Appl Biochem Biotechnol.* 77–79, 373–388.
115. Sedlak, M., and N. W. Y. Ho. 2004. "Production of ethanol from cellulosic biomass hydrolysates using genetically engineered *Saccharomyces* yeast capable of cofermenting glucose and xylose." *Appl Biochem Biotechnol.* 113–116, 403–416.
116. Kuyper, M., et al. 2004. "Minimal metabolic engineering of *Saccharomyces cerevisiae* for efficient anaerobic xylose fermentation: A proof of principle." *FEMS Yeast Res.* 4:655–664.
117. Op Den Camp, M. J. H., R. H. Harhangi, and C. Van Der Drift. 2003. "Fermentation of pentose sugars." Patent application WO 03/062430 A1.
118. Wisselink, H. M., et al. 2007. "Engineering of *Saccharomyces cerevisiae* for efficient anaerobic alcoholic fermentation of L-arabinose." *Appl Environ Microbiol.* Published online on June 1, 2007.
119. See Refs. 105, 237.
120. Ingram, L. O., et al. 1987. "Genetic engineering of ethanol production in *Escherichia coli*." *Appl Environ Microbiol.* 53:2420–2425.
121. Ohta, K., et al. 1991. "Genetic improvement of *Escherichia coli* for ethanol production: Chromosomal integration of *Zymomonas mobilis* genes encoding pyruvate decarboxylase and alcohol dehydrogenase II." *Appl Environ Microbiol.* 57:893–900.
122. De Carvalho Lima, K. G., C. M. Takahashi, and F. Alterthum. 2002. "Ethanol production from corn cob hydrolysates by *Escherichia coli* KO11." *J Ind Microbiol Biotechnol.* 29:124–128.
123. Asghari, A., et al. 1996. "Ethanol production from hemicellulose hydrolysates of agricultural residues using genetically engineered *Escherichia coli* KO11." *J Ind Microbiol.* 16:42–47.
124. Dumsday, G. J., et al. 1999. "Comparative stability of ethanol production by *Escherichia coli* KO11 in batch and chemostat culture." *J Ind Microbiol Biotechnol.* 23:701–708.
125. Takahashi, C. M., et al. 2000. "Fermentation of sugarcane bagasse hemicellulosis hydrolysate and sugar mixtures to ethanol by recombinant *Escherichia coli* KO11." *World J Microbiol Biotechnol.* 16:829–834.
126. Al-Zaag, A. 1989. "Molecular cloning of cellobiose and other β-glucosidase determinants from *Klebsiella oxytoca*." *J Biotechnol.* 12:79–86.
127. Doran, J. B., H. C. Aldrich, and L. O. Ingram. 1994. "Saccharification and fermentation of sugarcane bagasse by *Klebsiella oxytoca* P2 containing chromosomally integrated genes encoding the *Zymomonas mobilis* ethanol pathway." *Biotechnol Bioeng.* 44:240–247.
128. Ohta, K., et al. 1991. "Metabolic engineering of *Klebsiella oxytoca* M5A1 for ethanol production from xylose and glucose." *Appl Environ Microbiol.* 57:2810–2815.
129. Wood, B. E., and L. O. Ingram. 1992. "Ethanol production from cellobiose, amorphous cellulose, and crystalline cellulose by recombinant *Klebsiella oxytoca* containing chromosomally integrated *Zymomonas mobilis* genes for ethanol production and plasmids expressing thermostable cellulase genes from *Clostridium thermocellum*." *Appl Environ Microbiol.* 58:2103–2110.

130. Bothast, R. J., et al. 1994. "Fermentation of L-arabinose, D-xylose, and D-glucose by ethanologenic recombinant *Klebsiella oxytoca* strain P2." *Biotechnol Lett.* 16:401–406.
131. Zhou, S., F. C. Davis, and L. O. Ingram. 2001. "Gene integration and expression and extracellular secretion of *Erwinia chrysanthemi* endoglucanse CelY (*celY*) and CelZ (*celZ*) in ethanologenic *Klebsiella oxytoca* P2." *Appl Environ Microbiol.* 67:6–14.
132. Hartley, B. S., and G. Shama. 1987. "Novel ethanol fermentations from sugarcane and straw." *Philos Trans R Soc Lond* A. 321:555–568.
133. San Martin, R., et al. 1993. "Cultivation of an L-lactate dehydrogenase mutant of *Bacillus stearothermophilus* in continuous culture with cell recycle." *Biotechnol Bioeng.* 44:21–28.
134. Hartley, B. S. 1993. "Thermophilic ethanol production in a two-stage closed system." Patent U.S. 5.182,299.
135. Green, E., M. Javed, and N. Baghaei-Yazdi. 2004. "Ethanol production." Patent application U.S. 2004/0253713 A1.
136. Larsen, L., P. Nielsen, and B. K. Ahring. 1997. "*Thermoanaerobacter mathranii* spp. Nov, an ethanol-producing, extremely thermophilic anaerobic bacterium from a hot spring in Iceland." *Arch Microbiol.* 168:114–119.
137. Ahring, B. K., et al. 1996. "Pretreatment of wheat straw and conversion of xylose and xylan to ethanol by thermophilic anaerobic bacteria." *Bioresour Technol.* 58:107–113.
138. Georgieva, T. I., and B. K. Ahring. 2007. "Conversion of sugars in undetoxified lignocellulosic hydrolysates using thermophilic bacteria." *29th Symposium on Biotechnology for Fuels and Chemicals*, Denver, CO.
139. Fujita, Y., et al. 2004. "Synergistic saccharification, and direct fermentation to ethanol, of amorphous cellulose by use of an engineered yeast strain codisplaying three types of cellulolytic enzyme." *Appl Environ Microbiol.* 70:1207–1212.
140. Katahira, S., et al. 2004. "Construction of a xylan-fermenting yeast strain through codisplay of xylanolytic enzymes on the surface of xylose-utilizing *Saccharomyces cerevisiae* cells." *Appl Environ Microbiol.* 70:5407–5414.
141. Warnick, T., B.A. Methé, and S. B. Leschine. 2002. *Int J Syst Evol Microbiol.* 52:1155–1160.
142. Tolan, J. S. 1999. "Alcohol production from cellulosic biomass: The Iogen process, a model system in operation." In: *The Alcohol Textbook: A Reference for the Beverage, Fuel and Industrial Alcohol Industries.* K. A. Jacques, T. P. Lyons, and D. R. Kelsall (eds.). Nottingham University Press: Nottingham, UK, pp. 117–127.
143. Lawford, H. G., and J. D. Rousseau. 2003. "Cellulosic fuel ethanol." *Appl Biochem Biotechnol.* 105–108, 457–469.
144. Lynn, L. R., et al. 2005. "Consolidated bioprocessing of cellulosic biomass: An update." *Curr Opin Biotechnol.* 16:577–583.
145. Langvad, N. 2007. "BioGasol—biofuels for the future." California-Denmark Workshop on Clean Technology, San Francisco, California.
146. Mikkelsen, M. J., and B. K. Ahring. 2007. "Tailoring a thermophilic bacterium for optimal bioethanol production from lignocellulosic biomass." *29th Symposium on Biotechnology for Fuels and Chemicals*, Denver, CO.
147. Bohlmann, G. M. 2006. "Process economic considerations for production of ethanol from biomass feedstocks." *Ind Biotechnol.* Spring 2006, 14–20.
148. Solomon, B. D., J. R. Barnes, and K. E. Halvorsen. 2007. "Grain and cellulosic ethanol: History, economics, and energy policy." *Biomass Bioenerg.* 31:416–425.
149. Ciferno, J. P, and J. J. Marano. 2002. "Benchmarking biomass technologies for fuels, chemicals and hydrogen production." Report prepared for the National Renewable Energy Laboratory, U.S. Department of Energy.
150. Henstra, A. M., et al. 2007. "Microbiology of synthesis gas fermentation for biofuel production." *Curr Opin Biotechnol.* 18:1–7.
151. Hugenholtz, J., and L. G. Ljungdahl. 1990. "Metabolism and energy generation in homoacetogenic *clostridia*." *FEMS Microbiol Rev.* 87:383–390.

152. Phillips, J. R., E. C. Clausen, and J. L. Gaddy. 1994. "Synthesis gas as substrate for the biological production of fuels and chemicals." *Appl Biochem Biotechnol.* 45–46, 145–157.
153. Gaddy, J. L., and E. C. Clausen. 1992. "*Clostridium ljungdahlii*, an anaerobic ethanol and acetate producing microorganism." U.S. patent 5,173,429.
154. Gaddy, J. L. 1997. "*Clostridium* strain which produces acetic acid from waste gases." U.S. patent 5,593,886.
155. Gaddy, J. L. 2000. "Biological production of ethanol from waste gases with *Clostridium ljungdahlii.*" U.S. patent 6,136,577.
156. Gaddy, J. L., et al. 2003. "Methods for increasing the production of ethanol from microbial fermentation." U.S. patent application 2003/0211585.
157. Spivey, J. J., and A. Egbebi. 2007. "Heterogeneous catalytic synthesis of ethanol from biomass-derived syngas." *Chem Soc Rev.* Advance online publication; DOI: 1039/b414039g.
158. Spath, P. L., and D. C. Dayton. 2003. "Preliminary screening—technical and economic assessment of synthesis gas to fuels and chemicals with emphasis on the potential for biomass-derived syngas." National Renewable Energy Laboratory report number NREL/TP510–34929.
159. Anonymous. 2006. "BRI Energy seeking to build two gasification-fermentation ethanol plants." Available from: http://www.greencarcongress.com/2006/05/bri_energy_seek.html.
160. Bruce, W. F. 2006. "The co-production of ethanol and electricity from carbon-based wastes." A report from BRI Energy, Inc. Available from: http://www.magic-region.com/BRI_Energy_Backgrounder.pdf.
161. Anonymous. 2007. Syntec Biofuel Technology. Available from: http://www.syntecbiofuel.com/technology.html
162. Anonymous. 2006. "Syntec gearing up to commercialize its biomass-to-ethanol gasification/synthesis process." Available from: http://www.greencarcongress.com/2006/07/syntec_gearing_.html.
163. Van Kasteren, J. M. N., et al. 2005. "Bio-ethanol from bio-syngas." Technische Universiteit Eindhoven (TU/e) and Telos Ingenia Consultants & Engineers; Project number 0456.472; Document number 0456372-R02.
164. Bergman, P. C. A., S. V. B. van Paasen, and H. Boerrigter. 2002. "The novel "OLGA" technology for complete tar removal from biomass producer gas." Paper presented at the Pyrolysis and Gasification of Biomass and Waste, Expert Meeting, Strasbourg, France.
165. Anonymous. 2002. "Evaluation of the potential for the production of lignocellulosic based ethanol at existing corn ethanol facilities." NREL report number NREL/SR-510–32381.

| Company | Locations | Feedstocks | Current capacity (mg/yr) | Under construction & expansions (mg/yr) |
|---|---|---|---|---|
| Abengoa Bioenergy Corp. | York, NE | Corn/milo | 55 | |
| | Colwich, KS | | 25 | |
| | Portales, NM | | 30 | |
| | Ravenna, NE | | 88 | |
| AberdeenEnergy* | Mina, SD | Corn | | 100 |
| Absolute Energy, LLC* | St. Ansgar, IA | Corn | 100 | |
| ACE Ethanol, LLC | Stanley, WI | Corn | 41 | |
| Adkins Energy, LLC* | Lena, IL | Corn | 40 | |
| Advanced Bioenergy | Fairmont, NE | Corn | 100 | |
| AGP* | Hastings, NE | Corn | 52 | |
| Agri-Energy, LLC* | Luverne, MN | Corn | 21 | |
| Al-Corn Clean Fuel* | Claremont, MN | Corn | 35 | 15 |
| Amaizing Energy, LLC* | Denison, IA | Corn | 48 | |
| | Atlantic, IA | Corn | | 110 |
| Archer Daniels Midland | Decatur, IL | Corn | 1,070 | 550 |

(\*) Locally owned.
(#) Plant under construction
*Updated:* April 2, 2008.
*Source:* Renewable Fuels Association.

**TABLE 5A.1** Locations and Capacities of U.S. Ethanol Plants Using Corn and Other Minor Feedstocks

# Ethanol Production

| Company | Locations | Feedstocks | Current capacity (mg/yr) | Under construction & expansions (mg/yr) |
|---|---|---|---|---|
| | Cedar Rapids, IA | Corn | | |
| | Clinton, IA | Corn | | |
| | Columbus, NE | Corn | | |
| | Marshall, MN | Corn | | |
| | Peoria, IL | Corn | | |
| | Wallhalla, ND | Corn/barley | | |
| Arkalon Energy, LLC | Liberal, KS | Corn | 110 | |
| Aventine Renewable Energy, LLC | Pekin, IL | Corn | 207 | 226 |
| | Aurora, NE | Corn | | |
| | Mt. Vernon, IN | Corn | | |
| Badger State Ethanol, LLC* | Monroe, WI | Corn | 48 | |
| Big River Resources, LLC* | West Burlington, IA | Corn | 52 | |
| BioFuel Energy - Pioneer Trail Energy, LLC | Wood River, NE | Corn | | 115 |
| BioFuel Energy - Buffalo Lake Energy, LLC | Fairmont, MN | Corn | | 115 |
| BioFuel International | Clearfield, PA | Corn | | 110 |
| Blue Flint Ethanol | Underwood, ND | Corn | 50 | |

**TABLE 5A.1** *(Continued)*

# 184  Biofuels

| Company | Locations | Feedstocks | Current capacity (mg/yr) | Under construction & expansions (mg/yr) |
|---|---|---|---|---|
| Bonanza Energy, LLC | Garden City, KS | Corn/milo | 55 | |
| Bushmills Ethanol, Inc.* | Atwater, MN | Corn | 40 | |
| Calgren | Pixley, CA | Corn | | 55 |
| Cardinal Ethanol | Harrisville, IN | Corn | | 100 |
| Cargill, Inc. | Blair, NE | Corn | 85 | |
| | Eddyville, IA | Corn | 35 | |
| Cascade Grain | Clatskanie, OR | Corn | | 108 |
| Castle Rock Renewable Fuels, LLC | Necedah, WI | Corn | | 50 |
| Center Ethanol Company | Sauget, IL | Corn | | 54 |
| Central Indiana Ethanol, LLC | Marion, IN | Corn | 40 | |
| Central MN Ethanol Coop* | Little Falls, MN | Corn | 21.5 | |
| Chief Ethanol | Hastings, NE | Corn | 62 | |
| Chippewa Valley Ethanol Co.* | Benson, MN | Corn | 45 | |
| Cilion Ethanol | Keyes, CA | Corn | | 50 |
| Commonwealth Agri-Energy, LLC* | Hopkinsville, KY | Corn | 33 | |
| Corn, LP* | Goldfield, IA | Corn | 50 | |
| Cornhusker Energy Lexington, LLC | Lexington, NE | Corn | 40 | |
| Corn Plus, LLP* | Winnebago, MN | Corn | 44 | |
| Coshoctan Ethanol, OH | Coshoctan, OH | Corn | 60 | |

TABLE 5A.1  (Continued)

# Ethanol Production

| Company | Locations | Feedstocks | Current capacity (mg/yr) | Under construction & expansions (mg/yr) |
|---|---|---|---|---|
| Dakota Ethanol, LLC* | Wentworth, SD | Corn | 50 | |
| DENCO, LLC | Morris, MN | Corn | 21.5 | |
| E Energy Adams, LLC | Adams, NE | Corn | 50 | |
| E Caruso (Goodland Energy Center) | Goodland, KS | Corn | | 20 |
| East Kansas Agri-Energy, LLC* | Garnett, KS | Corn | 35 | |
| Elkhorn Valley Ethanol, LLC | Norfolk, NE | Corn | 40 | |
| ESE Alcohol Inc. | Leoti, KS | Seed corn | 1.5 | |
| Ethanol Grain Processors, LLC | Obion, TN | Corn | | 100 |
| First United Ethanol, LLC (FUEL) | Mitchell Co., GA | Corn | | 100 |
| Front Range Energy, LLC | Windsor, CO | Corn | 40 | |
| Gateway Ethanol | Pratt, KS | Corn | 55 | |
| Glacial Lakes Energy, LLC* | Watertown, SD | Corn | 100 | |
| Global Ethanol/ Midwest Grain Processor | Lakota, IA | Corn | 95 | |
| | Riga, MI | Corn | 57 | |
| Golden Cheese Company of California* | Corona, CA | Cheese whey | 5 | |
| Golden Grain Energy, LLC* | Mason City, IA | Corn | 110 | 50 |

TABLE 5A.1 (Continued)

# 186 Biofuels

| Company | Locations | Feedstocks | Current capacity (mg/yr) | Under construction & expansions (mg/yr) |
|---|---|---|---|---|
| Golden Triangle Energy, LLC* | Craig, MO | Corn | 20 | |
| Grand River Distribution | Cambria, WI | Corn | | 40 |
| Grain Processing Corp. | Muscatine, IA | Corn | 20 | |
| Granite Falls Energy, LLC* | Granite Falls, MN | Corn | 52 | |
| Greater Ohio Ethanol, LLC | Lima, OH | Corn | | 54 |
| Green Plains Renewable Energy | Shenadoah, IA | Corn | 50 | |
| | Superior, IA | Corn | | 50 |
| Hawkeye Renewables, LLC | Iowa Falls, IA | Corn | 105 | |
| | Fairbank, IA | Corn | 115 | |
| | Menlo, IA | Corn | | 100 |
| | Shell Rock, IA | Corn | | 110 |
| Heartland Corn Products* | Winthrop, MN | Corn | 100 | |
| Heartland Grain Fuelss, LP* | Aberdeen, SD | Corn | 9 | |
| | Huron, SD | Corn | 12 | 18 |
| Heron Lake BioEnergy, LLC | Heron Lake, MN | Corn | | 50 |
| Holt County Ethanol | O'Neill, NE | Corn | | 100 |
| Homeland Energy | New Hampton, IA | Corn | | 100 |
| Husker Ag, LLC* | Plainview, NE | Corn | 26.5 | |

TABLE 5A.1  (Continued)

# Ethanol Production

| Company | Locations | Feedstocks | Current capacity (mg/yr) | Under construction & expansions (mg/yr) |
|---|---|---|---|---|
| Idaho Ethanol Processing | Caldwell, ID | Potato wastes | 4 | |
| Illinois River Energy, LLC | Rochelle, IL | Corn | 50 | |
| Indiana Bio-Energy | Bluffton, IN | Corn | | 101 |
| Iroquois Bio-Energy Company, LLC | Rensselaer, IN | Corn | 40 | |
| KAAPA Ethanol, LLC* | Minden, NE | Corn | 40 | |
| Kansas Ethanol, LLC | Lyons, KS | Corn | | 55 |
| Land O'Lakes* | Melrose, MN | Cheese whey | 2.6 | |
| Levelland/Hockley County Ethanol, LLC | Levelland, TX | Corn | 40 | |
| Lifeline Foods, LLC | St. Joseph, MO | Corn | 40 | |
| Lincolnland Agri-Energy, LLC* | Palestine, IL | Corn | 48 | |
| Lincolnway Energy, LLC* | Nevada, IA | Corn | 50 | |
| Little Sioux Corn Processors, LP* | Marcus, IA | Corn | 52 | |
| Marquis Energy, LLC | Hennepin, IL | Corn | | 100 |
| Marysville Ethanol, LLC | Marysville, MI | Corn | | 50 |
| Merrick & Company | Golden, CO | Waste beer | 3 | |
| MGP Ingredients, Inc. | Pekin, IL | Corn/wheat starch | 78 | |
| | Atchison, KS | | | |

TABLE 5A.1 *(Continued)*

**188** Biofuels

| Company | Locations | Feedstocks | Current capacity (mg/yr) | Under construction & expansions (mg/yr) |
|---|---|---|---|---|
| Mid-America Agri Products/ Wheatland | Madrid, NE | Corn | | 44 |
| Mid America Agri Products/ Horizon | Cambridge, NE | Corn | | 44 |
| Mid-Missouri Energy, Inc.* | Malta Bend, MO | Corn | 45 | |
| Midwest Renewable Energy, LLC | Sutherland, NE | Corn | 25 | |
| Minnesota Energy* | Buffalo Lake, MN | Corn | 18 | |
| NEDAK Ethanol | Atkinson, NE | Corn | | 44 |
| New Energy Corp. | South Bend, IN | Corn | 102 | |
| North Country Ethanol, LLC* | Rosholt, SD | Corn | 20 | |
| Northeast Biofuels | Volney, NY | Corn | | 114 |
| Northwest Renewable, LLC | Longview, WA | Corn | | 55 |
| Otter Tail Ag Enterprises | Fergus Falls, MN | Corn | 57.5 | |
| Pacific Ethanol | Madera, CA | Corn | 40 | |
| | Boardman, OR | Corn | 40 | |
| | Burley, ID | Corn | | 50 |
| | Stockton, CA | Corn | | 50 |
| Panda Energy | Hereford, TX | Corn/milo | | 115 |
| Parallel Products | Louisville, KY | Beverage waste | 5.4 | |

TABLE **5A.1** (*Continued*)

# Ethanol Production

| Company | Locations | Feedstocks | Current capacity (mg/yr) | Under construction & expansions (mg/yr) |
|---|---|---|---|---|
| | R. Cucarmonga, CA | | | |
| Patriot Renewable Fuels, LLC | Annawan, IL | Corn | | 100 |
| Penford Products | Cedar Rapids, IA | Corn | | 45 |
| Phoenix Biofuels | Goshen, CA | Corn | 25 | |
| Pinal Energy, LLC | Maricopa, AZ | Corn | 55 | |
| Pine Lake Corn Processors, LLC* | Steamboat Rock, IA | Corn | 20 | |
| Platinum Ethanol, LLC* | Arthur, IA | Corn | | 110 |
| Plymouth Ethanol, LLC* | Merrill, IA | Corn | | 50 |
| Poet | Sioux Falls, SD | | 1253 | 282 |
| | Alexandria, IN | Corn | | # |
| | Ashton, IA | Corn | | |
| | Big Stone, SD | Corn | | |
| | Bingham Lake, MN | Corn | | |
| | Caro, MI | Corn | | |
| | Chancellor, SD | Corn | | |
| | Coon Rapids, IA | Corn | | |
| | Corning, IA | Corn | | |
| | Emmetsburg, IA | Corn | | |
| | Fostoria, OH | Corn | | # |

TABLE 5A.1  (*Continued*)

| Company | Locations | Feedstocks | Current capacity (mg/yr) | Under construction & expansions (mg/yr) |
|---|---|---|---|---|
| | Glenville, MN | Corn | | |
| | Gowrie, IA | Corn | | |
| | Groton, SD | Corn | | |
| | Hanlontown, IA | Corn | | |
| | Hudson, SD | Corn | | |
| | Jewell, IA | Corn | | |
| | Laddonia, MO | Corn | | |
| | Lake Crystal, MN | Corn | | |
| | Leipsic, OH | Corn | | |
| | Macon, MO | Corn | | |
| | Marion, OH | Corn | | # |
| | Mitchell, SD | Corn | | |
| | North Manchester, IN | Corn | | # |
| | Portland, IN | Corn | | |
| | Preston, MN | Corn | | |
| | Scotland, SD | Corn | | |
| Prarie Horizon Agri-Energy, LLC | Phillipsburg, KS | Corn | 40 | |
| Quad-County Corn Processors* | Galva, IA | Corn | 27 | |
| Red Trail Energy, LLC | Richardton, ND | Corn | 50 | |
| Redfield Energy, LLC* | Redfield, SD | Corn | 50 | |
| Reeve Agri-Energy | Garden City, KS | Corn/milo | 12 | |

TABLE 5A.1 *(Continued)*

# Ethanol Production

| Company | Locations | Feedstocks | Current capacity (mg/yr) | Under construction & expansions (mg/yr) |
|---|---|---|---|---|
| Renew Energy | Jefferson Junction, WI | Corn | 130 | |
| Siouxland Energy & Livestock Coop* | Sioux Center, IA | Corn | 60 | |
| Siouxland Ethanol, LLC | Jackson, NE | Corn | 50 | |
| Southwest Iowa Renewable Energy, LLC* | Council Bluffs, IA | Corn | | 110 |
| Sterling Ethanol, LLC | Sterling, CO | Corn | 42 | |
| Tate & Lyle | Loudon, TN | Corn | 67 | 38 |
| | Ft. Dodge, IA | Corn | | 105 |
| The Andersons Albion Ethanol, LLC | Albion, MI | Corn | 55 | |
| The Andersons Clymers Ethanol, LLC | Clymers, IN | Corn | 110 | |
| The Andersons Marathon Ethanol, LLC | Greenville, OH | Corn | 110 | |
| Tharaldson Ethanol | Casselton, ND | Corn | | 110 |
| Trenton Agri Products, LLC | Trenton, NE | Corn | 40 | |
| United Ethanol | Milton, WI | Corn | 52 | |
| United WI Grain Producers, LLC* | Friesland, WI | Corn | 49 | |
| Utica Energy, LLC | Oshkosh, WI | Corn | 48 | |

**TABLE 5A.1** (*Continued*)

## 192 Biofuels

| Company | Locations | Feedstocks | Current capacity (mg/yr) | Under construction & expansions (mg/yr) |
|---|---|---|---|---|
| VeraSun Energy Corporation | Aurora, SD | Corn | 1090 | 550 |
| | Ft. Dodge, IA | Corn | | |
| | Albion, NE | Corn | | |
| | Charles City, IA | Corn | | |
| | Linden, IN | Corn | | |
| | Welcome, MN | Corn | | # |
| | Hartely, IA | Corn | | # |
| | Bloomingburg, OH | Corn | | |
| | Albert City, IA | Corn | | |
| | Woodbury, MI | Corn | | |
| | Hankinson, ND | Corn | | # |
| | Central City, NE | Corn | | |
| | Ord, NE | Corn | | |
| | Dyersville, IA | Corn | | # |
| | Janesville, MN | Corn | | # |
| | Marison, SD | Corn | | |
| Western New York Energy, LLC | Shelby, NY | Corn | 50 | |
| Western Plains Energy, LLC* | Campus, KS | Corn | 45 | |
| Western Wisconsin Renewable Energy, LLC* | Boyceville, WI | Corn | 40 | |

TABLE 5A.1  (Continued)

# Ethanol Production

| Company | Locations | Feedstocks | Current capacity (mg/yr) | Under construction & expansions (mg/yr) |
|---|---|---|---|---|
| White Energy | Hereford, TX | Corn/milo | 100 | |
| | Russell, KS | Milo/wheat starch | 48 | 100 |
| | Plainview, TX | Corn | | |
| Wind Gap Farms | Baconton, GA | Brewery waste | 0.4 | |
| Renova Energy | Torrington, WY | Corn | 5 | |
| Xethanol BioFuels, LLC | Blairstown, IA | Corn | 5 | |
| Yuma Ethanol | Yuma, CO | Corn | 40 | |
| **Total current capacity at 114 ethanol plants** | | | **8,520.9** | |
| **Total under construction (80)/expansions (7)** | | | | **5,072.0** |
| **Total capacity** | | | **13,592.9** | |

TABLE 5A.1 *(Continued)*

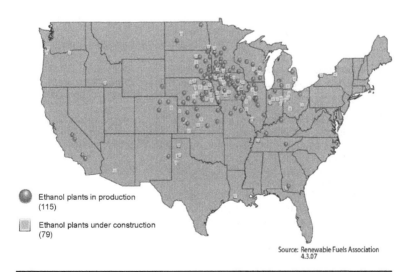

Ethanol plants in production (115)

Ethanol plants under construction (79)

Source: Renewable Fuels Association 4.3.07

**FIGURE 5A.1** Map of location of ethanol plants using corn and other minor feedstocks in operation and under construction in the United States.

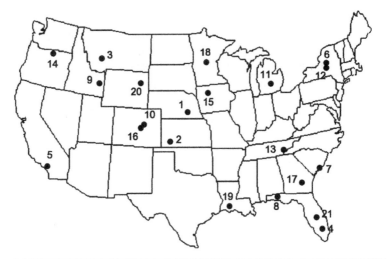

**FIGURE 5A.2** Map of location of pilot/demonstration and commercial ethanol plants using lignocellulosic biomass feedstocks in operation and under construction in the United States. (Source: *http://www.grainnet.com/pdf/cellulosemap.pdf.*)

*Legends:*
1. Abengoa Bioenergy, York, NE. Capacity: 10 mg/yr. Feedstock: Wheat straw. Plant type: Pilot.
2. Abengoa Bioenergy, Hugoton, KS. Capacity: 30 mg/yr. Feedstock: Wheat straw. Plant type: Commercial.
3. AE Biofuels, Butte, MT. Capacity: NA. Feedstock: Multiple sources. Plant type: Demonstration.
4. Alico, Inc., La Belle, FL. Capacity: NA. Feedstock: Multiple sources. Plant type: Commercial.
5. BlueFire Ethanol, Inc., Irvine, CA. Capacity: 17 mg/yr. Feedstock: Multiple sources. Plant type: Commercial.
6. Catalyst Renewables Corp., Lyonsdale, NY. Capacity: NA. Feedstock: Woodchips. Plant type: NA.
7. Clemson University Restoration Institute, North Charleston, SC. Capacity: 10 mg/yr. Feedstock: Multiple sources. Plant type: Pilot.
8. Gulf Coast Energy, Mossy Head, FL. Capacity: 70 mg/yr. Feedstock: Wood waste. Plant type: Commercial.
9. Iogen Biorefinery Partners, Inc., Shelley, ID. Capacity: 18 mg/yr. Feedstock: Multiple sources. Plant type: Demonstration/Commercial.
10. Lignol Innovations, Inc., Commerce City, CO. Capacity: 2 mg/yr. Feedstock: Wood. Plant type: Pilot.
11. Mascoma Corp., Lansing, MI. Capacity: 40 mg/yr. Feedstock: Wood. Plant type: Commercial.
12. Mascoma Corp., Rome, NY. Capacity: 0.5 mg/yr. Feedstock: Multiple sources. Plant type: Demonstration.
13. Mascoma Corp., Vonore, TN. Capacity: 5 mg/yr. Feedstock: Switchgrass. Plant type: Demonstration.
14. Pacific Ethanol, Boardman, OR. Capacity: 2.7 mg/yr. Feedstock: Mixed biomass. Plant type: Demonstration.
15. POET Biorefinery, Emmetsburg, IA. Capacity: 25 mg/yr. Feedstock: Corn cobs. Plant type: Commercial.

16. Pure Vision Technology, Ft. Lipton, CO. Capacity: 2 mg/yr. Feedstock: Corn stalks and grasses. Plant type: Pilot.
17. Range Fuels, Treutlen County, GA. Capacity: 20 mg/yr. Feedstock: Wood waste. Plant type: Commercial.
18. SunOpta Bioprocess LLC/Central Minnesota Ethanol Co-op, Little Falls, MN. Capacity: 10 mg/yr. Feedstock: Wood chips. Plant type: Commercial.
19. Verenium Energy, Jennings, LA. Capacity: 1.4 mg/yr. Feedstock: Wood waste. Plant type: Demonstration.
20. Western Biomass Energy (JL Process), Upton, WY. Capacity: 1.5 mg/yr. Feedstock: Wood waste. Plant type: Pilot.
21. Xethanol Corp./Southeast Biofuels. Auburndale, FL. Capacity: 8 mg/yr. Feedstock: Citrus peels. Plant type: Commercial.

*Notes:* mg/yr: million gallons per year. NA: Not available.

# CHAPTER 6
# Biodiesel

## 6.1 Introduction

Rudolf Diesel, pictured in Fig. 6.1, patented the first diesel engine in 1892 in Berlin, Germany, working for Linde Enterprises after moving from Paris. In 1894, he demonstrated a production scale engine nearly 3 m high, which exploded as he narrowly escaped with his life. In 1900, he demonstrated a working diesel engine using peanut oil as fuel at the World Exhibition in Paris.[1] Prior to his mysterious death in 1913, Rudolph Diesel stated that *"the use of vegetable oils as engine fuels may seem insignificant today but such oils may become, in the course of time, as important as petroleum and the coal tar products of the present time."*

The energy cycle founded by Rudolf Diesel enabled the production of an efficient, spark-free, and powerful engine compared to gasoline engines based on the Otto cycle that requires a spark for ignition of the fuel-air mixture. The main drawback at the time of invention of the diesel engine was the dramatically increased weight compared to a conventional gasoline engine. Therefore, the use of diesel engines was primarily to transport locomotives, trucks, and military submarines, while gasoline engines were primarily applied to automobiles and aircraft. Since then, high-performance, light-weight, turbo-diesel powered engines have come to the forefront primarily in Europe, where diesel fuel has become more cost effective for automobiles. Initially, Rudolf Diesel was interested in running his engine on either coal or vegetable-based fuels. Petroleum-based fuels became the main source due to lower cost over the past century. However, with the predictions based on Hubbert's curve that "peak" oil may occur within this decade, the price of oil would continue to rise until no longer affordable as world oil reserves approach depletion.

Recent incentives to reduce greenhouse gases, particularly carbon dioxide, have led to great interest in vegetable-based fuels because of a plant's inherent ability to capture solar energy through photosynthetic pigments (via light reactions) while efficiently sequestering carbon dioxide from the atmosphere as their primary carbon source (via dark reactions). This carbon is then biologically converted to high-energy starches, celluloses, proteins, and oils as storage and structural compounds. Some

FIGURE 6.1   Dr. Rudolf Diesel (1858–1913). (*Adapted from http://en.wikipedia.org/wiki/Biodiesel.*)

algae (e.g., diatoms) are known to efficiently convert carbon dioxide to nearly 60 to 70 percent of their dry weight in the form of storage oils. Theoretical yields of algal oils have been reported to produce 1000 to 10,000 (70 percent oil by weight in biomass) U.S. gal/acre/yr, which is greater than any other known plant source, including palm oil with 600 gal/acre/yr.[2] In the United States, soybeans are currently the largest source for biodiesel production, but only produces up to 50 gal/acre/yr. Total production of soybean oil for biodiesel is expected to hit peak levels by 2015, significantly raising the soybean prices.[3] The total production of oils and fats in the United States reached £35 billion in 2004. If this total amount were converted to biodiesel, only 5 billion gallons of biodiesel or about 8 percent of the diesel presently consumed would be produced. Therefore, other significant sources such as algal oils will be required to make a substantial contribution to supply for the increasing demand.

Soybean oil has a favorable energy balance ratio of about 3 energy units produced to each energy unit consumed, typically in the form of fossil-based fuels. Ethanol derived from corn starch is less favorable with a range between 0.8 and 1.7 energy ratios. Research on improving these ratios will be important to reduce dependency on foreign oil and other forms of fossil fuels. As costs for fossil-based fuels continue to escalate due to their inevitable depletion where supply no longer meets demand, biofuels will have to dramatically reduce dependency on fossil-fuels for their production.

Conventional diesel engines may operate on biodiesel (B100) or blends such as B20 (20 percent biodiesel blended with 80 percent no. 2 diesel) without major modifications and improve overall performance. The increasing solvent effect and lubrication properties with B20, B100, and E-biodiesel provide increased engine performance and life expectancy.

Injection timing may be adjusted to accommodate even better thermal efficiency when conventional diesel engines are operated using biodiesel. This results in significantly lower particulate emissions and carbon deposits in the engine and at the injector nozzles.[4] An initiative to replace the current kerosene-based JP-8 jet fuel with biofuels have also sparked great interest in the biodiesel sector.[5]

Business models for biodiesel production facilities typically follow two scales. Large-scale biodiesel companies (10 to 50 million gal/yr) typically require large corporate-owned farms to produce enough oils and fats to conform to a profitable economy of scale. Drawbacks include transportation costs and the ability to conduct sustainable agriculture through nitrogen inputs, run-off issues, and many other factors. Small-scale operations (0.1 to 2 million gal/yr) such as farm-based cooperatives like Piedmont Biofuels Coop have been successful with vertical integration strategies, utilizing all products and by-products within the farm region in partnership with local processing companies supplying oil-seed, animal fat, and waste greases.

### 6.1.1 Environmental Considerations

Emissions from the use of biodiesel in combustion engines are greatly reduced compared to conventional petroleum diesel fuels by up to 100 percent sulfur dioxide, 48 percent carbon monoxide, 47 percent particulate matter, 67 percent total unburned hydrocarbons, and up to 90 percent reduction in mutagenicity.[6] Figure 6.2 shows the emission

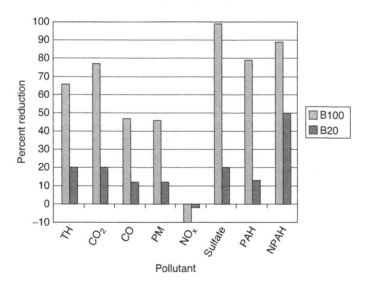

FIGURE 6.2  Percent reduction in emissions of pollutants in 100 percent biodiesel (B100) and 20 percent biodiesel blend (B20). TH is total hydrocarbons, $CO_2$ is carbon dioxide, CO is carbon monoxide, PM is particulate matter, $NO_x$ is nitrogen oxides, PAH is polycyclic hydrocarbons, and NPAH is nitrated PAHs. (*Adapted from http://www.biodiesel.org/pdf_files/fuelfactsheets/emissions.pdf.*)

reductions for biodiesel B100 and B20 compared to conventional no. 2 petroleum diesel fuels. Perhaps the most significant reduction based on life cycle analysis is the 78 percent reduction in carbon dioxide, which is considered the most important greenhouse gas in climatic models. Zhang et al.[7] showed that biodiesel has much higher biodegradability than low-sulfur diesel fuel and the addition of biodiesel to diesel fuels actually promotes the biodegradability of diesel fuel, making the blends more environmentally attractive.

### 6.1.1.1 Addressing the Nitrogen Oxide Problem

Although the environmental considerations for biodiesel are particularly favorable in terms of overall reduced emissions, one drawback is the potential relative increase in nitrogen oxides in the emissions of about 3 to 4 percent in B20, 4 to 6 percent in B40, and 6 to 9 percent in B100 over conventional petroleum diesel fuels.[9] However, reports show that actual reductions in $NO_x$ emissions are possible with corrections in injection timing and combustion temperatures.[10] The potential increase in nitrogen oxides is currently being addressed through research on engine thermodynamic mechanisms to improve emissions, catalytic conversion type technology to address the problem[9] or the additions of compounds such as antioxidants.[11] Nitrogen emissions in biodiesel may primarily occur through the Fenimore mechanisms where nitrogen in the air reacts during combustion of fuel to form $NO_x$ molecules[11] primarily as NO and $NO_2$. These oxidation mechanisms are shown in Fig. 6.3 where the rate of successive reactions forming radicals are highly temperature dependent. The first of these reactions was considered the rate-limiting step with the highest activation energy.[11] Nitrogen dioxide making up to 10 to 30 percent of the $NO_x$ emissions is formed from NO. The reaction takes place through peroxy radicals and may be reverted back to NO in the presence of free oxygen radicals.[11]

Some explanations for the reason why biodiesel increases $NO_x$ emissions over conventional diesel may be, in part, due to the physical property differences of the fuel's chemical structure. In particular, higher isentropic bulk modulus factors, which increase with increasing iodine values,[12,13] result in earlier injection rates at the fuel injectors in the diesel engine, leading to increased emissions.[14] Free radical formations are known to occur more prevalently at sights of higher degrees of unsaturation (double bonds). These shifts in dynamic starts of injection (SOI) timing due to increasing bulk modulus may be modified by

(a) $O^\bullet + N_2 \rightarrow NO + N^\bullet$
(b) $N^\bullet + O_2 \rightarrow NO + O^\bullet$
(c) $N^\bullet + {}^\bullet OH \rightarrow NO + H^\bullet$
(d) $NO + HO_2 \rightarrow NO_2 + {}^\bullet OH$
(e) $NO_2 + O^\bullet \rightarrow NO + O_2$

FIGURE 6.3 Oxidation mechanisms for formation of nitrogen oxides.

decreasing the timing for biodiesel to significantly decrease $NO_x$ emissions, but possibly not to levels of conventional diesel fuels.[9,15]

Addition of antioxidants to the fuel were proposed by Hess et al.[11] to scavenge these free radicals[16] during the combustion process to potentially decrease the rate of formation of $NO_x$ compounds. They found that butylated hydroxyanisole (BHA) and butylated hydroxytoluene (BHT) had a small, but significant effect on reduction of $NO_x$. Unfortunately, the natural antioxidants such as tocopherols normally present in fuels such as soy biodiesel did not have an effect in this study. If this technology is improved, additions of antioxidants are an attractive economic solution to the problem compared to many other strategies, while having additional advantages of preserving the fuel for extending the potential shelf life.[11]

### 6.1.1.2 Ethanol Fuel Additive

Ethanol contains only 66 percent of the energy value of gasoline and 60 percent of the energy value of diesel fuels resulting in a reduced driving range per equal volume of fuel; however, higher octane numbers for ethanol allow for greater compression ratios than for gasoline, thus increasing efficiencies by up to 30 percent and increasing the driving range to nearly 80 percent of gasoline vehicles.[17] Water-induced separation of gasoline-ethanol mixtures such as E10 or gasohol, E85 (85 percent ethanol), and E-diesel has posed problems for long-term storage of fuel mixtures and transportation through existing pipelines and is presently not permitted in the United States.[17] Methanol as a fuel contains only about 50 percent of the heating value of gasoline and poses much worse problems for water-induced separation and is highly toxic compared to ethanol. Recently, the Indy 500 racing organization has chosen to switch to ethanol from methanol high-octane value fuels. Methanol is derived from the fossil fuel industry and is not seen as sustainable in the foreseeable future. Biobutanol is a higher energy molecule than ethanol, but is more toxic. However, these longer carbon chain alcohols produced sustainably are a current topic with great interest.

Ethanol works well for the creation of biodiesel ethyl esters and may be used as an additive for fuel oxygenation with many advantages in terms of sustainable approaches to production and promising reductions in emissions. ETBE (ethyl tert butyl ether), another oxygenator compound derived from ethanol, has been used in the European Union (EU) as a replacement for MTBE. However, ETBE has not been proven to be any safer than MTBE and may face similar consequences in the near future.[18]

## 6.2 Biodiesel Production Chemistry and Thermodynamic Aspects

The basic chemistry for biodiesel production is relatively simple and occurs primarily through a transesterification step rendering this

$$\underset{\text{Triacylglycerol}}{\text{TAG}} + \underset{\text{Alcohol}}{\text{3R'OH}} \underset{}{\overset{\text{Catalyst}}{\longleftrightarrow}} \underset{\text{Fatty Acid Ester}}{\text{3R'COOR}} + \underset{\text{Glycerol}}{\text{C}_3\text{H}_5(\text{OH})_3}$$

FIGURE 6.4  Transesterification of triacylglycerol to fatty acid esters.

process to fairly straightforward commercialization.[19] The reactions associated most commonly to biodiesel production include transesterification and esterification, but with potential competing reactions including hydrolysis and saponification.[20]

### 6.2.1 Transesterification

The general transesterification reaction for acid or base-catalyzed conversion of oils consisting primarily of neutral triacylglycerols (TAG) in reaction with alcohols to form alkyl esters is shown by Fig. 6.4.

The alkyl esters produced depend on the alcohol used where methanol (R = $CH_3$) and ethanol (R = $CH_2CH_3$) are most common. The catalysts for transesterification include KOH, NaOH, or $H_2SO_4$. Excess alcohol with adequate catalyst generally forces the reaction equilibrium toward the products of biodiesel esters and glycerol.[7,21] With bio-based oils typically containing mostly TAG, the stoichiometric relationship requires 3 mol of alcohol per mole of TAG (3:1). The reaction usually requires excess amounts of alcohols ranging from 6:1 up to 20:1, depending on the reaction chemistry for base-catalyzed transesterification, and as high as 50:1 for acid transesterification.[7] The reaction for base-catalyzed systems will occur rapidly at room temperature, although higher temperatures of 50°C are often employed to reduced initial viscosity of oils while increasing reaction rates. Acid-catalyzed transesterification is often reacted at higher temperatures from just below the boiling point of the alcohol to 120°C in pressurized vessels. The transesterification reaction of TAGs takes place in three steps where TAG is first converted to a diacylglycerol (DAG) and one fatty acid ester. Then the DAG is converted to monoacylglycerol (MAG) liberating an additional fatty acid ester, and finally the MAG is converted to glycerol liberating the final fatty acid ester.

### 6.2.2 Esterification

The esterification process is a reversible reaction where free fatty acids (FFA) are converted to alkyl esters via acid catalysis (HCl or more commonly $H_2SO_4$). When oils are high in free fatty acids as common in waste cooking oils, the simultaneous esterification and transesterification reactions via acid catalysis is advantageous to potentially obtain nearly complete conversion to biodiesel. The esterification process follows a similar reaction mechanism of acid-catalyzed transesterification.[6] The reactants including FFA and alcohol are catalyzed by acid to create the alkyl ester and water as shown in Fig. 6.5. Wang et al.[22] used 2 percent ferric sulfate as a catalyst to conduct esterification as a pretreatment step before transesterification and obtained 97 percent conversion of waste cooking oil high in FFA. The

$$\text{RCOOH} + \text{R'OH} \underset{}{\overset{}{\longleftrightarrow}} \text{RCOOR'} + \text{H}_2\text{O}$$
FFA    Acid Catalyst   Alkyl Ester

**Figure 6.5** Acid-catalyzed esterification.

$$\text{TAG} + 3\text{CH}_2\text{CH}_3(\text{OC(O)CH}_3) \leftrightarrow 3\text{RCOOCH}_2\text{CH}_3 + \text{C}_3\text{H}_5(\text{OC(O)CH}_3)_3$$
Triacylglycerol   Ethyl Acetate   Lipase   Ethyl Ester   Triacetin

**Figure 6.6** Lipase-catalyzed interesterification.

advantages of this catalyst are no acidic wastewater and ease of recovery. Further exploration of this catalyst should be conducted for economic assessment.

### 6.2.3 Lipase-Catalyzed Interesterification and Transesterification

Interesterification and transesterification reactions via enzyme catalysis are biologically significant mechanisms for production of fatty acid esters and may be accomplished efficiently with enzyme catalysis using lipases secreted from various organisms such as *Candida antarctica*. For example, Novozyme-435 is an immobilized *C. antarctica* lipase B.[23] The reactant commonly used as the acyl acceptor for the interesterification of triglycerides is ethyl acetate rather than ethanol or methanol (as used in transesterification reactions). This reaction results in the production of triacetin and long-chain fatty acid methyl or ethyl esters, but not glycerol as in the case of esterification and transesterification with alcohols (see Fig. 6.6). Glycerol deactivates lipase mixtures[23] where the production of triacetin does not negatively affect the lipase activity presenting an advantage of this mechanism.[24]

### 6.2.4 Side Reactions: Saponification and Hydrolysis

Adverse reactions such as saponification and hydrolysis must be kept to a minimum. These reactions depend on the quality of the feedstock. Excessive amounts of free fatty acids are common in used cooking vegetable or animal-based oils due to oxidation and other reactions at excessive temperatures. With excessive free fatty acids present, the reaction mixture will be neutralized with excess base potentially resulting in two undesirable reactions.[6] Saponification creates soap and water as shown in Fig. 6.7a from excess free fatty acids in the mixtures. Also, a

(a) $\text{RCOOH} + \text{NaOH or KOH} \longleftrightarrow \text{R'COONa}^+ \text{ or } \text{R'COOK}^+ + \text{H}_2\text{O}$
    FFA    Base Catalyst    Soap

(b) $\text{RCOOR'} + \text{H}_2\text{O} \longleftrightarrow \text{RCOOH} + \text{R'OH}$
    Alkyl Ester   Base Catalyst   FFA

**Figure 6.7** Adverse reactions in biodiesel production including (a) saponification to create soaps and (b) FFA formation from hydrolysis.

second hydrolysis reaction causes conversion of biodiesel esters, via base catalysis, to free fatty acids as shown in Fig. 6.7b. Further reaction with catalyst will result in the formation of more soap and water until the catalyst is finally consumed and deactivated. An intermittent glycerolysis reaction is another approach to avoid soap formation by first converting free fatty acids to monoacylglycerols (MAGs) using high temperature (220 to 250°C) in the presence of alkaline catalyst or conducted at lower temperature with use of lipases.[25]

### 6.2.5 Alcohol Effect

The type of alcohol and its effect on biodiesel production has been studied little beyond the use of methanol and ethanol, which are the most widely used alcohols for biodiesel production.[26,27] Canakci and Van Gerpen[28] determined that increasing molar ratios of alcohol to TAG increased the rate of reaction and conversion yield as high as 98.4 percent at alcohol to TAG ratio of 30:1; but the effect decreased sharply beyond 6:1 ratios that resulted in nearly 90 percent conversion. The effect of increasing molecular weight and boiling points of alcohols used was studied by Nye et al.[29] They compared linear chains of alcohols, including methanol, ethanol, propanol, and butanol in acid-catalyzed (0.1 percent sulfuric acid) and base-catalyzed transesterification of waste frying oils to biodiesel esters. The longer chain alcohols showed the highest reaction rates for acid-catalyzed reactions; however, for base-catalyzed reactions, an opposite effect was observed. Therefore, alcohol solubility played a more important role for the acid-catalyzed reactions since longer chain alcohols are considerably more soluble in the initial oil than methanol. Behavior of different alcohols has a significant effect at supercritical conditions where no catalyst is required for both esterification and transesterification reactions.[30] This phenomena is discussed in further detail in Sec. 6.2.9.

### 6.2.6 Base or Alkali Catalysis

Base- or alkali-catalyzed reactions typically using sodium hydroxide or potassium hydroxide are most common when the oils are neutral or mostly free of fatty acids.[31-43] This is, in part, due to the low cost of the catalyst and mild temperatures, but primarily because the rate of reaction is up to three orders of magnitude greater than acid-catalyzed reactions.[6,27,44] The base-catalyzed reaction (shown in Fig. 6.3) is repeated twice per molecule of TAG until a molecule of glycerol is liberated from the three fatty acid moieties.[6] The specific transesterification occurs in four intermediate reversible steps (Fig. 6.8) that include: (1) producing an active species from the alcohol group in reaction with the base catalyst, (2) nucleophilic attack of the active species to the carbonyl group on the TAG to form a tetrahedral intermediate which is then broken down, (3) liberating the fatty acid ester compound and the active species is regenerated in the final step, and (4) allowing the process to continue for the remaining two attached fatty acid units.[6]

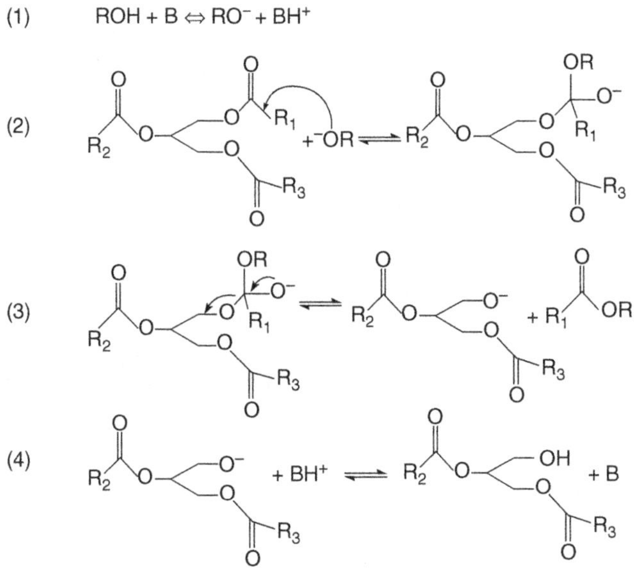

(1) ROH + B ⇔ RO⁻ + BH⁺

**FIGURE 6.8** Stepwise base-catalyzed transesterification of triacylglycerol to fatty acid esters. Steps 1 through 4 are repeated twice to yield three alkyl esters and glycerol.

### 6.2.6.1 Homogeneous Base Catalysts

Homogeneous base catalysts are mostly used in the production of biodiesel today primarily due to their low cost and high reaction rates at relatively low temperatures, which result in yields that often exceed 98 percent. The major drawbacks include difficulty in recovering the catalyst and their highly corrosive nature that is often detrimental to the useful life of equipment. Yields of methyl or ethyl esters from low-acid oil feedstocks often exceed 98 percent, with alkali catalysts typically in the concentration of nearly 1 percent for both sodium hydroxide and potassium hydroxide in dry conditions to avoid saponification side reactions.[45]

The temperature effect for base-catalyzed biodiesel conversion rates show increased reaction rates up to 55°C for methanol and 75°C for ethanol[46]; however, after 60 min of reaction time, very little effect can be seen. Therefore, ambient conditions are feasible for production of biodiesel using heterogeneous base catalyst in most cases.[41,44] Typical operating temperature ranges between 45 and 55°C to reduce viscosity of the reacting oils particularly when higher levels of saturated fats are present, while reaction time is reduced to less than 1 h in most cases.

The cost for the three most commonly used alkali catalysts sodium hydroxide, potassium hydroxide, and sodium methoxide reported in

2005 were $400, $770, and $2300/ton, respectively. Sodium hydroxide is the most commonly used base catalyst from the material cost perspective.[45] However, higher potential yields,[28] lower waste disposal cost, and better environmental characteristics are advantages of potassium hydroxide, making it more competitive with sodium hydroxide when considering the life cycle assessment of each catalyst. Leung and Guo[45] also noted that potassium hydroxide yielded better separation characteristics for oil feedstocks containing free fatty acids. The potassium soap and glycerol formed during the reaction were easier to remove than sodium soaps formed by sodium hydroxide or sodium methoxide. This was due to a softer potassium soap formation that did not settle into the glycerol layer, making a less viscous glycerol solution easier to remove. The soaps were then removed with a hot water washing step, making potassium hydroxide advantageous particularly when free fatty acids were present.

#### 6.2.6.2 Solid Base Catalysts

Heterogeneous solid base catalysts are advantageous due to ease of recovery and potential facilitation of continuous processing.[46–58] A summary of these catalysts given by Liu[56] includes hydrotalcites—for example, ETS-10 (Na, K), calcium oxides, quaternary ammonium silica gels, and other anionic exchange resins.

### 6.2.7 Acid Catalysis

Acid catalysts using hydrochloric or sulfuric acid typically result in substantially reduced transesterification rates by as much as 4000 times less than homogeneous base-catalyzed reactions and require higher reaction temperatures of nearly 100°C.[44,59–61] However, acid catalysts can be very effective in driving the transesterification process when oils contain large amounts of free fatty acids as in the case of some yellow and brown greases that are by-products of the food- and animal-rendering industries. Acid catalysts convert free fatty acids to biodiesel esters through esterification while simultaneously catalyzing the transesterification of triglycerides to biodiesel in a single step.[6,7] As Lotero et al.[6] reported, acid-catalyzed transesterification occurs in four intermediate reversible steps, including (1) protonation of the carbonyl group with acid catalyst making the adjacent carbon atom more prone to nucleophilic attack, (2) protonation of the alcohol group, (3) proton migration, and (4) breakdown of the intermediates freeing the alkyl ester from the intermediate DAG, which is then subject to further breakdown by repeating the series of intermediate reactions twice. The base-catalyzed transesterification reaction results in a more direct route, primarily resting on the first intermediate step producing a strong alkoxide ion that acts as a strong nucleophile (see Fig. 6.3). This is in contrast to the acid-catalyzed transesterification reaction that results in a more electrophilic species, but weaker nucleophile.[6]

### 6.2.7.1 Homogeneous Acid Catalysts

Relatively few studies exist on homogeneous acid-catalyzed transesterification.[28,41,62-65] Simultaneous esterification of the free fatty acids to alkyl esters occurs to increase biodiesel yields from lower quality feedstocks. The pioneering works of Liu,[66] Freedman et al.,[41] and Crabbe et al.[67] showed that increasing molar ratios to as high as 45:1 significantly increased the rate and percent conversion of acid-catalyzed reactions primarily with sulfuric acid, but also HCl, BF3, phosphoric, and sulfonic acids.[66] Three rate-limiting steps were determined by Liu[66] where initially a mass-transfer–controlled regime exists due to low solubility of alcohols and reagents. Then a kinetically-controlled regime occurs with rapid product formation, followed by a final regime represented by dramatically reduced rate of product formation near equilibrium.

Temperature also plays an important role in acid-catalyzed reactions compared to base-catalyzed reactions.[26,41,66] Freedman et al.[41] observed that 1 percent sulfuric acid catalyzed butanolysis of soybean oil to nearly complete conversion in 3 h at 117°C, but took significantly longer to obtain similar conversions at 77°C. Khan[68] noted that 90 percent ester conversions using 1.7 wt % sulfuric acid could occur in as little as 15 min, but not without significant side reactions such as alcohol etherification.[6] Al-Widyan and Al-Shyoukh[69] found that sulfuric acid with 100 percent excess ethanol performed better than similar concentrations of HCl in the conversion of used palm oil to biodiesel esters. The cost per ton of hydrochloric and sulfuric acid in U.S. $ are $120 and $60, respectively, giving an economic advantage to sulfuric acid use.[70,71]

### 6.2.7.2 Solid Acid Catalysts

Recently heterogeneous solid acid catalysis has been investigated.[63,64] Potential advantages include efficient conversion of waste cooking and rendered oils typically high in free fatty acids to biodiesel with simultaneous esterification and transesterification reactions with relative ease in the recovery and regeneration of catalyst. These are ambitious goals considering the challenges of creating such a catalyst with high turnover frequency, good absorption characteristics, low cost, and perhaps most important, reusability. Low-cost, stable materials with large interconnected pores saturated with a high concentration of stable acid sites are necessary for economic feasibility. A host of materials, including zeolites,[72] silica molecular sieves doped with metals (e.g., aluminum, titanium), sulfonated zirconia,[73] sulfonated tin oxide,[74] Amberlyst-15,[75] and Naphion NR50 have been examined.[64] A common problem with the solid catalysts currently being tested is the loss of activity after each use due to inhibitors that poison the catalyst with highly absorbed residues, and potential desorption of the acid active groups.[63] However, significant progress is being made to address these critical issues. Therefore, the future of solid acid catalysts is an important consideration of potential and dramatic process improvements, particularly for the use of waste cooking oils.

### 6.2.8 Enzyme Catalysis

Because interesterification and transesterification reactions occur naturally in biological systems, enzyme catalysts can increase the effectiveness of the reactions. The largest obstacle in most cases is cost for applying enzymes to industrial systems where the goal is to make a product in high volume and as low cost as possible. This is the case with both cellulosic ethanol production and enzyme-catalyzed biodiesel production.[23] Therefore, reusability of the enzyme in immobilized systems is critical for economics of these systems. Significant advances in enzyme biotechnology could provide a way to jump these hurdles. Modi et al.[23] discussed that potential advantages of lipase-catalyzed systems address problems associated with conventional homogeneous acid- and base-catalyzed systems such as glycerol recovery and removal of inorganic salts. Enzymes typically need small amounts of water for activity and constraints of water in conventional systems may be avoided, thus saving some energy in the removal of water from reacting alcohols. Immobilized lipase systems hold even greater promise for similar reasons to the solid acid and base catalysts in that they are more easily recovered and reused.[76] Immobilized systems are also subjected to diffusion limitations and potential deactivation by conventional linear alcohols such as methanol and ethanol[77] where the degree of deactivation has an inverse relation with the number of carbons on the alcohol chain.[78] Hsu et al.[79] developed an immobilized lipase from *Pseudomonas cepacia* in a sol-gel polymer matrix that achieved more than 95 percent conversion of tallow. Similar immobilized suspensions were prepared by Noureddini et al.[80] Lipase-catalyzed transesterification was carried out at 40°C for 1 h using mixtures of 10 g oil, 3 g immobilized lipase PS, 3 g methanol or 5 g ethanol, and 0.5 g water. Water concentrations below 0.2 percent substantially decreased conversion. Reusability of the immobilized lipase after 11 uses resulted in about 10 percent decrease in activity. Modi et al.[23] noted that *C. antarctica* lipase B immobilized on acrylic resin maintained most of its activity after 12 uses during interesterification, but then lost much of its activity after only six cycles during transesterification.

### 6.2.9 Supercritical Esterification and Transesterification

Supercritical and subcritical methanol transesterifications may be performed effectively without the need of a catalyst, thus omitting costly and time-consuming steps of removing catalyst as well as special equipment considerations for handling strong catalysts. The critical state of methanol is 239°C and 8.09 MPa. Table 6.1 shows the

TABLE 6.1 Critical Points and Reaction Pressures at 300°C for Different Alcohols and Potential Yields of Fatty Acid Esters after 10 min Esterification of Fatty Acids (Palmitic, Stearic, Oleic, Linoleic, Linolenic) and 30 min Transesterification of Rapeseed Oil

| Alcohol | Critical temperature (°C) | Critical pressure (MPa) | Reaction pressure at 300°C (MPa) | Esterification yield* (10 min) % | Transesterification yield (30 min) % |
|---|---|---|---|---|---|
| Methanol | 239 | 8.09 | 20 | 98 | 98 |
| Ethanol | 243 | 6.38 | 15 | 79 | 88 |
| 1-propanol | 264 | 5.06 | 10 | 81 | 85 |
| 1-butanol | 287 | 4.90 | 9 | 80 | 75 |
| 1-octanol | 385 | 2.86 | 6 | – | – |

*Average approximate value for all fatty acids tested.
Source: Adapted from Warabi et al., 2004.

critical states for alcohols ranging from methanol to 1-octanol.[30] They determined the reaction conditions for esterification and transesterification of different alcohols where percent conversion to fatty acid esters were shown for comparison at 10 min and 30 min reaction times at 300°C. In general the more unsaturated the fatty acid, the faster the rates of esterification occurred. Transesterification of rapeseed oil triglycerides exhibited slower rates of conversion than esterification reactions. However, the times to nearly complete conversion were close to that of alkali-catalyzed transesterification. Kusdiana and Saka[81] determined the reaction kinetics for subcritical and supercritical transesterification of rapeseed oil. They determined that supercritical methanol performed much better than subcritical conditions with a reaction temperature of 350°C and pressure of 19 MPa considered optimum. They determined that increasing molar ratio of methanol to oil dramatically increased conversion rates with a 42:1 ratio. Yields at these conditions were nearly 95 percent conversion compared to about 60 percent conversion for a 6:1 ratio at 350°C reacted for only 4 min. Esterification reactions of fatty acids and 1-propanol catalyzed in supercritical $CO_2$ by *C. antarctica* and *Mucor miehei* lipase encapsulated in lecithin water-in-oil microemulsions resulted in greater conversion rates. These suspension also used nearly threefold less enzyme[82] than other immobilized lipase systems.[83,84] A decrease in reaction rate with increasing pressure possibly occurred due to increased solubility of the reactants in supercritical $CO_2$, thereby partitioning the reactants away from the enzyme entrapped in the microemulsion-based organogel.[82]

### 6.2.10 Thermodynamics and Reaction Kinetics

Thermodynamic properties of biodiesel fuels and reactants play an important role in determining the reaction rate and phase separation kinetics, as with most basic chemical and biochemical reactions. The properties of the final biofuel product then determine important combustion, flow, and storage characteristics. The design of the diesel engine was based on thermodynamic concepts portrayed in the Diesel cycle (Fig. 6.9) where air-fuel mixtures are compressed adiabatically to a pressure high enough to raise the temperature to enable combustion, which typically requires a compression ratio of 15 to 20.

The physical and thermodynamic properties of biodiesel fuels compared to diesel fuels and alcohols are shown in Table 6.2. In general, the thermodynamic properties in terms of temperature (e.g., boiling point, flash point, pour point, cloud point) are typically higher for biodiesel than petroleum diesel, gasoline, and alcohols. Higher flash points result in a safer fuel for handling. Density and viscosity of biodiesel is typically higher than the petroleum fuels and alcohols. The heating values for biodiesel are comparable to petroleum fuels, which are nearly twice that of the alcohols.

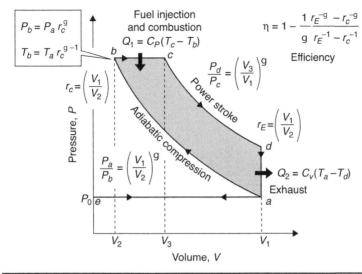

**FIGURE 6.9** Diesel cycle P-V diagram with efficiency ($\eta$) determined from compression ($r_c$) and expansion ($r_E$) ratios. Other variables: $\gamma$ is the ratio of specific heats ($C_p/C_v$) where $C_p$ and $C_v$ are the specific heats at constant pressure and volume, respectively. (*Adapted from http://hyperphysics. phy-astr.gsu.edu/hbase/thermo/diesel.html.*)

#### 6.2.10.1 Reaction Kinetics

Reaction kinetics determine the rate at which a given reaction will occur for the formation of biodiesel esters. The kinetics for biodiesel reactions depend on the physical state (solid, liquid, gas), concentration of reactants, temperature, and the types of catalysts used to lower the activation energy required for the reaction to occur. Phase and reaction equilibria gives a basis for the extent to which the phases and reactions will occur, respectively.

Reaction kinetics vary with system temperature and pressure. Freedman et al.[41,62] and Noureddini and Zhu[31] determined that transesterification follows pseudo first-order kinetics and the reverse reaction occurs as described by second-order kinetics. For conditions near the critical point of methanol, kinetics were considered first-order[30,81] represented by the following equation:

$$-\ln\frac{x_t}{x_0} = kt \tag{6.1}$$

where $x_t$ = the concentration of unreacted methyl esters at time $t$
$x_0$ = the initial methyl ester concentration
$k$ = the first-order reaction rate constant

A semilog plot of unmethylated esters content versus time yields the first-order rate constants by taking the slope. Typical values for

| Property (ASTM) | Units | Diesel (no. 2) | Biodiesel (general) | Sunflower ME | Rape ME | Soy ME | Sunflower EE | Rape EE | Soy EE | Gasoline | Ethanol | Methanol |
|---|---|---|---|---|---|---|---|---|---|---|---|---|
| Viscosity (D445) | v (40°C) (cSt) | 1.9–4.1 | 1.9–6.0 | 4.2–4.6 | 4.8 | 4.1–4.5 | 4.5 | – | 4.4 | 0.8 | 1.5 | 0.75 |
| Density (D4052) | R (25°C) (g/mL) | 0.85 | 0.88 | 0.86 | 0.89 | 0.89 | 0.87 | 0.90 | 0.88 | 0.72–0.78 | 0.794 | 0.796 |
| Boiling point | BP (°C) | 210–235 | – | – | – | 339 | – | – | – | 30–225 | 78 | 65 |
| Flash point (D93A) | FP (°C) | 60–80 (52) | 100–170 | 164–183 | 153 | 141–188 | 187 | 191 | 195 | –43 | 13 | 11 |
| Cloud point (D2500) | CP (°C) | –19 to 5 | –3 to 12 | –3 | –3 | 2 | –1 | 8 | 1 | – | – | – |
| Pour point | PP (°C) | –35 to –15 | –15 to 16 | –6 | – | –2 | –8 | – | –4 | – | – | – |
| Combustion point | CBP (°C) | 92[2] | – | 183 | – | 171 | 192 | – | – | – | – | – |
| Autoignition temperature | AIT (°C) | 254 | – | – | – | – | – | – | – | 370 | 423 | 464 |
| Distillation (D86) | 50% (°C) | – | – | – | – | – | 353 | – | 344 | – | – | – |

| Property | Units | | | | | | | | | | |
|---|---|---|---|---|---|---|---|---|---|---|---|
| Cetane index (D4737) | CI | 40–55 | 48–60 | 49 | 52 | 45–47 | 49.4 | 48.2 | <15 | <15 | <15 |
| Octane number | – | – | – | – | – | – | – | – | 82–92 | 90–102 | 89–102 |
| Lubricity (D6079) | HFRR (μm) | 685 | 314 | – | – | – | – | – | – | – | – |
| Lubricity | BOCLE (g) | 3600 | >7000 | – | – | – | – | – | – | – | – |
| Higher heating value | HHV (MJ/kg) | 45.2 | – | 40.1 | 40.0 | 39.8 | 38.9 | 40.0 | 43.5 | 27 | 20.1 |
| Heat of vaporization | HV (kJ/kg) | 375 | – | – | – | – | – | – | 380 | 920 | 1185 |
| Water | % vol | 0.05 | 0.05 | – | – | – | – | – | – | – | – |
| Carbon (D189) | wt % | 87 | 77 | – | – | – | – | – | – | – | – |
| Hydrogen | wt % | 13 | 12 | – | – | – | – | – | – | – | – |
| Oxygen | wt % | 0 | 11 | – | – | – | – | – | – | – | – |
| Sulfur (D4294) | wt % | 0.05 | 0.05 | 0.01 | <0.01 | 0.01 | <0.01 | 0.1 | – | – | – |
| Ramsbottom residue | % | 0.14 | – | – | – | – | 0.30 | 0.69 | – | – | – |

*Source:* Adapted from Lotero et al., 2005; Brown, 2003; ASTM, 2005; Encinar et al., 2005; Encinar et al., 2002; and Barnwal and Sharma, 2005.

**TABLE 6.2** Physical Properties of Diesel and Biodiesel Fuels

first-order reaction rates for esterification of fatty acids and trans-methylation of vegetable oils in supercritical methanol range from 0.0002 to 0.1 s$^{-1}$ in the temperature range of 240 to 490°C.[81,88] Once the reaction constants were determined, the Arrhenius equation was applied to determine the activation energy as given in the following equation:

$$\ln k = \ln A - \frac{E_a}{RT} \qquad (6.2)$$

where $E_a$ = activation energy, kcal/gmol or kJ/gmol
$R$ = Universal gas constant
$T$ = temperature, K

The activation energy is defined as the minimum amount of energy required for a reaction to occur. He et al.[88] showed that activation energy at supercritical conditions was dramatically higher than at conditions less than the critical point.

**Example 6.1** Determine the activation energy from the following subcritical and supercritical methanol rate data for transmethylation of soybean oil. Data has been taken from Kusdiana and Saka.[81]

**Solution** From Fig. 6.10, the activation energy is determined from the slopes of the individual regions of supercritical and near to subcritical. For supercritical conditions, the slope of the linear fit is $-E_a/R = -5.6643$; therefore $E_a = 5.6643 \times 1.9872 = 11.2$ cal/gmol. For the near and subcritical conditions, we take the slope

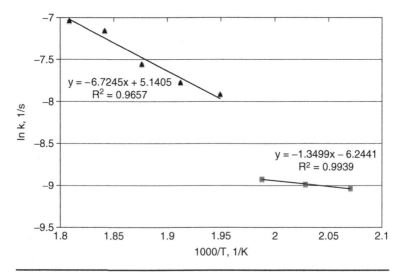

**Figure 6.10** Arrhenius plot example of transesterification of rapeseed oil with sub- and supercritical methanol for soybean oil. (*Adapted from He et al., 2007.*)

of these points, which are separated by the clear break due to very different reaction rates. From this slope, we obtain $E_a = 4.62883 \times 1.9872 = 9.2$ cal/gmol, which is a lower activation energy that explains the difference in reaction rates due to temperature (and subsequentially different pressures in this case).

The kinetic analysis of enzyme-catalyzed reactions typically follows Michaelis-Menton kinetics represented by Eq. (6.1).

$$v = \frac{v_{max} S}{K_s + S} \qquad (6.3)$$

where  $v$ = the reaction velocity
$v_{max}$ = maximum reaction velocity
$K_s$ = half-saturation constant (equal to the substrate concentration at ½ $v_{max}$)
$S$ = substrate (reactant) concentration

The Lineweaver-Burk double reciprocal plot of the $1/v$ versus $1/S$ yields the kinetic coefficients $v_{max}$ and $K_s$ after linear regression as shown in Eq. (6.2).

$$\frac{1}{v} = \frac{K_s}{v_{max}} \frac{1}{S} + \frac{1}{v_{max}} \qquad (6.4)$$

The Ping Pong Bi Bi inhibition model is another mechanistic model that more accurately describes reaction kinetics for esterification and transesterification reactions catalyzed by immobilized lipases in supercritical carbon dioxide and other organic solvents.[82,84,89–91] This model considers inhibitory effects by excess alcohols or other possible inhibitors. The alcohol participates in the nucleophilic attack on the carboxylic lipase intermediate, resulting in the release of an ester. Inhibition may occur if the alcohol reacts with the free enzyme described by the constant, $K_i^B$ (mM) in the following equation:

$$v_0 = \frac{v_{max}[A][B]}{K_m^A[B]\left(1+([B]/K_i^B)\right) + K_m^B[A] + [A][B]} \qquad (6.5)$$

where  $v_0$ (mM/min) = initial reaction velocity
$v_{max}$ (mM/min) = maximum reaction velocity
$[A], [B], K_m^A,$ and $K_m^B$ = concentrations (mM) and Michaelis-Menten constants (mM) for the catalyst and alcohol, respectively

### 6.2.10.2 Thermogravimetric Kinetics

When oils are subjected to a constant rate temperature change over time, the weight and composition of the material is measured by generating a thermogravimetric curve. Mass loss steps are typically attributed to volatilization and/or combustion of triglycerides. Thermogravimetric analysis of biodiesel esters typically results in two significant mass loss steps, including evaporation of methyl esters between 180 and 220°C and combustion and further volatilization above 300°C for castor oil methyl and ethyl esters with over 95 percent mass loss attributed to evaporation and oxidation of the esters.[92] Stability and combustion characteristics of oils and biodiesel esters determined through thermogravimetric analysis yield activation energy values through the following equation:

$$g = \frac{A}{\phi}\int_0^T e^{-E_a/RT} dT \qquad (6.6)$$

where  $g$ = kinetic thermogravimetric rate
$A$ = pre-exponential factor
$\phi$ = heating rate
$E_a$ = activation energy
$R$ = Universal gas constant
$T$ = temperature

From an analysis of castor oil and its methyl and ethyl esters, activation energies averaged 120, 82, and 87 kJ/gmol, respectively. Stability decreased in the order of castor oil, ethanol-biodiesel, and methanol-biodiesel as expected based on decreasing volatility, while methanol-biodiesel would exhibit higher degree of combustion as indicated by the lowest value of activation energy.[92] Thermogravimetric analysis (TGA) may be applied to determine boiling point data for mixtures of fatty acid esters in biodiesel, but the technique is expensive and decomposition of some compounds at high temperature reduces the reliability of this method, making the prediction of boiling points highly desirable to the industry[93] as discussed in the next section.

### 6.2.10.3 Vapor Pressure and Boiling Point

Yuan et al.[93] determined vapor pressure and boiling point data of pure methyl esters and biodiesel fuels and then used the Antoine equation and a group contribution method to predict the temperature-dependent behavior of vapor pressures. To ensure quality control measures in the rapidly growing bioenergy sector, reliability in the predictions of thermodynamic properties are necessary, particularly with the diverse feedstock oils implemented for biodiesel production. Boiling point is an important parameter in that it provides the basis for the oil feedstock and biodiesel critical properties as well as vapor pressure, density, surface tension, viscosity, and many other important parameters.[93] Reid et al.[94] is an excellent resource for providing the foundational equations for predictions of thermodynamic properties. Boiling points for some fatty acid esters are listed, but are not readily available for the wide range oils and esters in the production of biodiesel. The relationship of vapor pressure to boiling point is often represented by the Antoine equation[94] as shown in equation:

$$\ln P = A - \frac{B}{C+T} \tag{6.7}$$

where $T$ = temperature, °C
A, B, and C = Antoine constants with C being a function of the boiling point temperature $T_b$ as given in equation:

$$C = 239 - 0.19 T_b \tag{6.8}$$

If the composition of fatty acid esters is known in the biodiesel, the vapor pressure in biodiesel mixtures may be approximated by the following equation:

$$P_{vmix} = \sum_i P_{vi} x_i \tag{6.9}$$

Correlations for boiling points for fatty acid methyl esters (FAME) were determined with reasonable success by Yuan et al.[93] When compared to data from Graboski and McCormick[95] and Goodrum[96] with the following equation and are shown in Table 6.3.

$$T_b = 218.49 \ln(CN) - 6.933 \tag{6.10}$$

where CN is the carbon number of the fatty acid.

| FAME | $T_b$ (K) | $T_b$ (predicted) | A | B | C |
|---|---|---|---|---|---|
| C14:0 | 568.2 | 569.7 | 9.6258 | 2194.36 | −95.50 |
| C16:0 | 611.2 | 598.8 | 9.5714 | 2229.94 | −111.01 |
| C18:0 | 625.2 | 624.6 | 9.3746 | 2174.39 | −131.23 |
| C18:1 | 622.2 | 622.2 | 9.9155 | 2583.52 | −96.15 |
| C18:2 | 639.2 | 639.2 | 8.2175 | 1450.62 | −188.03 |
| C18:3 |  | 639.2 | 8.1397 | 1387.93 | −196.16 |
| C22:1 |  | 665.4 | 10.7518 | 3423.99 | −69.43 |
| SME | 642.1 | 642.5 |  |  |  |
| RME | 620.8 | 627.0 |  |  |  |
| TME | 611.3 | 612.4 |  |  |  |

SME: soybean oil methyl esters, RME: rapeseed oil methyl esters, TME: tallow methyl esters.
*Source*: Adapted from Yuan et al., 2005.

TABLE 6.3  Normal Boiling Points and Antoine Constants for Pure FAMEs Typically Found in Biodiesel Fuels

### 6.2.10.4 Phase Separation and Solubility

Phases represented in biodiesel reactions include the alcohol, oil, fats, soaps (emulsions), biodiesel esters, and glycerol components, each separating in part to different liquid or semisolid phases. Saturated fats tend to form a solid phase at room temperature and are often heated to create a continuous liquid phase reaction mixture. Also supercritical and subcritical alcohol phases occur in high temperature reactions. Alcohols at atmospheric conditions occur in the gaseous phase in equilibrium with the liquid phase. Figure 6.11 shows a ternary equilibrium phase diagram for FAME, methanol, and glycerol. The tie lines represent a family of curves obtained from experimental data where equilibrium phase concentrations were determined for each component. For solutions lying within the boundaries of the solubility curve, the components will separate into two phases and the amounts of each component may be precisely determined. For the mixture represented by $M$, the compositions in the upper lighter biodiesel rich layer will contain about 70 percent FAMEs, 29 percent methanol, and 1 percent glycerol and the lower glycerol-rich phase will contain nearly 5 percent FAMEs 65 percent methanol, and 30 percent glycerol.

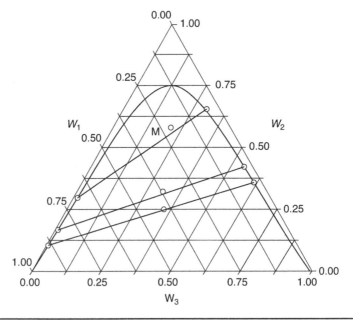

**Figure 6.11** Ternary phase diagram for ($w_1$) FAME, ($w_2$) methanol, and ($w_3$) glycerol showing tie lines for system at 298.15K. (*Adapted from Zhou et al., 2006.*)

## 6.3  Oil Sources and Production

Plant- and algal-derived oils make up the majority of potential sources for biodiesel production primarily for their direct utilization of our most powerful energy source, the sun. Animal and fungal sources are also important, but derive their energy primarily from other carbon sources and would provide to a lesser extent to biodiesel feedstocks. Table 6.4 lists the oil content fatty acid compositions of most promising current plant and animal sources for biodiesel production. Some species of rapeseed and cottonseed oils contain unusually high degrees of the erucic acid (C22:1) with compositions as high as 50.9 and 58.5 wt %, respectively.[93] Currently, soybean, canola, and palm oils are the primary sources for biodiesel production with canola oil leading worldwide and soybean oil leading in the United States. Algal oils are another potential source in that it may produce up to theoretical yields of 10,000 gal/acre/yr.[2]

### 6.3.1  Plant Oils

Most plant oils have a promising fatty acid ester profiles and produce cetane values near 60. The most prevalent plant oil for production of biodiesel worldwide is from rapeseed or commonly called *canola* in the United States, which yields oil as much as 127 gal/acre/yr. Rapeseed is a mustard-like plant that is typically grown in cooler summer climates such as Northern United States, Canada, and Germany, producing magnificent fields of colorful yellow flowers (Fig. 6.12) containing the seeds with high oil content and unique fatty acid composition. Disadvantages of vegetable oils for biodiesel are their relative lower energy content (8 percent less than petroleum diesel), higher viscosity, lower volatility, reactivity of unsaturated hydrocarbons (oxidative stability), and potential presence of unreacted glycerols.[26] The current industry has many innovative approaches to mitigate these potential drawbacks such as blending fuels with additives to increase oxidative stability or reducing viscosity with lower viscosity feedstocks such as rapeseed oil.

Soybeans are grown for their high protein and lipid content, and being a legume, they have the advantage of introducing some nitrogen back into the soil naturally, making this plant more sustainable than intense nitrogen consumers. Because of the value of the products and ability to be used in crop rotation with nitrogen-intensive crops such as corn, soybean oil has become the largest source of food oils and consequently biodiesel, accounting for more than 50 percent of all bio-based oils in the United States, where 6.8 billion kg (2 billion gallons) were harvested from more than 30 million ha[103] in 2002. In the United States, soybean oil is the most prevalent form of oil used in the production of biodiesel even though oil yields are less than 50 gal/acre/yr, but with the wide-scale production of soybeans used in important crop rotations with corn, the total oil yields are significant

| Source | Oil content | Iodine value[4] (oil) | Cetane index (esters) | Palmitic 16:0 | Palmitoleic 16:1 | Stearic 18:0 | Oleic 18:1 | Linoleic 18:2 | Linolenic 18:3 |
|---|---|---|---|---|---|---|---|---|---|
| **Vegetable-based oils** | | | | | | | | | |
| Rape (canola) oil[6] | 30 | 98 | 55 | 3.5 | – | 0.9 | 64.4 | 22.3 | 8.2 |
| Olive oil[6] | 20 | 81 | 60 | 9.2 | 0.8 | 3.4 | 80.4 | 4.5 | 0.6 |
| Sunflower oil[6] | 47 | 125 | 52 | 6.0 | – | 4.2 | 18.7 | 69.3 | – |
| Safflower oil[6] | 60 | | | 5.2 | – | 2.2 | 76.3 | 16.2 | – |
| Soybean oil[6] | 18 | 130 | 53 | 10.6 | – | 4.8 | 22.5 | 52.3 | 8.2 |
| Palm oil[6] | 35 | 54 | 65 | 47.9 | – | 4.2 | 37.0 | 9.1 | 0.3 |
| Cottonseed oil[26] | 40 | 105 | 55 | 28.7 | – | 0.9 | 13.0 | 57.4 | – |
| Poppyseed oil[26] | – | – | – | 12.6 | 0.1 | 4.0 | 22.3 | 60.2 | 0.5 |
| Sesameseed oil[26] | 49 | – | – | 13.1 | – | 3.9 | 52.8 | 30.2 | – |
| Linseed (flax) oil[26] | 35 | 178 | – | 5.1 | 0.3 | 2.5 | 18.9 | 18.1 | 55.1 |
| Wheat grain oil[26] | 11 | – | – | 20.6 | 1.0 | 1.1 | 16.6 | 56.0 | 2.9 |
| Corn oil[26] | – | 120 | 53 | 11.8 | – | 2.0 | 24.8 | 61.3 | – |
| Castor oil[26] | – | 85 | – | 1.1 | – | 3.1 | 4.9 | 1.3 | – |
| Laurel leaf oil[26] | – | – | – | 25.9 | 0.3 | 3.1 | 10.8 | 11.3 | 17.6 |
| Peanut oil[26] | 48 | 93 | – | 11.4 | – | 2.4 | 48.3 | 32.0 | 0.9 |
| Hazelnut oil[26] | 62 | – | – | 4.9 | 0.2 | 2.6 | 83.6 | 8.5 | 0.2 |

| Oil | | | | | | | | | |
|---|---|---|---|---|---|---|---|---|---|
| Walnut oil[26] | 60 | – | – | 7.2 | 0.2 | 1.9 | 18.5 | 56.0 | 16.2 |
| Almond oil[26] | 54 | – | – | 6.5 | 0.5 | 1.4 | 70.7 | 20.0 | – |
| Coconut oil[26] | 35 | 10 | 70 | 9.7 | 0.1 | 3.0 | 6.9 | 2.2 | – |
| Jatropha curcas[98] | | 185 | – | 13.3 | 1.0 | 4.9 | 32.0 | 45.0 | 0.2 |
| Hempseed oil[99] | 35 | – | – | 6.0 | – | 2.0 | 12 | 60 | 20.0 |
| Rice bran oil[100] | 10 | – | – | 21.5 | – | 2.9 | 38.4 | 34.4 | 2.2 |
| Camelina oil[26] | | 155 | – | 5.4 | – | 2.6 | 14.3 | 14.3 | 38.4 |
| Seashore Mallow[101] | 22 | 102 | – | 24.1 | 0.6 | 1.0 | 13.7 | 55.2 | 0.8 |
| Evening primrose[99] | 17 | – | – | 6.0 | – | 2.0 | 11.0 | 81.0 | – |
| Pumpkin seed[99] | 47 | – | – | 9.0 | – | – | 34.0 | 50.0 | 8.0 |
| **Animal-based oils** | | | | | | | | | |
| Poultry fat[6] | – | – | – | 22.2 | 8.4 | 5.1 | 42.3 | 19.3 | 1.0 |
| Lard[6] | – | 65 | 65 | 17.3 | 1.9 | 15.6 | 42.5 | 9.2 | 0.4 |
| Tallow[6] | – | 50 | 75 | 28.4 | – | 14.8 | 44.6 | 2.7 | – |
| **Waste oils** | | | | | | | | | |
| Yellow grease[6] | – | – | – | 23.2 | 3.8 | 13.0 | 44.3 | 7.0 | 0.7 |
| Brown grease[6] | – | – | – | 22.8 | 3.1 | 12.5 | 42.4 | 12.1 | 0.8 |
| Brown grease ME[6] | – | – | – | | 23.0 | 12.9 | 42.5 | 11.6 | 0.8 |
| White grease[6] | – | – | – | 23.3 | 3.5 | 11.0 | 47.1 | 11.0 | 1.0 |

**TABLE 6.4** Typical Oil Content and Fatty Acid Compositions (% by wt of Total Lipids) of Plant and Animal Oils. Cetane and Iodine Values are also Reported

FIGURE 6.12 Rapeseed crops grown for biodiesel and canola food oils. (*Adapted from http://www.dupontelastomers.com/autofocus/a2/af2. asp?article=af2_biofuel.*)

to make this source economically attractive for fuel production in addition to food-grade oil production. Because of increased use for food oils for biodiesel production, prices for oil have dramatically increased. In many cases the prices have more than doubled in the past 3 years as in the case of soybean oil and poultry fat.

As shown in Table 6.5, palm oil would require about 22 percent of the U.S. cropping area to displace petroleum diesel for transportation demands in the United States. However, palm production has already shown great ecological damage from planting in the tropical regions of Indonesia due to burning of rainforests creating one of the largest emissions of carbon dioxide worldwide. The present climate in the United States would not support palm mass production, except for in the subtropical regions of Florida and Hawaii, making this source unrealistic except for possibly supporting some fuel to the subtropical regions. Microalgae with 30 to 50 percent oil (by wt) in its biomass would require only 2 percent of the existing U.S. cropping area to potentially replace U.S. petroleum diesel needs for transportation fuels (about 60 billion gal/yr) and could potentially replace all transportation fuels (180 billion gallons of gasoline and diesel fuel) currently consumed in the United States with less than 5 percent of the existing U.S. cropping area.[2] The DOE NREL report[104] cited that 0.1 percent of the climatically suitable land (200,000 ha) would be required to produce 1 quad of fuel. Therefore, microalgae appears to be a viable plant oil source for this type of mass production. Microalgal production also typically requires considerably less water than other conventional oilseed crops such as soybeans and canola.[104]

| Oil source | Production (gal/acre/yr) | Land area* (M acre) | Percent of U.S. cropping area* |
|---|---|---|---|
| Microalgae | 6280 | 10 | 2 |
| Chinese Tallow tree | 700 | 85 | 20 |
| Palm oil | 635 | 94 | 22 |
| Coconut oil | 287 | 209 | 50 |
| Jatropha | 202 | 297 | 71 |
| Castor oil | 151 | 397 | 95 |
| Olive oil | 129 | 465 | 111 |
| Rapeseed oil | 127 | 472 | 112 |
| Poppy seed oil | 124 | 484 | 115 |
| Peanut oil | 113 | 531 | 126 |
| Sunflower oil | 102 | 588 | 140 |
| Tung oil tree | 100 | 600 | 143 |
| Rice bran oil | 88 | 682 | 162 |
| Safflower oil | 83 | 723 | 172 |
| Sesame seed oil | 74 | 811 | 193 |
| Linseed (flax) oil | 51 | 1176 | 280 |
| Hazelnut oil | 51 | 1176 | 280 |
| Soybean oil | 48 | 1250 | 298 |
| Hempseed oil | 39 | 1538 | 366 |
| Cottonseed oil | 35 | 1714 | 408 |
| Kenaf | 29 | 2069 | 493 |
| Corn oil | 18 | 3333 | 794 |

*To displace all petroleum transport fuel needs in the United States.
Source: Adapted from Christi, 2007.

TABLE 6.5 Production of Oil from Plant Sources, Land Area, and Percent Cropping Area Required to Displace Transportation Fuel in the United States in Biodiesel Equivalents

### 6.3.2 Microbial and Algal Oils

Considerable work toward production of oils by algal and microbial strains with oleaginous traits has been conducted with main goals of producing biodiesel[104] as well as for the recently growing nutraceutical market.[105] Table 6.6 shows the fatty acid compositions of several algal and fungal species. These species not only produce the typical fatty acids produced by higher order plants, but typically also produce long-chain polyunsaturated fatty acids (LC-PUFA), important to metabolism in animals, including nerve cell function and brain cell development.

| Microbial-based oils | Oil content (% dw) | Lauric 14:0 | Palmitic 16:0 | 16:1 | 18:0 | Oleic 18:1 | Linoleic 18:2 | Linolenic 18:3 | ArA 20:4 ω-6 | EpA 20:5 ω-3 | DhA 22:6 ω-3 |
|---|---|---|---|---|---|---|---|---|---|---|---|
| **Algae** | | | | | | | | | | | |
| *Amphidinium carterae*[20] | – | 8.0 | 15.0 | 5.0 | – | 5.0 | 6.0 | 17.0 | – | 4.0 | 2.0 |
| *Botryococcus braunii*[106] | 50 | – | 15.4 | 10.6 | 28.2 | 13.3 | 22.12 | – | – | – | – |
| *Chlorella* spp.[2,107] | 30 | – | 25.0 | 2.0 | 0.9 | 5.0 | 20 | 19.0 | – | – | – |
| *Chlorella pyrenoidosa*[107,114] | 45 | – | 22.0 | 3.0 | 0.9 | 6.5 | 18 | 27.0 | – | – | – |
| *Chlorella pyrenoidosa*[115] | 13 | – | 14.6 | 3.7 | – | 18.1 | 12.3 | 15.8 | – | – | – |
| *Chlorella vulgaris*[107,114,120] | 38 | – | 26.0 | 2.0 | 0.8 | 16.0 | 24 | 20.0 | – | – | – |
| *Chlorella protothecoides*[122,129,130] | 55 | 1.3 | 12.9 | – | 2.7 | 60.8 | 17.3 | – | – | – | – |
| *Crypthecodinium cohnii*[108,124] | 15 | 17.0 | 17.0 | 1.0 | 3.0 | 10.0 | – | – | – | – | 44.0 |
| *Cylindrotheca fusiformis*[109] | 14 | – | 25.1 | 27.0 | 1.6 | 3.8 | 0.5 | 0.5 | 9.1 | 12 | 0.7 |
| *Dunaliella salina*[119] | 55 | 0.4 | 48.9 | – | 0.5 | 17.5 | 8.1 | 2.3 | – | – | – |
| *Dunaliella bardawil*[13] | 8 | – | 1.7 | 1.3 | – | 2.5 | 2.5 | 0.5 | 1.5 | 1.02 | – |
| *Isochrysis* spp.[110,124] | – | – | 14.5 | 4.5 | – | 21.4 | 2.5 | 6.0 | – | 0.2 | 8.3 |
| *Isochrysis galbana*[124] | 29 | 12.0 | 10.0 | 11.0 | 0.7 | 3.0 | 2.0 | – | – | 25.0 | 11.0 |
| *Nannochloropsis* spp.[17,112] | 50 | – | 27.9 | 32.4 | 2.1 | 10.4 | 1.9 | – | 2.1 | 20.1 | 0.5 |
| *Neochloris oleoabundans*[2] | 45 | – | – | – | – | – | – | – | – | – | – |

| Organism | Oil content (%) | | | | | | | | | | |
|---|---|---|---|---|---|---|---|---|---|---|---|
| Nitzschia closterium[125] | 21 | 17.1 | 30.7 | 0.3 | – | 0.9 | 1.9 | 7.1 | 2.3 | 0.2 | – |
| Nitzschia laevis[2,116] | 46 | – | 15.4 | 43.9 | 1.3 | 7.9 | 5.0 | 0.8 | 3.7 | 11.9 | – |
| Nitzschia cf. ovalis[2,117] | 46 | – | 18.8 | 28.2 | – | 0.7 | 0.2 | 0.4 | 2.6 | 24.0 | 4.0 |
| Nitzschia paleacea[118] | 20 | 11.5 | 29.5 | 1.3 | – | 1.2 | 0.8 | 18.1 | 1.2 | 0.6 | – |
| Pavlova lutheri[121] | 15 | – | 26 | 26.0 | 0.4 | 3.0 | 2.0 | 0.6 | – | 15.5 | 7.5 |
| Phaeodactylum tricornutum[111] | 25 | 5.9 | 14.5 | 8.8 | 0.3 | 0.7 | 0.8 | 2.7 | 3.2 | 24.3 | 1.0 |
| Schizochytrium spp.[2,123,124] | 77 | 4.0 | 55.0 | – | – | 1.0 | – | – | 13.0 | – | 30.0 |
| Spirulina platensis[115] | 10 | – | 35.8 | 0.9 | – | 5.0 | 16.3 | 18.2 | – | – | – |
| Tetraselmis spp.[2,117] | 18 | – | 17.5 | 3.3 | 1.2 | 17.9 | 6.0 | 17.0 | 1.3 | 4.2 | – |
| Tetraselmis sueica[110] | 19 | – | 20.3 | 2.7 | 1.2 | 12.4 | 13.8 | 11.8 | 1.8 | 4.3 | – |
| Thalassiosira spp.[117] | – | – | 20.7 | 42.0 | – | 0.7 | 1.8 | 1.5 | 0.2 | 11.3 | 0.8 |
| **Filamentous fungi** | | | | | | | | | | | |
| Aspergillus terreus[127] | 57 | 2 | 23 | – | – | 14 | 40 | 21 | – | – | – |
| Cunninghamella japonica[127] | 60 | – | 16 | – | 14 | 48 | 4 | 8 | – | – | – |

**TABLE 6.6** Typical Oil Content and Fatty Acid Content in Oils from Microbial Sources

| Microbial-based oils | Oil content (% dw) | Lauric 14:0 | Palmitic 16:0 | 16:1 | 18:0 | Oleic 18:1 | Linoleic 18:2 | Linolenic 18:3 | ArA 20:4 ω-6 | EpA 20:5 ω-3 | DhA 22:6 ω-3 |
|---|---|---|---|---|---|---|---|---|---|---|---|
| Mortierella alpina 1S-4[126] | 56 | 2.4 | 15.0 | – | 2.3 | 10.0 | 7.2 | 4.0 | 40.3 | – | – |
| Mortierella elongata[118,128] | 43 | – | 9.4 | – | 3.5 | 50.9 | 8.2 | 3.5 | 16.5 | – | – |
| Mucor circinelloides[127] | 20 | – | 23 | 1 | 12 | 40 | 11 | 16 | – | – | – |
| Penicillium spinulosum[127] | 64 | – | 18 | – | 12 | 12 | 43 | 21 | – | – | – |
| Pythium irregulare[118,138] | 43 | 16.8 | 18.6 | 4.1 | 2.8 | 17.3 | 16.0 | 1.2 | 8.2 | 10.5 | – |
| Rhizopus arrhizus[127] | 57 | | | | 6 | | | | | | |
| **Yeast** | | | | | | | | | | | |
| Cryptococcus albidus[127] | 65 | – | 12.0 | 1.0 | 3 | 73.0 | 12.0 | – | – | – | – |
| Lipomyces starkeyi[127] | 63 | – | 34 | 6 | 5 | 51 | 3 | – | – | – | – |
| Rhodotorula glutinis[127] | 72 | – | 37 | 1 | 3 | 47 | 8 | – | – | – | – |
| Trichosporon pullulans[127] | 65 | – | 15 | – | 2 | 57 | 24 | 1 | – | – | – |

TABLE 6.6 (Continued)

### 6.3.2.1 Algal Oils

Microalgae exhibit versatile growing conditions established over billions of years as one of the more primitive and well-established eukaryotic organisms on earth.[131] With the power of both mitochondria and chloroplasts, algae are able to inhabit many environments as long as water and micronutrients are present. Slime molds living in dark septic systems or dinoflagellates growing in extreme saltwater conditions are examples of the adaptability of algae to live in most environments. Their versatility can serve many advantages with overproduction of oils for potential use in foods and biofuels. Algae have been shown to accumulate an impressive amount of lipids of over 80 percent of their dry weight[132,133] and commonly produce levels of 20 to 50 percent[2] as shown in Table 6.4.

The three most prevalent groups of algae targeted for biodiesel production include the diatoms (Bacillariophyceae) containing nearly 100,000 species that make up a majority of phytoplankton in salt and brackish waters, green algae (Chlorophyceae) common in many freshwater systems, blue-green algae (Cyanophyceae), which are actually bacteria that contain chloroplasts and are important to nitrogen fixation in aquatic systems, and finally the golden algae (Chrysophyceae) with about 1000 known species able to store carbon as oil and complex carbohydrates.[104]

Unicellular green algae, dinoflagellates, and diatoms have unique capabilities of growth in a wide variety of environmental conditions, including salinities to near saturation.[119] One such organism to receive great attention is the microalgae *Dunaliella salina*, which is able to grow in high salinity environments by ionic homeostasis without the need for a rigid cell wall with self-regulating adjustment of intracellular glycerol metabolism, thus balancing external osmotic pressure.[119] Cells grown in high salt concentrations tend to enhance production of C18 unsaturated fatty acids in the form of storage lipids potentially through salt-induced desaturase metabolism. Another organism that currently is the major source for single-cell oils (SCO) is *Crypthecodinium cohnii*, which has been intensively studied.[108,124,134] This organism is currently the only commercial source of SCO containing large amounts of docosahexaenoic acid (DHA), nearly 60 percent of the fatty acids, primarily used as in infant formulas.

Algae have been suggested as good candidates for fuel production because of their higher photosynthetic efficiency, higher biomass production, and faster growth compared to other energy crops.[135] Most of the microalgae that are grown in aquaculture systems are typically grown under autotrophic conditions. In autotrophic growth, algae utilize sunlight as their energy source and carbon dioxide as their carbon source. Certain species of algae can also grow under heterotrophic conditions. In heterotrophic growth, algae utilize a reduced organic compound as their carbon and energy source. Miao and Wu[135] concluded in a research paper that green algae, *Chlorella*

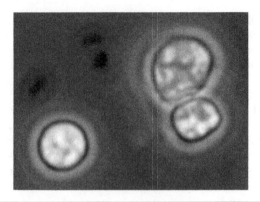

FIGURE 6.13 Photomicrograph of *Chlorella* spp. (*Photo by Scott Davis, Clemson University Aquaculture Center, Clemson, SC.*)

*protothecoides,* produced a higher lipid to biomass ratio when grown under heterotrophic conditions rather than autotrophic conditions (approximately 55 percent to 15 percent lipid to biomass). *Chlorella vulgaris* showed an increase in biomass production when the organism was grown under mixotrophic conditions.[135] Mixotrophic growth is when the organism utilizes both the energy from the sun and the energy in glucose.[136] *Chlorella* spp. is pictured in Fig. 6.13.

Accumulation of oils in microorganisms is typically affected by C:N ratios where N depletion results in accelerated triacylglycerol (TAG) formation.[137] Temperature has also shown significant effect toward the fatty acid composition of oil where lower temperatures may shift toward accumulation of long-chain polyunsaturated fatty acids, but with overall decreased growth rates and subsequently compromised total lipid accumulation.[138] Industrial bioreactor processes for microbial lipid production are typically operated in fed-batch mode where the first phase of operation concentrates on achieving high cell density without nitrogen limitation, and the culture media is then shifted to much higher C:N ratios or nitrogen depletion to achieve oil accumulation. As the culture shifts from active-growth to a nitrogen-depleted idiophase, the cells lose their ability to divide and also lose their flagella due to decreased essential protein and nucleic acids required for cell division, and the metabolism shifts primarily to triacylglycerol (TAG) accumulation as the cells become "cyst-like."[137]

Light saturation for microalgae is an important variable for production of oils by phototrophic mechanisms, and typical saturation constants range from 185 $\mu Em^{-2}s^{-1}$ for *P. tricornutum*[139] to 200 $\mu Em^{-2}s^{-1}$ for *P. cruentum*,[140] only a fraction of typical midday equatorial light intensities of nearly 2000 $\mu Em^{-2}s^{-1}$. Above the saturation constants for these organisms, reversible ultraviolet (UV) damage to the photosynthetic mechanism occurs causing photoinhibition resulting in significantly reduced growth rates.[2]

The most common means for propagation of algae occur in lakes, ponds, constructed raceway ponds, bioreactors, and photobioreactors. The disadvantages of growing algae in natural lakes and ponds are primarily due to the methods for harvesting algae from natural settings and the risk of contamination of species that would ultimately overtake the oleaginous cultures of microalgae due to many uncontrollable environmental factors. For these reasons, commercial application for production of algae primarily utilize raceway ponds or photobioreactor technology.[133] Christi[2] and Richmond[131] have written reviews about mass production of algal oils for foods, chemicals, and biodiesel fuels.

### 6.3.2.2 Raceway Ponds

Raceway ponds have been constructed for many purposes and include production of oils for food and fuel purposes and the use for partitioned aquaculture systems (PAS). The PAS system is primarily used for growing fish such as farm-raised catfish or tilapia as part of a complex ecosystem of algae and fish working together in high concentrations typically maintained within partitioned raceways mixed with a paddlewheel to maintain sufficient oxygen and mixing to increase concentrations of both algae and fish by an order of magnitude over natural systems.[141] Similar disadvantages of raceway ponds exist in the potential for contamination as in natural systems, but could be better controlled by manipulating the environments to select for oleaginous algal species. Also, covers could be placed to potentially keep the system more stable, but would increase maintenance cost of the conventional systems.

The raceway pond system is shown in Fig. 6.14 and has been utilized for many purposes since the late 1950s. The raceways have been modeled

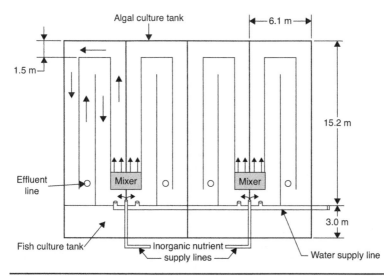

FIGURE 6.14  Raceway pond design. (*Adapted from Drapcho and Bruna, 2006.*)

for implementation near a coal-powered plant to utilize the massive amounts of carbon dioxide emitted for potential cleanup. Currently, the largest raceway system is used for food purposes[133] and is 440,000 m². The system consists of water aquaduct raceways about 0.3 m in depth typically made of concrete and sometimes lined with white plastic. The water is circulated with a paddlewheel to maintain adequate mixing and reduced sedimentation. Feed is introduced daily near the front of the paddlewheel primarily during sunlight hours when the algae is most active. Algae is harvested near the end of the channels just before the paddlewheel.[2] Raceways are significantly less costly to build than photobioreactors, but are far less productive and have considerable losses associated making current bioreactor technologies more competitive, particularly when tighter control parameters are desired[2] such as the case of coproduction of food and biofuel systems.

### 6.3.2.3 Bioreactors

Figure 6.15 shows the concept of recirculation scheme to address mixed culture conditions common in raceway pond designs. Species common to this system would include fast growing unicellular green algae (e.g., *Chlorella* spp.) with slow growth filamentous blue-green algae (e.g., *Spirulina* spp., *Oscillatoria* spp.). The separator proposed contains a 26 μm wire mesh for selective capture and recycle of the slower growing filamentous cultures.[104,142] The mass balances for substrate and biomass are given in Eqs. (6.7) and (6.8), respectively, with the growth rate modeled through Monod kinetics.[104]

$$\frac{dS}{dt} = D(S_0 - S) - \frac{X^a \mu^a}{Y^a} - \frac{X^b \mu^b}{Y^b} \tag{6.11}$$

$$\frac{dX^i}{dt} = X^i \mu^i - X^i D(1 + \alpha - \alpha \beta^i) \tag{6.12}$$

$$\mu^i = \frac{\mu_m^i S}{K_s^i + S} \tag{6.13}$$

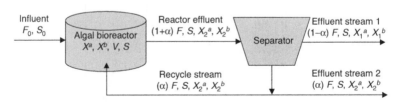

**FIGURE 6.15** Proposed continuous mixed bioreactor system with recycle for single limiting substrate and two competing algal species.

where  $D$ = dilution rate $(F/V)$
 $S_0$ = influent substrate concentration
 $S$ = bioreactor substrate concentration
 $X$ = bioreactor biomass concentration
 $X^{a,b}$ = biomass species a or b
 $Y^{a,b}$ = yield coefficient for species a or b
 $\mu^i$ = specific growth rate of species $i$
 $\mu^t_m$ = maximum specific growth rate of species $i$
 $\alpha$ = ratio of recycle flow to influent flow
 $\delta$ = ratio of effluent stream 2 to influent flow
 $\beta = X_2^i / X^i$

Photobioreactors have been proposed with literature that focus on scale-up of these devices.[140,143–145] These systems would maintain the greatest control of the system, but would be more expensive to build than conventional raceway pond systems. However, the need for sustained quality in production must be considered. Photobioreactors with tighter control over environmental factors ultimately outperform the pond systems.[2] Also, photobioreactors have been adapted to exhaust systems from coal-power plants and demonstrated at the pilot scale as a means of improving coal emissions while creating biodiesel powered by sunlight (see Fig. 6.16).

Typical photobioreactor configurations include an array of glass tubular reactors exposed to either natural light with arrays oriented from north to south or artificial light[146] with a reflective surface underneath to optimize light capture. Artificial light has been applied successfully for production of high-value products, but would be prohibitively expensive for biodiesel production alone.[2] An airlift pump is used in place of mechanical pumps due to potential disruption of algal cells[141] to move the algal biomass through the array of phototubes from a reservoir reactor used also for degassing excess oxygen created during

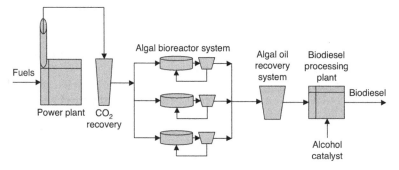

FIGURE 6.16  Algal bioreactor system in conjunction with fossil-fuel power plant to capture carbon dioxide and other emissions for potential use in biodiesel production.

FIGURE **6.17** Photobioreactor for production of algal biodiesel from coal power plant effluent gases. (*Courtesy of GreenFuel, http://news.cnet.com/photos-betting-big-on-biodiesel/2009-1043_3-5714336.html.*)

photosynthesis.[2] Cooling coils are placed within the degassing reservoir to maintain temperature control. Helical coil photobioreactors and other geometric configurations have been proposed,[145,146] some of which include the use of fiber optics for light concentration. Figure 6.17 shows a 1000 L helical tubular design bioreactor system.[2] A system like this could be optimized to coincide with raceway designs for continuous inoculum of species to potentially push the more economical raceway designs toward optimal production of oils.

Other bioreactors have been successfully commercialized for production of algal and fungal lipids primarily for the production of high-value nutraceutical and functional food-based lipids taking advantage of high-lipid yields (>50 percent dw) under heterotrophic conditions operated in fed-batch conditions.[137] The fed-batch bioreactor method would be the most expensive option for the production of biodiesel, but could potentially capitalize on the use of cellulosic-based sugars to replace corn sugars currently used to make the process more feasible. Perhaps a better method may combine a closed-system pond method operating mixotrophically in fed-batch mode to potentially exploit the best characteristics of the bioreactor methods with pond systems, taking advantage of both energy provided by sunlight and cellulosic sugars (also indirectly derived from sunlight) to close the loop of efficiency.

Christi[2] compared photobioreactor systems to raceway ponds and found that for a facility producing 100,000 kg algal biomass utilizing 183,333 kg carbon dioxide annually, photobioreactors could produce about one-third more oil in an area nearly one-third less than raceway ponds primarily due to increased volumetric productivity (1.5 compared to 0.12 kg/m$^3$/day) and more efficient use of carbon dioxide since significant losses occur in raceway systems. Further advantages were cited for separation costs in bioreactors by filtration and centrifugation, and sedimentation would be greatly reduced due to a nearly thirty-fold increase in cell densities achieved in the bioreactor compared to the raceway system.[2,142]

### 6.3.3 Used Cooking Oils

Cooking oils contain many types of vegetable-based oils as well as rendered animal oils. There are enough used cooking oils and fats generated in the United States annually, including 18 billion pounds of soybean oil and 11 billion pounds of animal fat, to produce an estimated 5 billion gallons of biodiesel.[147] Fats, moisture, proteins, and animal fragments extracted from meats during the frying process could become an important component of waste cooking oils particularly when they affect the biodiesel production process in terms of filtration requirements, potential poisoning of the catalyst, and altered fatty acid ester composition in the final product. Waste cooking oils having less than 15 percent free fatty acids (FFA) as a by-product of oxidation are considered yellow grease and when oils exceed 15 percent FFA, as might occur particularly in the summer months during storage of waste grease, they are considered a lower value brown grease.[148] Yellow grease may be combined with lower FFA yellow grease to be sold at a higher price.[148] Zhang et al.[7] proposed an acid-catalyzed esterification and transesterification process for waste cooking oil conversion to biodiesel.

### 6.3.4 Straight Vegetable Oil

Straight vegetable oil (SVO) for use as fuel in diesel engines is another potentially viable use for vegetable-based hydrocarbons. The advantages include bypassing the processing of oils to biodiesel esters substantially reducing the cost of fuel. However, several drawbacks exist, including the general requirement that the engine must be hot for adequate combustion. A common approach is to simply switch from biodiesel to SVO during operation once adequate temperatures are reached to handle the higher viscosity SVO. Also, with current precision injection systems designed for lower viscosity diesel fuel, SVO tends to cause buildup of free glycerin residues on the injectors requiring periodic maintenance. Some older diesel engine designs tend to run fairly well on SVO and are commonly operated in this fashion in many developing countries. These older designs may have increased emissions due to the lack of advanced precision systems in place, resulting in less efficient combustion characteristics. High viscosity characteristics of SVO, particularly in the case of oils containing primarily highly saturated 16 to 18 carbon fatty acids, may also lead to engine degradation.[103] Use of oils containing short-chain unsaturated fatty acids could help curtail this problem. However, these oils are only found in rare exotic plants in nature, such as those in the genus *Cuphea* containing about 37 wt % of capric acid and 40 wt % caprylic acid.[149] A blend of these fatty acids was tested in a diesel engine yielding similar properties to biodiesel, but with reduced cold-flow properties due to capric acid.[150] Some attempts to genetically modify plants such as soybeans to produce high degrees of shorter chain fatty acids have shown some promise, but have not yet been considered economically feasible. Continued research for the potential use of straight algal oils, perhaps in combination with the addition of ethanol to reduce viscosity, should be conducted due to their unique diversity in fatty acid composition and potential high oil yields. Avoiding the transesterification step would significantly decrease the production cost with only degumming steps possibly being necessary.

### 6.3.5 Biosynthesis of Oils and Modification

The biosynthetic machinery in plants and microbial sources for oil accumulation and fatty acid modification giving the species oleaginous traits may result in an impressive 70 percent oils by weight in biomass, typically in the form of triacylglycerols in most species. Ratledge[151] outlined specific biosynthetic mechanisms, including the fatty acid synthase (FAS) system and polyketide synthase mechanisms. As fatty acids are synthesized, a series of desaturases and elongases catalyze the modifications in production of the final lipid compounds. Commercially, three processes for production of microbial oils include the Martek (DHASCOTM), OmegaTech (DHASCO-STM), and Nutrinova processes primarily for production of high-DhA algal oils and the DSM and Wuhan Alking processes for production of high-ArA *Mortierella alpina* strains.[151] The commercial successes for mass production of single-cell oils show promise for biodiesel production, particularly when combined with high-value coproduction of nutraceuticals, precursors for pharmaceuticals, and components for potential biomaterials.

Most species cannot store lipids in excess of 20 percent of their dry weight. A few oleaginous species exist, however, that are capable of storing lipids to as much as 70 to 80 percent of their dry weight, mostly in higher plants, yeast, filamentous fungi, and algal species. These lipids are primarily stored in the form of neutral triacylglycerols. Bacteria typically store lipids in the form of poly-$\beta$-hydroxy-butyrates or poly-$\beta$-hydroxy-alanaoates.[151] For organisms to store excess lipids, excess carbon sources with limited nitrogen is typically required. Nitrogen is utilized primarily during growth of the organism, but on becoming limited allows the carbon to funnel into pathways for energy storage in the form of lipids. Nonoleaginous organisms typically store energy in the form of complex carbohydrates such as glycogen.[151] The precursor for lipid accumulation is the sufficient supply of both acetyl-CoA in the cytosol and NADPH, needed as a reducing agent during fatty acid biosynthesis. Acetyl CoA is directly formed from citric acid as an offshoot from the citric acid cycle, thus funneling carbon toward lipid production primarily during nitrogen limitation in oleaginous organisms. NADPH production is primarily attributed to the malic enzyme catalyzing malate to pyruvate and forming NADPH occurring cytosol at the mitochondrial membrane. The schematic in Fig. 6.18 shows the flux of carbon toward

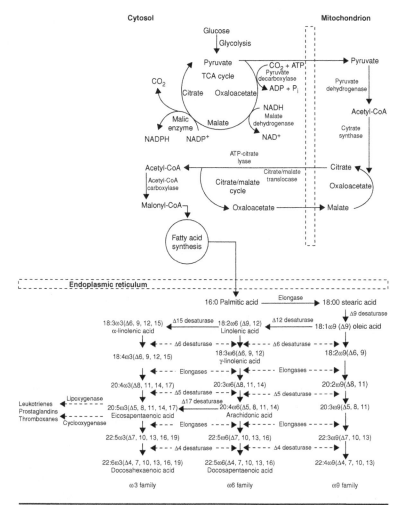

FIGURE 6.18 Biosynthetic pathways for fatty acid metabolism. (*Adapted from Ratledge, 2004.*)

lipid biosynthesis. Once the fatty acids are produced primarily in the form of C16 and C18 saturated fatty acids, a series of desaturases and elongases modify the fatty acids into longer chain polyunsaturated fatty acids. After synthesis is complete the fatty acids are then esterified with glycerol to triacylglycerol droplets via the endoplasmic reticulum.[151]

Due to the demand for soybean oil for biodiesel purposes, research in genetic modification for enhanced performance has been conducted, including alteration of the fatty acid profiles to increase stability of the fuel in warmer climates or to enhance lubrication (cold flow) properties in cooler climates.[103] Algae also holds greater promise for improved production through gene-expression.[131] Enclosed bioreactor configurations enable prevention of potential contamination of transgenic variants to the environment. Molecular-level or metabolic engineering may enhance growth rate, oil production rate, photosynthetic rates while reducing effects of temperature dependence, inhibition from light saturation, and susceptibility to photooxidation,[2] but has thus far received little attention.[152] Genetic-level expression systems are often kept within the chloroplast environment separated from the cytoplasm of the plant system, protecting these expression systems from potential transfer to other systems. However, transgenes have crossed species requiring greater caution in handling these biosystems with respect to the environment.[103]

## 6.4 Coproducts

High-value pigments containing unique metabolic properties are a benefit of cottonseed oil when compared to many other oils. Algal oils may also contain high proportions of valuable omega-3 fatty acids, which have shown substantial health benefits. Capitalizing on both biodiesel production and fractionation of the valuable pigments and/or long-chain polyunsaturated fatty acids may lend to substantial profits, similar to the cocoa industry with nearly all of the seed material utilized for valuable products. The biodiesel produced has additional benefits to the farmer and processor in that it may be used directly back into the equipment for harvesting, transportation, and processing.

Another way of increasing value of oil feedstocks is with bioconversion methods that also aid in utilization of the feedstock material. Promising soil-borne fungi, *Pythium irregulare* and *Mortierella alpina*,[153] and microalgae, including *Schizochytrium* spp. and *Crypthecodinium cohnii*[151] are capable of adding important long-chain omega-3 fatty acid components, including eicosapentaenoic acid (EpA) and docosapentaenoic acid (DhA), and omega-6 fatty acids, including arachidonic acid (ArA). An added benefit to fungal species is their enhanced ability to enzymatically digest complex substrates (e.g., hemicelluloses, lignocellulosics) associated with plant biomass. To date, fish oil is the main supplement for eicosapentaenoic acid with a market-value of about $500/lb.

However, fish acquires EpA from microbial sources such as algae or fungi and the extracted oil often contains undesirable odor compounds and accumulated mercury. Through extensive research EpA and ArA have shown to play an important metabolic role in preventing some of the most problematic diseases, namely atherosclerosis, cancer, and arthritis. Plant oils do not contain EpA and ArA, but plant omega-3 and omega-6 fatty acids must be converted to EpA and ArA through further metabolism.[138] However, research has shown that the plant-based oils high in omega-3 fatty acids are not as effective as fish oil in prevention of the major diseases and human infant brain development. The fungal bioconversion by *P. irregulare* has shown that much of the oil present in the plant material is unaltered, but most of the complex carbohydrate material is utilized in the production of EpA and ArA.[118,138] Figure 6.19 shows fungal oil after extraction from *P. irregulare* by supercritical carbon dioxide.[138] One added advantage occurs in that the oil does not contain odor and off-flavor compounds often associated with many fish oils in the present market.

Polar lipids represent an important class of unique lipids with therapeutic value.[154] Plant and algal species produce phospho- and sulfolipids, including phosphatidylcholine, phosphatidylinositol, phosphatidylethanolamine, phosphatidylglycerol, monogalactosyldiacylglycerol, digalactosyldiacylglycerol, and sulfoquinovosyl diacylglycerols (SQDG).[155] SQDG are important structural components of chloroplast membranes in eukaryotic algae and lamellas in prokaryotic cyanobacteria,[156] and appear to be the only natural lipid with sulfonic acid linkage.[157] SQDG extracted from cyanobacteria have been shown to inhibit HIV-1 in cultured human lymphoblastoid T-cell lines and thus were placed as high priority for further research by

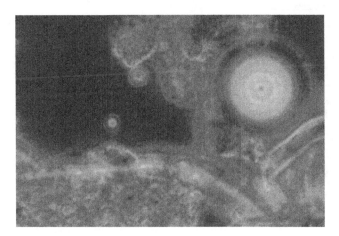

FIGURE **6.19** Photomicrograph of *Pythium irregulare* with coalesced droplet of fungal oil after extraction by supercritical carbon dioxide.

the National Cancer Institute.[158] The cyanobacteria evaluated included species of *Lyngbya*, *Phormidium*, *Oscillatoria*, *Scytonema*, *Calothrix*, and *Anabaeana*.[159,160] SQDG content in the *Anabaena* biomass increased throughout the exponential and decreasing exponential growth phases to values of nearly 11 mg SQDG/g algal dry weight, indicating that SQDG may be characterized as a mixed primary and secondary product and relatively unaffected by the surface light intensities tested.[161] By comparison to higher plants, sulfolipid yield was found to be 0.18 mg SQDG/g fresh spinach.[162]

Another innovative approach is engineered bioplastics that may have properties making them available to degrade to biodiesel.[163] The researchers claim that the biopolyester material is stronger than polyethylene, but may be broken down into biodiesel in the field to serve as a source of electricity. The U.S. armed forces currently dispose of more than 30,000 tons of plastics per year, which could be diverted to energy in the field—if easily converted to biodiesel esters. The glycerol by-product holds promise to increase value of the biodiesel industry particularly through utilization of this reduced substrated through anaerobic fermentations that result in a wide array of value-added products, including organic acids, biohydrogen, methane, ethanol, and many others.[20]

## 6.5 Methods of Biodiesel Production

Preferred methods of production of biodiesel typically consist of reaction of oil sources with alcohols with aid of either acid or base

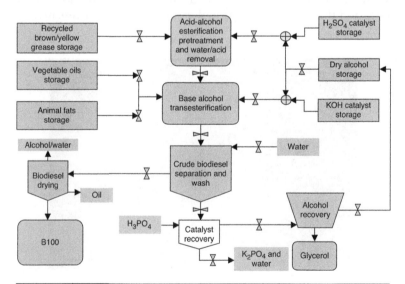

**FIGURE 6.20** General biodiesel production process flow diagram for varied oil sources, including waste cooking oils.

catalysis. Noncatalyzed reactions have also been reported with most recent approaches using supercritical methanol.[81,164,165] The preferred catalyzed methods are somewhat time consuming and require the removal and handling of catalyst.[81] Solid-acid catalysts have recently been proposed as a method to easily remove and recycle the catalytic material.[56] Once the method of production has been determined, quality control characterization must be applied to assess the process in terms of final product quality.

### 6.5.1 General Biodiesel Production Procedures

The simple procedures outlined for production of fuel-grade biodiesel from various plant-based oil sources is given in this section and should be performed before any large-scale batches are attempted to ensure confidence in the level of completion of the reaction, which is heavily reliant on the quality of the feedstock oil. Similar procedures exist using the same type of chemistry for quantification of lipid esters for determination of fatty acid composition of oils by gas or liquid chromatography in the food and pharmaceutical industries.[157] In most studies for production of biodiesel with ethanol and methanol using NaOH or KOH homogeneous catalysts, high yields of biodiesel may be obtained within 1 h of reaction time using 1 percent by weight catalyst concentration and between a 4:1 and 9:1 molar ratio of alcohol to oil. Additional catalyst would be needed to neutralize free fatty acids, if they exist, based on titration. The quality of oil feedstock is vital to the success in production. Feedstocks obtained from oil pressing (e.g., screw or hydraulic presses) and extraction (e.g., hexane) should be degummed by treating the oil for 4 to 8 hours with 300 to 3000 ppm phosphoric acid (depending on the natural levels of gums present) followed by washing with water.

Safety plays an important role for biodiesel production as the reactants are either strongly corrosive as in the case of alkali and acid catalysts and alcohols, toxic when using methanol, or highly flammable as in the case of alcohols. Once biodiesel is produced, this end product is nearly nontoxic and much less flammable than the alcohols, but more flammable than the starting oils. However, unrecovered catalyst and unreacted alcohols will still remain and must be treated as potentially toxic, corrosive, and flammable materials. The by-products of glycerin and wash water must not be poured down the drain—even if neutralized, but handled according to strict waste management laws usually requiring disposal in approved containers through certified waste treatment facilities. Glycerin is often recovered from this process and may be converted to soaps that may be used to clean the production facility, but need to follow necessary safety procedures for further use. When methanol is used, extra care should be taken to avoid any contact. In all cases, adequate ventilation is extremely important with care taken to avoid any electrical sparks in the areas where vapors may reach concentrations near the critical value of the flash point, which is the lowest temperature at which combustion of vapors may occur—only 11 to 13°C for alcohols compared to nearly 150°C for biodiesel.

The first basic steps for producing a batch of biodiesel is titration and determination of the presence of water in the feedstock. Titration is conducted to determine the amount of catalyst needed and for choosing the best route for transesterification, whether acid- or base-catalyzed or a combination of processes. Determination of water may be quickly accomplished by boiling a small sample of oil and checking for spattering, which will occur violently even in the presence of very small amounts of water due to rapid vaporization caused by the extreme difference in boiling points (over 300°C for most oils). Most stored oils will contain some water that will settle out over time, but could be trapped in the oil with the presence of surfactants found in rancid oils such as soaps created by ambient saponification.[6]

### 6.5.1.1 Titration

Titration is necessary to first determine the best route for biodiesel production whether using the conventional alkali-catalyzed approach, acid-catalyzed approach, or acid pretreatment followed by the alkali approach. If the titration shows an excess of 15 percent or greater free fatty acid concentration as in the case of brown greases, typically an acid-catalyzed esterification and transesterification procedure is implemented. If the concentration of free fatty acids are between 5 and 15 percent as in the case with yellow greases, then an acid pretreatment followed by base-catalyzed procedure may be chosen. Typically for oils less than 5 percent free fatty acids, the alkali-catalyzed procedure will be implemented since the cost of an additional acid pretreatment could be cost-prohibitive. However, concentrations of free fatty acids greater than 1 percent will contribute to soap production, making the downstream wash steps more difficult. The soap reaction occurs more rapidly than biodiesel formation, thus requiring an excess of catalyst normally consumed in the soap reaction. The intermittent glycerolysis reaction using glycerol to convert free fatty acids to MAGs typically at high temperatures is often implemented to remove the fatty acids prior to transesterification.

The titration determines the amount of excess catalyst needed in the chosen reaction mechanism. For alkali-catalyst procedure an excess amount of catalyst may be required to first neutralize the free fatty acids before transesterification of oil triglycerides to ethyl esters can occur. Typical concentrations of catalyst to achieve efficient transesterification range around 1 percent by weight and any excess catalyst is added beyond 1 percent based on the titration values. A simple titration procedure is outlined below.

1. Measure 10 mL of ethanol in a test tube.
2. Measure 1 mL of oil and mix with ethanol.
3. Add about 0.5 mL of phenolphthalein solution (10 to 20 drops).

4. Titrate the oil/ethanol solution with 1 g/L KOH or NaOH solution in distilled or deionized water using a burette until the color begins to turn pink after adequate mixing (near pH 8.0).

5. Estimate amount of KOH or NaOH needed for transesterification from the following formula:

$$y = 9 + x$$

where $y$ is grams of KOH or NaOH catalyst (usually 9 to 15 g) to use in 1 L of oil (about 900 g) and $x$ is the number of milliliters of KOH or NaOH solution used in titration. The 9 g of catalyst gives you a starting value of about 1 wt % in 1 L of oil and the additional catalyst designated by $x$ is required to first neutralize any free fatty acids before rapid transesterification may take place.

Following titration, a decision should be made toward the best method for production of biodiesel, depending on the amount of free fatty acids and relative acidity of the oil feedstock.

1. When only 9 to 15 g of catalyst are required per liter of oil, then the "Alkali Transesterification" procedure, described next, may be followed with reasonable success.

2. If after titration, the feedstock sample oil requires more than 15 g catalyst per liter of oil to complete transesterification, then an "Acid Esterification Pretreatment" with 1 wt % sulfuric acid or conducting a glycerolysis reaction using glycerol and base catalyst at 220°C should first be considered. These intermittent pretreatment steps should be considered because values greater than 1.5 wt % base catalyst will cause excessive saponification, resulting in significant loss in biodiesel yields as well as making the separation steps for glycerol and the water wash solutions much more difficult. Acid esterification may be conducted at temperatures slightly less than the boiling point of the alcohol (or up to 120°C if pressure vessels are available).

3. If after titration, the feedstock sample requires more than 25 g catalyst per liter of oil, the oil is highly rancid with more than 15 percent free fatty acids and is considered brown grease. Therefore, acid-catalyzed esterification of free fatty acids with simultaneous acid transesterification reactions should be considered using 1 wt % sulfuric acid conducted at higher temperatures [ranges typically from slightly less than the boiling point temperature of the alcohol (70°C for methanol and 80°C for ethanol) to 120°C in appropriate pressure vessels]. See the procedure for "Acid Esterification and Transesterification."

### 6.5.1.2 Alkali Transesterification

The transesterification reaction is shown where no more than 0.5 percent water can be present for the reaction to take place and to avoid significant saponification. (Note: Any volume may be used keeping the same ratios of reactants.) This procedure is conducted in 50 mL centrifuge tubes for reduced transfer steps, but other volumes are typically used such as 15 mL centrifuge tubes or 500 mL flasks.

1. Add 30 mL of oil to a 50 mL centrifuge tube (or 10 mL to a 15 mL centrifuge tube or 300 mL to a 500 mL flask).
2. Add the appropriate weight of catalyst determined in the titration procedure to the alcohol and blend to completely dissolve in alcohol. For example, if the oil were primarily free of fatty acids as in the case with most refined oils, then you will need 9 g of catalyst per liter of oil. Therefore, the amount of catalyst for the 30 mL sample would be 9/1000 × 30 mL = 0.27 g of catalyst per sample.
3. Mix 9 mL catalyst and alcohol solution with the oil (or 3 mL to the 15 mL centrifuge tube or 90 mL to the 500 mL flask). This will produce approximately 6 to 1 molar ratio of ethanol to oil.
4. Heat at 55°C and mix (or vortex) continuously or intermittently every 5 min for 1 h. Two layers should begin to form after 5 min with the darker glycerol layer sinking to bottom. If using ethanol and dark crude oil, the two layers will be initially difficult to distinguish due to solubilization of pigments into the upper layer by ethanol.
5. Centrifuge for 3 min at 3000 rpm (or let settle for several hours). Phase separation is an important step dependent on several factors, including the amount of excess alcohol and the amount and type of catalyst chosen. Korus et al.[166] determined that the use of potassium hydroxide yielded better separation than potassium ethoxide.
6. Estimate the conversion produced by measuring the volume of glycerol and biodiesel layers.

**Example 6.2** To produce biodiesel ethyl or methyl esters from rapeseed oil, ethanol or methanol, and potassium hydroxide are used in the reaction:

a. Determine the theoretical conversion efficiency of the biodiesel reaction and the amount of ethanol required to convert 1 L of rapeseed oil reacted with 1 percent potassium hydroxide.
b. Determine the amount of ethanol required to make a molar ratio of 6 mol of ethanol to 1 mol of oil.

Base calculations on triolean since this is often the predominant triglyceride in vegetable-based oils: $C_{57}H_{98}O_6$, MW 878 g/gmol, density 0.88 kg/L
Ethanol: $C_2H_5OH$, MW: 46 g/gmol, density 0.794 kg/L

Methanol: CH$_3$OH, MW: 32 g/gmol, density: 0.796 kg/L
Potassium hydroxide in alcohol (1 wt %/vol):

**Solution** With the assumption that the oil may be represented well by triolean, the following calculations are made:

  a. For complete conversion of one triglyceride molecule to three alkyl esters, the theoretical molar ratio for alcohol to triglyceride molecule is 3 to 1; therefore, 3 gmol EtOH × 46 g/gmol = 138 g of ethanol per gmol of triglyceride are needed. 1 gmol of triolean × 878 g/gmol = 878 g triolean; 878 g triolean/138 g EtOH = 6.4 g oil/1 g EtOH needed for complete conversion.
  b. For a molar ratio of 6:1 only 3 mol of ethanol per mole of triglyceride are utilized in the reaction and the other 3 mol are in excess, but shift the reaction equilibrium toward the products.

### 6.5.1.3 Acid Esterification Pretreatment

Two-stage reaction processes incorporating an acid esterification pretreatment step followed by an alkali transesterification step have been proposed to increase yields of biodiesel from free fatty acid (FFA) containing oils.[7,70,148,167–169] When oils exceed FFA values of 5 percent, the following procedure should be considered.

1. Add 30 mL of oil to a 50 mL centrifuge tube (or 10 mL to a 15 mL centrifuge tube or 300 mL to a 500 mL flask).
2. Make a 1 wt % acid catalyst solution in dry alcohol by using sulfuric acid.
3. Mix 9 mL catalyst solution to the oil (or 3 mL to the 15 mL centrifuge tube or 90 mL to the 500 mL flask).
4. Heat at 50 to 60°C and mix (or vortex) continuously or intermittently every 15 min for 2 h. A water layer will begin to develop, depending on the amount of FFA initially present. Some glycerol will also be liberated by acid-catalyzed transesterification.
5. Centrifuge for 3 min at 3000 rpm (or let settle for several hours) and remove water layer.
6. Add 5 mL of glycerol to wash the mixture by inverting slowly.
7. Titrate the biodiesel mixture to test for residual free fatty acids or acid catalyst.
8. Centrifuge for 3 min at 3000 rpm to remove wash layer and measure pH.
9. Titrate the sample and repeat the pretreatment procedure if necessary until less than 5 wt % of free fatty acids remain.
10. Proceed to the "Alkali Transesterification" procedure to convert the remaining oil to biodiesel.

### 6.5.1.4 Acid Esterification and Transesterification

Acid esterification and transesterification using sulfuric or hydrochloric acids are another convenient way to produce biodiesel, but often takes 1 to 2 days to complete the reaction compared to only 1 hour for the alkali-catalyzed techniques. When the FFA have exceeded 15 wt % creating brown grease, the pretreatment step outlined about earlier may be applied followed by alkali transesterification. To achieve adequate conversion of FFA to alkyl esters, allowing the reactions to proceed to completion with acid catalyst is recommended to avoid excessive and time-consuming wash steps.

1. Add 30 mL of oil to a 50 mL centrifuge tube (or 10 mL to a 15 mL centrifuge tube or 300 mL to a 500 mL flask).
2. Make a 1 wt % acid catalyst solution in dry alcohol with sulfuric acid.
3. Mix 9 mL catalyst solution to the oil (or 3 mL to the 15 mL centrifuge tube or 90 mL to the 500 mL flask).
4. Heat at 50 to 60°C and allow to react overnight in a heated shaker bath. The free fatty acids present will react to create alkyl esters and a water layer. Some glycerol will also be liberated by acid-catalyzed transesterification. (Note: As water is liberated during esterification of FFA, this will further slow the transesterification step.) Intermittent centrifugation to remove water may be necessary to complete the conversion process.
5. Centrifuge for 3 min at 3000 rpm (or let settle for several hours) and remove water layer.
6. Estimate the conversion produced by measuring the volume of glycerol and biodiesel layers.

### 6.5.1.5 Biodiesel Water Wash

Washing the biodiesel is necessary to remove excess catalyst, glycerol, and glycerides not fully reacted that would be detrimental to an engine. If an excessive amount of MAG, DAG, or TAGs as well as FFAs exist, then further esterification or transesterification may be necessary to achieve ASTM biodiesel standards for free and total glycerine.[19]

**Laboratory Scale**

1. Pour the top layer (biodiesel) into a second centrifuge tube.
2. Add about 20 mL of water and invert tube gently several times.
3. Allow to settle in 50°C bath for 5 min.
4. Centrifuge 3 min at 3000 rpm (if separation does not occur, add several drops of acetic acid to help break emulsion).

5. Pour top layer into third centrifuge tube.
6. Measure pH of bottom wash layer; initially the wash water will be cloudy due to catalyst, glycerol, and soap formation.
7. Repeat washing process until pH drops to 7.0 and specific gravity is between 0.85 and 0.9.
8. Repeat the washing process as needed to obtain the desired pH. The wash water removed should become clear as another indicator of adequate washing.
9. Test for glycerine in the washed biodiesel[170] using AOCS CA 14-56.

**AOCS Method for Determination of Glycerine**   American Oil Chemists' Society (AOCS) Official Method Ca 14-56 testing for Total, Free, and Combined Glycerol using Iodometric-Periodic Acid.[170]

## 6.5.2   Pilot and Commercial Scale

The pilot and commercial scale continuous or semi-continuous biodiesel operations for Alkali Transesterification, Acid Esterification Pretreatment, and Acid Esterification and Transesterification are described by Zhang et al.[7] Batch operation or semicontinuous operations are typically implemented for smaller scale. The process design for the Alkali Transesterification process occurs in eight basic steps:

1. Storage and transfer: Oil is typically heated from the storage vessel to operating temperatures to reduce viscosity caused primarily by the presence of saturated fatty acids in the oil for easy transfer by pump to the reactor vessel.
2. Transesterification: Based on the kinetics of transesterification reactions typically yielding 90 to 98 percent oil conversion to alkyl esters, the reaction conducted at ambient pressure with temperatures ranging from room temperature to slightly below the boiling point of the alcohol was considered the best for economical conversion efficiencies.
3. Alcohol recovery: The excess alcohol from the transesterification reaction should be recovered to the greatest extent possible and recycled back to the transesterification step, requiring a complete distillation system and recycle system. Up to five theoretical stages with a reflux ratio of 2 should be employed to achieve efficient alcohol recovery.
4. Water washing: Multiple-stage water washing steps (up to four theoretical stages) are typically employed to purify the biodiesel from catalyst, free glycerol, and alcohols.
5. Biodiesel ester purification: To obtain ASTM-certified biodiesel at 99.6 percent purity, further purification of the

esters must be employed with ester distillation with four theoretical stages where water and alcohols are removed as vent gases, esters removed as liquid distillate, and residual oils remain in the bottoms, which are recycled back to the transesterification step.

6. Alkali removal: The removal of alkali may be accomplished with the addition of phosphoric acid, resulting in a salting out of $Na_3PO_4$ or $K_3PO_4$. Potassium phosphate is much more desirable in that it may be further used as a fertilizer.

7. Glycerine purification: The glycerine should be purified into two grades to be sold commercially with the higher grade for food and pharmaceutical uses and lower grade for industrial or land application uses. With the use of UNIQUAC model in the simulations run by Zhang et al.,[7] glycerin of 92 percent high purity was obtained through four theoretical stages with a reflux ratio of 2 through the bottoms with water and alcohol removed in the vent gases.

8. Waste treatment: The compositions of waste streams, the alcohol recovery stages, ester purification, and glycerol purification consist primarily of water and alcohol and may be reused in the washing steps. Recovery of solid waste should be utilized as fertilizer to reduce waste treatment costs and water use.

The Acid Esterification Pretreatment requires an additional three steps prior to the transesterification step.[7] These steps include:

1. Esterification: The recommended approach for the esterification reaction requires a temperature of 70°C, resulting in a pressure of about 400 kPa when a 6:1 molar ratio methanol is used.[171] A 1 wt % sulfuric acid to oil ratio is used to completely convert free fatty acids to biodiesel esters with formation of water.

2. Glycerine washing: Sulfuric acid and water must be completely removed before alkali transesterification may be accomplished. The addition of 10 wt % ratio of glycerin to oil with mixing will accomplish complete recovery of alcohol (about 60 percent of wash layer or stream), glycerol (33 percent of wash), sulfuric acid (3 percent sulfuric acid) and water (1 percent), and traces of esters and residual oil.

3. Alcohol recovery: The alcohol recovery is similar to that described previously with the exception that five theoretical stages were used with a reflux ratio of 5.

4. Acid neutralization: Sulfuric acid is recovered from the glycerol phase with calcium oxide to produce calcium sulfate, $CaSO_4$, and water then separated by gravity settler.

Glycerolysis pretreatment is easier to implement than acid esterification since the reaction only requires the addition of glycerol, base catalyst conducted at high temperature. This avoids the need for neutralization and removal steps since the free fatty acids are simply converted to MAGs in the presence of catalyst.

For pilot scale operation, the wash step often results in a bottleneck of the process due to a potential of significant emulsification and degree of washing required to remove free glycerin, excess alcohol, and catalyst from the newly-formed biodiesel esters. More advanced operations will incorporate a pilot-scale continuous centrifuge to first separate most of the free glycerin from the biodiesel layer before being sent to a separate conical washing tank. However, overnight gravity settling will accomplish nearly the same results. The wash tanks typically contain a misting spray nozzle that disperses fine droplets of water to the surface of the biodiesel and allows to slowly settle to the bottom of the conical vessel until the wash water becomes clear and pH is around neutral. Addition of dispersed air at the bottom of the vessel is commonly practiced particularly for more difficult emulsions that may result. Small air bubbles are trickled through the settled water layer carrying moisture back through the biodiesel layer for more complete washing. Low airflow rates are typically required to avoid potential foaming of residual soaps during this gentle mixing process. Addition of a 5 percent acetic acid solution may be incorporated to aid in the breaking of significant emulsions.

### 6.5.3 Quality Control Analytical Technique

Quality control standards for biodiesel production have been formalized since the late 1990s in the United States and the EU. The National Biodiesel Board helped to establish the current ASTM standards as well as Environmental Protection Agency (EPA) health data. The current standards for pure (100 percent) biodiesel or B100 are outlined in ASTM standard D6751-07 for the U.S. standards and similar EU standards outlined in EN 14214.[172] This standard is for B100 before use in on-road compression-ignition diesel engines or prior to use in blends with petroleum diesel fuel. Certified laboratories such as Intertek, Inc. that comply with the current standards offer services to producers for testing and sample management for the producers to remain in compliance as outlined for both ASTM D6751 and EU specifications. Quality control methods and analytical techniques are outlined in this section. Specific details for ASTM methods must be purchased from ASTM and only general concepts are outlined here.

#### 6.5.3.1 Chromatography Methods

Gas chromatography (GC) is the preferred method for determination of total and free glycerin as well as the fatty acid ester composition within the biodiesel. GC methods will also determine the amount of

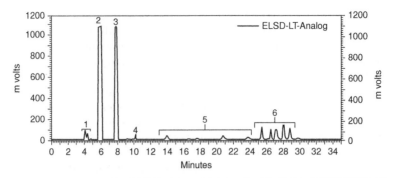

**FIGURE 6.21** Monoglycerides (MAGs), 2-ethyl linoleate (FAEE), 3-ethyl oleate (FAEE), 4-ethyl palmitate (FAEE), 5-diglycerides (DAGs), 6-unreacted triglycerides present in the biodiesel. The system used included Shimadzu HPLC system equipped with Alltech HP prevail C18 column of length 150 mm and ID 4.6 mm. The mobile phase was acetonitrile and dichloromethane, with a gradient of dichloromethane maintained to separate the biodiesel sample. (*Adapted from Joshi et al., 2008.*)

alcohols present within the biodiesel. Free and total glycerol methods are outlined in ASTM 6584. High pressure liquid chromatography (HPLC) techniques have also been conveniently developed not only for testing fatty acid alkyl esters, but also the mono-, di-, and triacylglycerol components to aid in the study of the conversion process kinetics, whether accomplished by esterification or transesterification reactions.[173,174] The method outlined consists of an HPLC system with reverse phase column (C18) for separation using acetonitrile:water mobile phase and evaporative light scattering detector (ELSD). An attached chromatogram is shown in Fig. 6.21 that outlines the conversion of cottonseed oil to ethyl esters. HPLC may be employed to obtain a complete lipid profile, but typically, separation characteristics make analysis difficult to distinguish the many variations of lipids that may contain more than 100 combinations of lipid compounds. The need for two separation columns usually involving reverse phase and silver-ion column separations would be necessary to separate and detect each lipid compound.[157]

### 6.5.3.2 Viscosity Testing

*Viscosity* is the rheological property of fluids that gives a measure of the resistance of a fluid to flow. The flow parameters of biodiesel are particularly important to ensure that adequate fuel is pumped to injectors for complete combustion over wide temperature ranges that exist for start-up and operation of diesel engines dependent on climate conditions. Viscosity is also important to the producers and distributors for handling, storage, and mixing of the biofuels. Kinematic viscosity measured in units of centistokes is determined according to ASTM D445 procedures conducted at 40°C.

### 6.5.3.3 Cetane Values

The cetane index helps to determine combustion performance, cold start performance, and emissions.[40] Higher cetane values are desired. Table 6.2 shows standard characteristics of diesel and biodiesel fuels. Biodiesel esters tend to show relatively higher cetane values averaging near 50 compared to No. 2 diesel,[40] which averages near 46. Cetane values are obtained by ASTM D976 as calculated from the API specific gravity (ASTM D1298) and simulated distillation at 50 percent (ASTM D2887) values. *Distillation* is the progressive boiling off of individual components based on differences in boiling points, resulting in a distillation curve of compounds. Because biodiesel contains a fairly homogenous quantity of straight-chain hydrocarbons with lengths of 16 and 18 carbons, a single boiling point is typically obtained rather than a defined distillation curve. High distillation values typically result in fuels of high boiling point and flash point (determined by ASTM D93) and serves to provide evidence that the transesterification has not gone to completion, resulting in poor cold-starting characteristics and increased viscosity. Low distillation values are typical of alcohol carryover that may contribute to poor timing and seal material failure (primarily for older engines that may not contain Viton tubing and seals). DuPont Performance Elastomers (DPE) Viton fluoroelastomer were developed in the 1950s and have since been widely applied to vehicles to handle extreme variations in temperature and fuel types.[175]

### 6.5.3.4 Total Acid Number

Titrations are performed for both the feedstock before biodiesel production and after for the determination of total acid number (TAN) by ASTM D664 methods. The titration conducted before conducting the biodiesel reactions are required to determine the amount of catalyst needed for transesterification, and results from this simple test may help the producer to determine whether to initially pretreat the feedstock to first accomplish acid-catalyzed esterification before conducting the much faster base-catalyzed transesterification procedure.

### 6.5.3.5 Iodine Values

Iodine values (IV) are useful for determination of the overall degree of saturation of the oil, which is important for viscosity, cloud point, and reactivity characteristics. At lower temperatures the oil becomes solid when saturated, but remains liquid with higher degrees of unsaturation. This is important for biodiesel characteristics where ideally the fuel will remain more liquid at lower temperatures, but remains somewhat stable from oxidation or hydrogenation reactions. Iodine is introduced and reacts with the double bonds within the fatty acid structure, thus saturating the oil with iodine only at the double-bond sites. The amount of iodine absorbed in grams per 100 mL of oil determines the iodine value.

High degrees of unsaturation are generally not desirable for fuels because their oxidation reactions generally found at high temperatures during combustion may result in irreversible polymerization to plastic-like substances. Iodine values greater than 50 may result in decreased engine life, but give better viscosity characteristics in cooler conditions. Therefore coconut, palm, tallow, and lard will have low iodine numbers indicating a stable oil, but viscosity of these oils at room temperature and low temperature conditions will cause the oils to become solid, indicating viscosity problems in the resulting biodiesel.

#### 6.5.3.6 Free and Total Glycerin

One of the more important quality parameters for B100 testing is the glycerin in the free form and the bonded mono-, di-, or triglycerol form, indicating to producers an incomplete transesterification process or incomplete washing of the final product. For distributors, presence of glycerin in the final fuel may result in the fouling of pumps and filters or separation during storage of the fuel. The end-users are affected by the potential fouling of injectors resulting in poor combustion of the fuel. This test is typically conducted using gas chromatography methods according to ASTM D6584 methods where results are determined in percent by weight of the individual glycerin components. Free and total fatty acids and fatty acid esters may also be determined. HPLC methods without the need for derivatization could be employed as described earlier.[157] Also, AOCS Official Method Ca 14-56 testing for Total, Free, and Combined Glycerol using Iodometric-Periodic Acid is often applied.[170]

## 6.6 Economics

The cost of production of biodiesel fuel relies on a complex system of economics that begins with the infrastructure of agricultural processing of oilseed crops or other promising sources such as algae, which will be specific to regions optimal for the crop and further dependent on tolerance to variable climate conditions. Zhang et al.[71] presented an economic assessment and sensitivity analysis of biodiesel production from waste cooking oils via acid-catalyzed esterification and transesterification. The authors presented an economic comparison of four continuous processes of acid and base catalysts on virgin canola oil and waste cooking oils high in free fatty acids. A sensitivity analysis was then utilized to assess individual factors on the final analysis and they concluded that plant capacity, raw material cost, and final biodiesel price were the main factors.[71] Based on after-tax rate of return, Zhang et al.[71] determined that the alkali treatment worked best for the virgin canola feedstock, but the acid-catalyzed process, even with much slower process time, worked best for the waste cooking oil. The alkali transesterification process for waste cooking oils was not as cost effective primarily because an esterification pretreatment step was required to first reduce the free fatty acid content. Haas et al.[70] presented a full working process model in ASPEN PLUS9 to determine biodiesel production costs. Both papers concluded that the liberated glycerol component could also play an important role in the final economics.[70,71] Haas et al.[70] showed a linear impact of crude glycerin (80 wt %) on the biodiesel production cost. However, Zhang et al.[71] showed that the yearly change in glycerol prices for two different levels of purity (85 vs. 92 wt %) had a relatively small impact on break-even price of biodiesel. The break-even cost is estimated by the annual sales value of the product required to operate the facility at zero profit, where the annual sales value equals the annual cost.[166] The break-even cost is important for economic assessment when comparing different processes. Korus et al.[166] determined that break-even cost for methyl esters and ethyl esters from rapeseed oil catalyzed by 1 percent potassium hydroxide was $1.85 and $2.08 per U.S. gallon, respectively.

The remaining steps considered in the final economic analysis include transportation of the biomass to pressing and oil extraction facilities, processing the oil through refining, transesterification, and further cleaning steps followed by transportation, distribution, and storage considerations. Further utilization of by-products in each of these steps is critical to creating an efficient biorefinery, much like the designs of petroleum refineries of today. Finally, tax incentives, insurance issues, risk assessment, and many other factors may significantly influence the outcome of the analysis.[17]

With any process economic analysis in a biorefinery setting, the cost estimation of raw materials and manufacturing capital and operating cost may be accomplished to aid in first simulating a production scheme before putting into place at the commercial level. Supply and demand economics control the cost of feedstock materials and are markedly different, depending on collection and transportation cost. Agricultural milling residues account for the lowest cost feedstock due to availability and lower transport cost, but due to limited supply, the cost increases dramatically at the upper limit of the supply curve. Agricultural harvesting residue costs are typically higher due to increased transport cost. Dedicated crops will incur the highest initial cost, but increased supply compared to residues potentially makes the market more feasible for high-volume energy production when combined with available residue markets. Table 6.7 shows feedstock yield and cost.

| Feedstock | U.S. production (millions MT) | Yield (MT/ha/yr) | Cost (U.S. $/MT) |
|---|---|---|---|
| Corn | 225 | 6.3 | 83 |
| Sugarcane | 25 | – | 30 |
| Sorghum | 16 | 16 | 63 |
| Potato | 17 | – | 160 |
| Switchgrass | – | 8 | 104 |
| Hybrid poplar | – | 10 | 66–114 |
| Napier grass | – | 14 | 64 |
| Sycamore | – | 8 | 89 |
| Agricultural residues | 50 | – | 13–47 |
| Forest residues | 6 | – | 13–43 |
| MSW (lignocellulosic) | 55 | – | 0–25 |

*Source*: Adapted from Brown, 2003.

**TABLE 6.7** Typical Cost for Manufacturing Several Bio-Based Products

## 6.6.1 Feedstock Cost

To estimate the feedstock crop production cost, both fixed and variable costs are associated. Fixed cost includes payment of loan principle and interest for crop production equipment such as machinery for plowing, disking, and fertilizing as well as postharvest machinery for cutting, baling, forage chopping, collecting, and combining. Variable costs for these machinery include fuel and maintenance. Greater detail for individual equipment and labor cost may be found in Brown (2003).[17] Typical fixed and variable machinery production costs range from around $2 and $1/ha, respectively, for machinery like rotary hoe equipment to over $50 and $30/ha, respectively, for equipment like silage harvesters. Labor costs and rates associated with crop production are considered separately and typically range from 1 to 13 h/ha with harvesting soybeans followed by corn crops averaging 6 h/ha.[17] Material expenses for seed, cuttings, fertilizer, and pesticides create costs that may exceed $70/ha each for nitrogen and herbicides, depending on the needs of the crop. Other significant costs include insurance, land rental, and transportation. From the analysis of corn production,[176] the breakdown for total fixed and variable costs were nearly $450 and $350/ha, respectively. Material (seeds, fertilizers, and pesticides) and machinery (preharvest and postharvest) costs accounted for about 60 percent and 20 percent, respectively of the total variable costs, while land costs accounted for nearly 70 percent of the fixed costs. However, in this study, transportation costs were not included when estimating *farm-gate price*. The total unit production for corn was nearly \$130/MT with crop yield of 6.3 MT/ha/yr calculated as:

$$\text{Unit production cost (\$/MT)} = \text{Total expenses (\$/ha)} / \text{Crop yield (MT/ha)}$$

An important note stated by Brown[17] was that overproduction of many of these crops has resulted in the cost of production exceeding the market price by as much as 50 percent in the United States, creating an obvious competitive disadvantage facing growers and producers in the twenty-first century.

A simplified cash flow analysis is convenient for determining the time value of money over an investment period for crop production. The cash flow process is typically divided into annual net cash flows by subtracting annual revenues from the sale of products, coproducts, and by-products from outflows, including operating expenses, capital invested, and income taxes. The net present value (NPV) is then obtained by summing all annual discounted cash flows, $C_D$, where discounts adjust for inflation and real rate of return.

$$NPV = \sum_n C_D = \sum_n \frac{C_T}{(1+i)^n} \qquad (6.14)$$

where $C_T = C_R(1-x) - C_O(1-x) - C_I - xC_P + x(C_F - C_S)/t_d$ (6.15)

where $C_D$ = total discounted cash flow
$C_T$ = total cash flow
$C_R$ = total revenue = market price (\$/MT) × yield (MT/ha)
$C_O$ = total operating cost
$C_I$ = total capital cost invested
$C_P$ = loan principle cost
$C_F$ = fixed capital cost
$C_S$ = salvage cost
$x$ = tax rate
$t_d$ = time of depreciation
$i$ = nominal rate of return $\approx f + r$
$f$ = inflation rate
$r$ = real rate of return
$n$ = years of investment

An example cash flow sheet for production of feedstock crops such as soybeans for production of soy protein isolate and oil for biodiesel is given in Table 6.8.

| Crop: | | Market: | | Yield: | | Tax: | | Salvage: $c_s =$ | | Fixed: $c_f =$ | | | | Date: | |
|---|---|---|---|---|---|---|---|---|---|---|---|---|---|---|---|
| | | Year 1 | | Year 2 | | Year 3 | | | | Year 4 | | Year 5 | | Total | |
| Expenses | Fixed | Variable | Fixed | Variable | Fixed | Variable | Fixed | Variable | Fixed | Variable | | | | | |
| $C_I$ | | | | | | | | | | | | | | | |
| Machinery | | | | | | | | | | | | | | | |
| Seeds | | | | | | | | | | | | | | | |
| Fertilizers | | | | | | | | | | | | | | | |
| Pesticides | | | | | | | | | | | | | | | |
| Insurance | | | | | | | | | | | | | | | |
| Interest | | | | | | | | | | | | | | | |
| Labor | | | | | | | | | | | | | | | |
| Land | | | | | | | | | | | | | | | |
| $C_O$ | | | | | | | | | | | | | | | |
| Total cost | | | | | | | | | | | | | | | |
| Income tax | | | | | | | | | | | | | | | |
| $C_T$ | | | | | | | | | | | | | | | |
| $C_D$ | | | | | | | | | | | | | | | |
| NPV | | | | | | | | | | | | | | | |

**TABLE 6.8** Example Cash Flow Sheet Template for Feedstock Production

## 6.6.2 Manufacturing Cost

Several texts outlining manufacturing costs are available for further details for chemical and bioprocessing facilities,[177] and more specifically bioenergy plants.[17,70] Manufacturing costs consist of direct costs, indirect costs, and overhead expenses.[177] Direct costs include the components of raw materials, equipment, utility costs, supplies, and labor. Indirect costs include depreciation, taxes and exemptions, rental fees, and insurance. Overhead expenses consist of administration, supervision, medical costs, regulatory affairs, and procurement. To determine the economic analysis of a manufacturing facility, this may be divided into capital and operating cost. Capital cost includes all cost of the facility, equipment, installation, and labor for installation. Operating cost is the required cost for operation of the plant, including labor for operation, cost of feedstock raw materials, and payment of principal and interest on loans.[17] Table 6.9 describes the differences in key feedstock, operating, and capital parameters for energy from biomass sources.

## 6.6.3 Capital Cost

Direct fixed capital cost includes equipment, materials, and installation. A scaling relationship for determination of the economy of scale for typical processing equipment required is expressed by:

$$C = C_b (S/S_b)^n (I/I_b) \qquad (6.16)$$

where $C$ and $S$ are the cost and size of equipment, respectively, as a function of the base cost, $C_b$, and base size, $S_b$. The economy of scale is estimated by this power relationship where $n$ is the sizing exponent ($n < 1$). The cost must be adjusted for inflation if the base cost inflation index, $I_b$, is different than the inflation index of the current year, $I$. Table 6.10 shows the inflation index factors from 1955 to adjusted inflation factors[17] through 2014. Material (fittings, and so on) and installation labor factor into total equipment capital cost, which are often significant for processing equipment. Compensating for these factors, the direct unit cost, $C_{dc}$, is estimated by:

$$C_{dc} = C(1 + x_m + x_l + x_m x_l) \qquad (6.17)$$

where $x_m$ and $x_l$ are material and labor module factors, respectively.

The indirect capital costs include freight, engineering, overhead, insurance, and taxes associated with the purchased equipment, which account for nearly 25 percent of the equipment and material costs and 70 percent of the installation labor cost.[17] The indirect cost, $C_i$, may

| Bioenergy product | Characteristics | Feedstock value | Production cost | Total capital cost (U.S. $) |
|---|---|---|---|---|
| Electricity from combustion of biomass | 50 MW biomass plant $1.90/GJ biomass | – | $0.06/kWh (1990 $) | 80 M |
| Electricity from gasification of biomass | 50 MW plant | – | $0.05/kWh | 75–138 M |
| Biogas from anaerobic digestion | 200 ton/day plant (28 M m³ methane/yr) | – | $5–6/GJ (1987 $) $2–4/GJ (natural gas) | 8.8–26.4 M |
| Ethanol from corn | New plant (5000 bbl/day) Existing plant 265 M L/yr | $2–5/bu | $0.53–0.79/L $0.26–0.40/L | 140 M (1987 $) |
| Ethanol from lignocellulosics | New plant (5000 bbl/day) Projected improvements | $42/dry ton of wood | $0.13–0.31/L $0.08–0.11/L | 175 M (1994 $) |
| Methanol from Syngas | 7500 bbl/day plant Projected goal | $40/dry ton of wood | $0.27/L $0.15/L | 280 M (1991 $) |
| Bio-oil from fast pyrolysis | 5000 bbl/day | $1.70/GJ | $0.18/L | 63 M (1987 $) |
| Biodiesel from vegetable oil | 2.9 M L/yr 37.9 M L/yr | $0.53–0.79/L refined $0.16–0.26/L waste | $0.30–0.69/L $0.25/L (No. 2 diesel) | 0.25 M 11.3 M |

**TABLE 6.9** Characteristics, Feedstock, Production, and Capital Costs for Energy Products from Biomass Sources

then be estimated from the material and labor module factors according to the following relation:

$$C_{ic} = C(0.25 + 0.15 x_m + 0.7 x_l + 0.7 x_m x_l) \qquad (6.18)$$

Finally, contingency and profit fees are typically added as well as auxiliary facilities cost that may be necessary, depending on the size

| Year | Inflation index | Year | Inflation index | Year | Inflation index | Year | Inflation index |
|---|---|---|---|---|---|---|---|
| 1955 | 0.151 | 1970 | 0.219 | 1985 | 0.608 | 2000 | 0.972 |
| 1956 | 0.154 | 1971 | 0.229 | 1986 | 0.619 | 2001 | 1.000 |
| 1957 | 0.159 | 1972 | 0.236 | 1987 | 0.641 | 2002 | 1.018 |
| 1958 | 0.163 | 1973 | 0.251 | 1988 | 0.668 | 2003 | 1.042 |
| 1959 | 0.164 | 1974 | 0.278 | 1989 | 0.700 | 2004 | 1.067 |
| 1960 | 0.167 | 1975 | 0.304 | 1990 | 0.738 | 2005 | 1.093 |
| 1961 | 0.169 | 1976 | 0.321 | 1991 | 0.769 | 2006 | 1.119 |
| 1962 | 0.171 | 1977 | 0.342 | 1992 | 0.792 | 2007 | 1.147 |
| 1963 | 0.173 | 1978 | 0.368 | 1993 | 0.816 | 2008 | 1.175 |
| 1964 | 0.175 | 1979 | 0.410 | 1994 | 0.837 | 2009 | 1.204 |
| 1965 | 0.178 | 1980 | 0.465 | 1995 | 0.861 | 2010 | 1.233 |
| 1966 | 0.183 | 1981 | 0.513 | 1996 | 0.886 | 2011 | 1.262 |
| 1967 | 0.189 | 1982 | 0.545 | 1997 | 0.906 | 2012 | 1.293 |
| 1968 | 0.197 | 1983 | 0.562 | 1998 | 0.920 | 2013* | 1.323 |
| 1969 | 0.207 | 1984 | 0.587 | 1999 | 0.941 | 2014* | 1.353 |

*Estimated from previous years with increase of 0.030; factors are normalized to year 2001 with projections beyond this year.
*Source*: Adapted from Brown, 2003.

**TABLE 6.10** Inflation Index Values from 1955 to Projected Year of 2014

of the project.[17] The contingency and fee cost are typically around 20 percent of the total indirect and direct costs and if auxiliary facilities are required, these costs may be estimated to as much as 30 percent of the total module cost. Therefore, the total capital cost, $C_t$, may be approximated as:

$$C_{tc} \approx 1.5(C_{dc} + C_{ic}) \tag{6.19}$$

### 6.6.4 Operating Cost

Operating costs after the plant is constructed include both direct costs such as raw materials and operating labor and indirect cost, including overhead and insurance. Payment of the capital charges or loan and interest on capital during the life of the plant is also included in the operating cost. The raw material cost is determined from throughput in the plant calculated by:

$$C_R = C_r \dot{m} f_c \tag{6.20}$$

where $C_R$ = cost of raw materials, \$/kg
$\dot{m}$ = plant capacity, kg/s
$f_c$ = capacity factor, which is the fraction of time during a year that the plant is in actual operation

The total direct cost includes the cost of raw materials, operating labor, supervisory labor, utilities, operating supplies, and so on. An approximate direct cost based on these typical expenses is given by:

$$C_{do} = C_R + 1.3C_o + 0.2C_{tc} \tag{6.21}$$

Indirect operating expenses are represented by overhead, taxes, insurance, and other expenses and may be estimated by:

$$C_{io} = 0.95C_o + 0.05C_{tc} \tag{6.22}$$

Finally, the annual capital charges are estimated by the annual interest rate on the loan over the number of years of the loan, $n$, by:

$$C_{co} = \frac{iC_{tc}(1+i)^n}{(1+i)^n - 1} \tag{6.23}$$

The product cost is the total annual operating costs divided by the annual production in total energy, volume of product, and so on:

$$C_{po} = (C_{do} + C_{io} + C_{co})/\text{production} = C_{to}/\text{production} \tag{6.24}$$

## 6.7 Summary and Conclusions

Biodiesel production entails a large number of variables to determine the life-cycle assessment for potential implementation of technologies to make the processes practicable while taking into account the environmental consequences. The factors examined in this chapter include:

- Environmental considerations: nitrogen oxides, emissions, and feedstock considerations
- Potential feedstocks and oil sources: plant, animal, and microbial sources
- Production chemistry: transesterification, esterification, interesterification, and side reactions with emphasis on catalyst type or supercritical conditions

- Processing considerations: transesterification, esterification pretreatment, and esterification processing
- Economics: feedstock and manufacturing (capital and operating) economic analysis

For biodiesel production to accomplish the goals of displacement of petroleum resources and aiding in our current national security to break dependency on the middle east imports, a complex understanding of the environmental impacts must be considered. Some of these considerations include the enormous amount of inputs such as fertilizers and their project runoff into rivers, energy input to obtain sufficient feedstock, and transportation concerns for shipping vast amounts of materials from relatively dilute biomass energy sources to bioenergy facilities to concentrate the energy source providing services to customers. This chapter explains the chemical and biosystems engineering approaches to manufacturing of biodiesel fuels when applied to design of facilities relative to the types of feedstock oils used as raw materials. A basic understanding of the processing mechanisms involved tied with economic, social, and policy concerns is instrumental for incorporation of the best practices in the implementation of biodiesel fuels as a potential source for displacement of current petroleum-based transportation fuels.

## Acknowledgments

The authors would like to acknowledge those who helped to influence and provide advice for this chapter, including the Clemson University Bioenergy Creative Inquiry Research Group. Special thanks goes to undergraduate research students Justin Montanti and Lauren Staples for their valuable input and patience with this project, and to graduate students Hem Joshi, Derek Little, Arpan Jain, Meidui Dong, Cheng-Yi Kuan, Yen-Hui Chen, Holly Garrett, and others who have made direct contributions in the pilot and research labs to gather information for this text. Our thanks goes to funding organizations for research toward biodiesel and construction of the pilot facility at Clemson University, including USDA and Tom Wedegaertner of Cotton Incorporated. Special thanks goes to Clemson University Environmental Health and Safety, including Philip Carroll and Robin Newberry for their invaluable support of this project. We express our appreciation to Drs. Feng Chen, Xi Wang, Jim Frederick, Shelie Miller, Nick Rigas, Rob Leitner, David Brune, Harold Allen, and others for their involvement with biodiesel research and development. Finally, thanks goes to Dr. George Chumanov for being next door stimulating further discussion toward our renewable energy future.

## Problems

1. Write the mass balance of a continuous transesterification process for production of crude biodiesel from 250,000 kg/h rapeseed oil reacted with 75,000 kg/h ethanol and 250 kg/h potassium hydroxide with 1 h residence time. Determine the production yield of biodiesel assuming 95 percent conversion rate making use of:
   a. Microsoft Excel or similar spreadsheet to compute the mass balance for each stream.
   b. Intelligen SuperPro Designer (if available) to model the conversion system.

2. If ethanol were used in the formation of biodiesel, a typical process to remove water is necessary since ethanol naturally forms an azeotrope with water at approximately 95.6 percent ethanol and 4.4 percent water by weight. An azeotrope is a mixture of two or more liquids that cannot be separated further by distillation. In other words, the azeotrope of ethanol and water forms a constant boiling solution at 78.2°C where the composition of the liquid phase is the same as that of the vapor phase in equilibrium. Determine the energy required to dry ethanol to produce 99.5 percent ethanol feasible for biodiesel production using the following methods:
   a. Azeotropic distillation—Ethanol may be separated by forming a ternary azeotrope with benzene to displace water content typically bound tightly to ethanol in the binary mixture.
   b. Absorption using zeolite.
   c. Use of corn grits has shown to be a low-cost alternative for removal of water from ethanol as demonstrated by Ladisch et al.[177]

3. To produce biodiesel ethyl esters from rapeseed oil, ethanol and potassium hydroxide are used in the reaction:
   a. Determine the theoretical conversion efficiency of the biodiesel reaction and the amount of ethanol required to convert 1 L of rapeseed oil reacted with 1 percent potassium hydroxide.
   b. Determine the amount of ethanol required to make a molar ratio of 6 mol of ethanol to 1 mol of oil.
   c. Determine the amount of ethanol not reacted in the 6:1 molar mixture.
   d. Determine the amount of methanol required in this reaction for a molar ratio of 6:1.

4. Determine the activation energy from the following subcritical and supercritical methanol rate data for transmethylation of rapeseed oil (data taken from Kusdiana and Saka[81]; *data extrapolated to complete problem).

| T (°C) | k (s−1) |
|---|---|
| 487 | 0.0803 |
| 431 | 0.0503 |
| 385 | 0.0249 |
| 350 | 0.0178 |
| 300 | 0.0071 |
| 270 | 0.0007 |
| 230 | 0.0003 |
| 220* | 0.00027 |
| 210* | 0.00025 |
| 200* | 0.00022 |

5. Based on the ternary diagram of methanol, glycerol, and biodiesel esters (see Fig. 6.11) determine the amount of methanol present in the upper and lower layers of the final product for the mixtures shown by the three tie lines.
6. A biodiesel process utilizes two reactor vessels to complete transesterification. Soybean oil and sodium ethoxide are fed continuously at 60°C to the first reactor at rates of 5000 kg/h and 900 kg/h, respectively. These reactors are to be purchased and incorporated into a 10-million-gallon-per-year (MGY) facility.
   a. Determine the volume of both reactors if 100 kg/h sodium ethoxide is added to the second reactor.
   b. Determine the direct capital cost of each reactor.
7. Estimate the continuous centrifuge capital and operational cost for separating the glycerol from biodiesel esters in a 50 MGY plant.
8. A concentration of 103 mM of lauric acid is esterified to biodiesel propyl esters in supercritical carbon dioxide (35°C and 110 bar) catalyzed by 1.2 mg of immobilized *C. antarctica* lipase in a microemulsion-based organogel (MBG). From the following data, determine the kinetic constants from the Lineweaver-Burk double reciprocal plot.

| v (mM/min) | 1-propanol (mM) |
|---|---|
| 0.182 | 200 |
| 0.153 | 100 |
| 0.137 | 50 |
| 0.111 | 33 |
| 0.098 | 25 |

# References

1. http://en.wikipedia.org/wiki/Biodiesel.
2. Christi, Y. 2007. "Biodiesel from microalgae." *Biotechnol Adv.* 25:294–306.
3. Collins, K. 2007. "A Status Report on Biofuels, Food & Feed Tradeoffs." U.S. Department of Agriculture, Office of the Chief Economist. USDA Agricultural Projections to 2016, OCE-2007-1, February 2007, available at: www.ers.usda.gov/publications/oce071/.
4. Rao, P. S., and K. V. Gopalakrishnan. 1991. "Vegetable oils and their methyl esters as fuels for diesel engines." *Indian J Technol.* 29(6):292–297.
5. Arkoudeas, P., et al. 2003. "Study of using JP-8 aviation fuel and biodiesel in CI engines." *Energy Conversion and Management.* 44(7):1013–1025.
6. Lotero, E., et al. 2005. "Synthesis of biodiesel via acid catalysis." *Ind Eng Chem Res.* 44:5353–5363.
7. Zhang, Y., et al. 2003. "Biodiesel production from waste cooking oil: 1. Process design and technological assessment." *Bioresour Technol.* 89:1–16.
8. National Biodiesel Board. 2007. Biodiesel Emissions Fact Sheet. http://www.biodiesel.org/pdf_files/fuelfactsheets/emissions.pdf.
9. Szybist, J. P., S. R. Kirby, and A. L. Boehman. 2005. "$NO_x$ emissions of alternative diesel fuels: A comparative analysis of biodiesel and FT diesel." *Energy Fuels.* 19:1484–1492.
10. Walker, K. 1994. "Biodiesel from rapeseed." *J R Agric Soc Engl.* 155:43–44.
11. Hess, M. A., et al. 2005. "Effect of antioxidant addition on $NO_x$ emissions from biodiesel." *Energy Fuels.* 19:1749–1754.
12. Boheman, A. L., E. Morris, and J. Szybist. 2004. *Energy Fuels.* 18:1877–1882.
13. McCormick, R. L., et al. 2001. "Impact of biodiesel source material and chemical structure on emissions of criteria pollutants from a heavy-duty engine." *Environ Sci Technol.* 35:1742–1747.
14. Heywood, J. R. 1988. *Internal Combustion Engine Fundamentals.* McGraw-Hill: New York.
15. Cheng, A. S., A. Upatnieks, and C. Mueller. 2005. "Investigation of the impact of biodiesel fueling on $NO_x$ emission using an optical direct-injection diesel engine." Presented at the Biodiesel Technical Workshop.
16. Frankel, E. N. 1998. *Lipid Oxidation.* The Oily Press: Dundee, Scotland.
17. Brown, R. C. 2003. *Biorenewable Resources.* Iowa State Press, Blackwell Publishing Co.: Iowa.
18. Tenenbaum, D. J. 2000. "A burning question—Do farmer-set fires endanger health?" *Environ Health Perspect.* 108:A351.
19. ASTM. 2005. "ASTM Standards Related to Biodiesel Fuel Blend Stock (B100) for Middle Distillate Fuels."
20. Yazdani, S. S., and R. Gonzalez. 2007. "Anaerobic fermentation of glycerol: A path to economic viability for the biofuels industry." *Curr Opin Biotechnol.* 18:213–219.
21. Meher, L. C., D. Vidya Sagar, and S. N. Naik. 2006. "Technical aspects of biodiesel production by transesterification—a review." *Renew Sustain Energy Rev.* 10:248–268.
22. Wang, Y. K., et al. 2007. "Preparation of biodiesel from waste cooking oil via two-step catalyzed process." *Energy Conversion and Management.* 48:184–188.
23. Modi, M. K., et al. 2007. "Lipase-mediated conversion of vegetable oils into biodiesel using ethyl acetate as acyl acceptor." *Bioresour Technol.* 98(6):1260–1264.
24. Du, W., et al. 2004. "Comparative study on lipase-catalyzed transformation of soybean oil for biodiesel production with different acyl acceptors." *J Mol Catal, B Enzym.* 30(3–4):125–129.
25. Yang, T., et al. 2005. "Enzymatic production of monoacylglycerols containing polyunsaturated fatty acids through an efficient glycerolysis system." *J Agric Food Chem.* 53:1475–1481.
26. Demirbas, A. 2003. "Biodiesel fuels from vegetable oils via catalytic and non-catalytic supercritical alcohol transesterifications and other methods: A survey." *Energy Conversion and Management.* 44:2093–2109.

27. Fukuda, H., A. Kondo, and H. Noda. 2001. "Biodiesel fuel production by transesterification of oils." *J Biosci Bioeng*. 92(5):405–416.
28. Canakci, M., and J. Van Gerpen. 1999. "Biodiesel production via acid catalysis." *Trans ASAE*. 42:1203–1210.
29. Nye, M. J., et al. 1983. "Conversion of used frying oil to diesel fuel by transesterification: preliminary test." *J Am Oil Chem Soc*. 60:1598–1601.
30. Warabi, Y., D. Kusdiana, and S. Saka. 2004. "Biodiesel fuel from vegetable oil by various supercritical alcohols." *Appl Biochem Biotechnol*. 113–116, 793–801.
31. Noureddini, H., and D. Zhu. 1997. "Kinetics of transesterification of soybean oil." *J Am Oil Chem Soc*. 74:1457–1463.
32. Ataya, F., M. A. Dube, and M. Ternan. 2006. "Single-phase and two-phase base-catalyzed transesterification of canola oil to fatty acid methyl esters at ambient conditions." *Ind Eng Chem Res*. 45:5411–5417.
33. Boocock, D. G. B., et al. 1998. "Fast formation of high purity methyl esters from vegetable oil." *JAOCS*. 75:1167–1172.
34. Mao, V., S. K. Konar, and D. G. B. Boocock. 2004. "The pseudo-single-phase, base-catalyzed transmethylation of soybean oil." *JAOCS*. 81:803–808.
35. Komers, K., et al. 2002. "Kinetics and mechanism of the KOH-catalyzed methanolysis of rapeseed oil for biodiesel production." *Eur J Lipid Sci Technol*. 104, 728–737.
36. Stepan, E., et al. 2006. "Kinetics of transesterification reaction of the sunflower oil." *Rev Chim*. 57, 693–698.
37. Vicente, G., M. Martínez, and J. Aracil. 2005. "Kinetics of sunflower oil methanolysis." *Ind Eng Chem Res*. 44:5447–5454.
38. Vicente, G., M. Martínez, and J. Aracil. 2004. "Integrated biodiesel production: A comparison of different homogeneous catalysts systems." *Bioresour Technol*. 92:297–305.
39. Franceschini, G., and S. Macchieto. 2007. "Validation of a model for biodiesel production through model-based experiment design." *Ind Eng Chem Res*. 46:220–232.
40. Encinar, J. M., J. F. Gonzalez, and A. Rodriguez-Reinares. 2005. "Biodiesel from used frying oil: Variables affecting the yields and characteristics of the biodiesel." 44:5491–5499.
41. Freedman, B., R. O. Butterfield, and E. H. Pryde. 1986. "Transesterification kinetics of soybean oil." *JAOCS*. 63:1375–1380.
42. Kulkarni, M. G., K. Dalai, and N. N. Bakhshi. 2006. "Utilization of green seed canola oil for biodiesel production." *J Chem Technol Biotechnol*. 81:1886–1893.
43. Mittelbach, M., and H. Enzelsberger. 1999. "Transesterification of heated rapeseed oil for extending diesel fuel." *JAOCS*. 76:545–550.
44. Srivastava, A., and R. Prasad. 2000. "Triglycerides-based diesel fuels." *Renewable Sustainable Energy Rev*. 4:111–133.
45. Leung, D. Y. C., and Y. Guo. 2006. "Transesterification of neat and used frying oil: Optimization for biodiesel production." *Fuel Process Technol*. 87:883–890.
46. Bournay, L., et al. 2005. "New heterogeneous process for biodiesel production: A way to improve the quality and the value of the crude glycerin produced by biodiesel plants." *Catal Today*. 106:190–192.
47. Bournay, L., et al. 2005. "Process of producing alkyl esters from vegetable or animal oil and aliphatic monoalcohol." *US Patent*. 6:878, 937.
48. Stern, R., et al. 1999. "Process for the production of esters from vegetable oils or animal oils and alcohol." *US Patent*. 5:908, 946.
49. Cantrell, D. G., et al. 2005. "Structure-reactivity correlations in MgAl hydrotalcite catalyst for biodiesel synthesis." *Appl Catal A-Gen*. 287:183–190.
50. Suppes, G. J., et al. 2004. "Transesterification of soybean oil with zeolite and metal catalysts." *Appl Catal A-Gen*. 257:213.
51. Watkins, R. S., A. F. Lee, and K. Wilson. 2004. "Physicochemical properties of Pt-SO$_4$/Al$_2$O$_3$ alkane oxidation catalysts." *Green Chem*. 6:335–340.
52. Kim, H. J., et al. 2004. "Preparation of biodiesel catalyzed by solid super base of calcium oxide and its refining process." *Catal Today*. 93–95, 315.
53. Ebiura, T., et al. 2005. "Selective transesterification of triolein with methanol to methyl oleate and glycerol using alumina loaded with alkali metal salt as a solid-base catalyst." *Appl Catal A-Gen*. 283:111–116.

54. Di Serio, M., et al. 2006. "Transesterification of soybean oil to biodiesel by using heterogeneous basic catalysts." *Ind Eng Chem Res.* 45:3009–3014.
55. Reddy, C., et al. 2006. "Room temperature conversion of soybean oil and poultry fat to biodiesel catalyzed by nanocrystalline calcium oxides." *Energy Fuels.* 20:1310–1314.
56. Liu, Y., E. Lotero, and J. G. Goodwin. 2006. "Effect of carbon chain length on esterification of carboxylic acids with methanol using acid catalysis." *J Catalysis.* 243:221–228.
57. Liu, Y. 2007. "Biodiesel Forming Reactions and Heterogeneous Catalysis." PhD Dissertation, Clemson University, Clemson, SC.
58. Shibasaki-Kitakawa, N., et al. 2007. "Biodiesel production using anionic ion-exchange resin as heterogeneous catalyst." *Bioresource Technol.* 98:416.
59. Ma, F., and M. A. Hanna. 1999. "Biodiesel production: A review." *Bioresour Technol.* 70:1.
60. Nye, M. J., and P. H Southwell. 1983. "Esters from rapeseed oil as diesel fuel." In Vegetable Oils Diesel Fuel: Seminar III; ARM-NC-Agriculture: Peoria, IL, p. 78.
61. Harrington, K. J., and D'Arcy-Evans, C. 1985. "Transesterification in situ of sunflower seed oil." *Ind Eng Chem Prod Res Dev.* 24:314.
62. Freedman, B., E. H. Pryde, T. L. Mounts. 1984. "Variable affecting the yields of fatty esters from transesterified vegetable oils." *JAOCS.* 61:1638–1643.
63. López, D. E., J. G. Goodwin, and D. A. Bruce. 2007. "Transesterification of Triacetin with Methanol on Nafion Acid Resins." *J Catal.* 245, 379.
64. López, D. E., et al. 2005. "Transesterification of triacetin with methanol on solid acid and base catalysts." *Appl Catal A Gen.* 295:97–105.
65. Mittelbach, M., and H. Enzelsberger. 1999. "Transesterification of heated rapeseed oil for extending diesel fuel." *JAOCS.* 76:545–550.
66. Liu, K. S. 1994. "Preparation of fatty acid methyl esters for gas chromatographic analysis of lipids in biological materials." *JAOCS.* 71:1179–1187.
67. Crabbe, E., et al. 2001. "Biodiesel production from crude palm oil and evaluation of butanol extraction and fuel properties." *Process Biochem.* 37:65–71.
68. Khan, A. K. 2002. Research into Biodiesel Kinetics & Catalyst Development. Brisbane, Australia.
69. Al-Widyan, M. I., and A. O. Al-Shyoukh. 2002. "Experimental evaluation of the transesterification of waste palm oil into biodiesel." *Bioresour Technol.* 85(3):253–256.
70. Haas, M. J., et al. 2006. "A process model to estimate biodiesel production cost." *Bioresour Technol.* 97:671–678.
71. Zhang Y., et al. 2003. "Biodiesel production from waste cooking oil: 2. Economic assessment and sensitivity analysis." *Bioresour Technol.* 90:229–240.
72. Sadidharan, M., and R. Kumar. 2004. "Transesterification over various zeolites under liquid-phase conditions." *J Mol Catal, A Chem.* 210:93–98.
73. Yadav, G. D., and J. J. Nair. 1999. "Sulfated zirconia and its modified versions as promising catalysts for industrial processes." *Microporous Mesoporous Mater.* 33:1–48.
74. Chavan, S. P., et al. 1996. "Use of solid superacid (sulfonated $SnO_2$) as efficient catalyst in facile transesterification of ketoesters." *Tetrahedron Lett.* 37:233–236.
75. Chavan, S. P., et al. 2001. "Transesterification of ketoesters using Amberlyst-15." *Synth Commun.* 31:289–294.
76. Zhao, X., et al. 2007. "An organic soluble lipase for water-free synthesis of biodiesel." *Appl Biochem Biotechnol.* 143(3):236–243.
77. Samukawa, T., et al. 2000. "Pretreatment of immobilized *Candida antarctica* lipase for biodiesel fuel production from plant oil." *J Biosci Bioeng.* 20(2):180–183.
78. Chen, J. W., and W. T. Wu. 2003. "Regeneration of immobilized candida antarctica lipase for transesterification." *J Biosci Bioeng.* 95(5):466–469.
79. Hsu, A. F., et al. 2002. "Immobilized lipase-catalyzed production of alkyl esters of restaurant grease as biodiesel." *Biotechnol Appl Biochem.* 36(3):181–186.

80. Noureddini, H., X. Gao, and R. S. Philkana. 2005. "Immobilized *Pseudomonas cepacia* lipase for biodiesel fuel production from soybean oil." *Bioresour Technol.* 96(7):769–777.
81. Kusdiana, D., and S. Saka. 2001. "Kinetics of transesterification in rapeseed oil to biodiesel fuels as treated in supercritical methanol." *Fuel.* 80:693–698.
82. Blattner, C., et al. 2006. "Biocatalysis using lipase encapsulated in microemulsion-based organogels in supercritical carbon dioxide." *J Supercritical Fluids.* 36:182–193.
83. Rantakyla, M., and O. Aaltonen. 1994. "Enantioselective esterification of ibuprofen in supercritical carbon dioxide by immobilized lipase." *Biotechnol Lett.* 16:825–830.
84. Marty, A., D. Combes, and J. S. Condoret. 1992. "Fatty acid esterification in supercritical carbon dioxide." *Progress Biotechnol.* 8:425–432.
85. http://hyperphysics.phy-astr.gsu.edu/hbase/thermo/diesel.html.
86. Encinar, J. M, et al. 2002. "Biodiesel fuels from vegetable oils: Transesterification of *Cynara cardunculus* L. oils with ethanol." *Energy Fuels.* 16:443–450.
87. Barnwal, B.K., and M. P. Sharma. 2005. "Prospects of biodiesel production from vegetable oils in India." *Renew Sustain Energy Rev.* 9:363–378.
88. He, H., S. Sun, and T. Wang. 2007. "Transesterification kinetics of soybean oil for production of biodiesel in supercritical methanol." *JAOCS.* 84:399–404.
89. Stamatis, H., et al. 1993. "Kinetic study of lipase catalyzed esterification reactions in water-in-oil microemulsions." *Biotechnol Bioeng.* 42:931–937.
90. Chulalaksananukul, W., J. S. Condoret, and D. Combes. 1993. "Geranyl acetate synthesis by lipase-catalyzed transesterification in supercritical carbon dioxide." *Enzyme Microb Technol.* 15:691–698.
91. Delimitsou, C., et al. 2002. "Activity and stability studies of Mucor miehei lipase immobilized in novel microemulsion-based organogels." *Biocatal Biotransform.* 20:319–327.
92. Conceição, M. M., et al. 2007. "Dynamic kinetic calculation of castor oil biodiesel." *J Therm Anal Calorim.* 87:865–869.
93. Yuan, W., A. C. Hansen, and Q. Zhang. 2005. "Vapor pressure and normal boiling point predictions for pure methyl ester and biodiesel fuels." *Fuel.* 84:943–950.
94. Reid, R. C., J. M. Prausnitz, and T. K. Sherwood. 1987. "The Properties of Gases and Liquids." 4th ed. McGraw Hill, Inc.: New York.
95. Graboski, M. S., and R. L. McCormick. 1998. "Combustion of fat and vegetable oil derived fuels in diesel engines." *Prog Energy Combust Sci.* 24:125–164.
96. Goodrum, J. W. 2002. "Volatility and boiling points of biodiesel from vegetable oils and tallow." *Biomass Bioenergy.* 22:205–211.
97. Zhou, H., H. Lu, and B. Liang. 2006. "Solubility of multicomponent systems in the biodiesel production by transesterification of *Jatropha curcas* L. oil with methanol." *J Chem Eng Data.* 51:1130–1135.
98. Tiwari, A K., A. Kumar, and H. Raheman. 2007. "Biodiesel production from Jatropha oil (*Jatropha curcas*) with high free fatty acids: An optimized process." *Biomass Bioenergy.* 31:569–575.
99. Erasmus, U. 2007. *Fats That Heal, Fats That Kill: The Complete Guide to Fats, Oils, Cholesterol and Human Health.* Alive Books: New York.
100. Xu, Z., and J. S. Godber. 1999. "Purification and identification of components of γ-oryzanol in rice bran oil." *J Agric Food Chem.* 47(7):2724–2728.
101. Gallagher, J. 2007. "Seashore Mallow Potential Biodiesel Source." http://www.evworld.com/news.cfm?newsid=15682.
102. http://www.dupontelastomers.com/autofocus/a2/af2.asp?article=af2_biofuel.
103. Kinney, A. J., and T. E. Clemente. 2005. "Modifying soybean oil for enhanced performance in biodiesel blends." *Fuel Processing Technol.* 86:1137–1147.
104. Sheehan, J., et al. 1998. "A Look Back at the U.S. Department of Energy's Aquatic Species Program-Biodiesel from Algae." U.S. DOE.
105. Van Elswyk, M. 2003. "Products containing highly unsaturated fatty acids for use by women and their children during stages of preconception, pregnancy and lactation/post-partum." Patent WO03017945.

106. Rao, A. R., et al. 2007. "Effect of salinity on growth of green alga *Botryococcus braunii* and its constituents." *Bioresour Technol.* 98:560–564.
107. Petkov, G., and G. Garcia. 2007. "Which are fatty acids of the green alga *Chlorella*?" *Biochem Syst Ecol.* 35:281–285.
108. De Swaaf, M. E., et al. 2003. "Analysis of docosahexaenoic acid biosynthesis in *Crypthecodinium Cohnii* by c13 labeling and desaturase inhibitor experiments." *J Biotechnol.* 103:21–29.
109. Liang, Y., K. Mai, and S. Sun. 2005. "Differences in growth, total lipid content and fatty acid composition among 60 clones of *Cylindrotheca fusiformis*." *J Appl Phycol.* 171:61–65.
110. Volkman, J., et al. 1989. "Fatty acid and lipid composition of 10 species of microalgae used in mariculture." *J Exp Mar Biol Ecol.* 128:219–240.
111. Robles, M A., et al. 1999. "Lipase-catalyzed esterification of glycerol and polyunsaturated fatty acids from fish and microalgae oils." *J Biotechnol.* 701:379–391.
112. Xu, F., et al. 2004. "Growth and fatty acid composition of *Nannochloropsis* spp. grown mixotrophically in fed-batch culture." *Biotechnol Lett.* 26:1319–1322.
113. Vanitha, A., et al. 2007. "Comparative study of lipid composition of two halotolerant alga, *Dunaliella bardawil* and *Dunaliella salina*." *Int J Food Sci Nutr.* 581:373–382.
114. Pratt, R., and E. Johnson. 1963. "Production of protein and lipid by *Chlorella vulgaris* and *Chlorella pyrenoidosa*." *J Pharm Sci.* 521:979–984.
115. Oiles, S., and R. Pire. 2001. "Fatty acid composition of *Chlorella* and *Spirulina* microalgae species." *JAOAC Int.* 841:1708–1714.
116. Wen, Z. Y., and F. Chen. 2002. "Continuous cultivation of the diatom *Nitzschia aevis* for eicosapentaenoic acid production: Physiological study and process optimization." *Biotechnol Prog.* 181:21–28.
117. Pratoomyot, J., P. Srivilas, and T. Noiraksr. 2005. "Fatty acids composition of 10 microalgal species." *Songklanakarin J Sci Technol.* 271:1180–1187.
118. Cheng, M. H., T. H. Walker, and G. J. Hulbert. 1998. "Fungal production of eicosapentaenoic acid from soybean oil in an external-loop airlift bioreactor." *Bioresour Technol.* 67:101–110.
119. Azachi, M., et al. 2002. "Salt induction of fatty acid elongas and membrane lipid modifications in the extreme halotolerant alga *Dunaliella salina*." *Plant Physiol.* 129:1320–1329.
120. Tokusoglu, O., and M. K. Unal. 2003. "Biomass nutrient profiles of three microalgae: *Spirulina platensis, Chlorella vulgaris, and Isochrisis galbana*." *J Food Sci.* 68:1114–1148.
121. Carvalho, A. P., and F. X. Malcata. 2005. "Optimization of x-3 fatty acid production by microalgae: Crossover effects of $CO_2$ and light intensity under batch and continuous cultivation modes." *Marine Biotechnol.* 7:381–388.
122. Xu, H., X. Miao, and Q. Wu. 2006. "High quality biodiesel production from a microalga *Chlorella prototheocoides* by heterotrophic growth in fermenters." *J Biotechnol.* 126(4):499–507.
123. Wu, S. T., S. T. Yu, and L. P. Lin. 2005. "Effect of culture conditions on docosahexaenoic acid production by *Schizochytrium* spp. S31." *Process Biochem.* 40:3103–3108.
124. Sijtsma, L., and M. E. de Swaaf. 2004. "Biotechnological production and applications of the omega-3 polyunsaturated fatty acid docosahexaenoic acid." *Appl Microbiol Biotechnol.* 64:146–153.
125. Renaud, S. M., et al. 1995. "Effect of temperature on the growth, total lipid content and fatty acid composition of recently isolated tropical microalgae *Isochrysisi* spp., *Nitzschia closterium, Nitzschia paleacea*, and commercial species *Isochrysis* spp. (clone T.ISO)." *J Appl Phycol.* 71:595–602.
126. Ladisch, M. R., et al. 1984. "Cornmeal adsorber for dehydrating ethanol vapors." *I&EC Process Design Development.* 23:437.
127. Higashiyama, K., et al. 1999. "Effects of dissolved oxygen on the morphology of an arachidonic acid production by *Mortierella alpina* 1S-4." *Biotechnol Bioeng.* 63:442–448.
128. Ratledge, C. 1993. "Single cell oils—have they a biotechnological future?" *Trends Biotechnol.* 11:278–284.

129. Yamada, H., S. Shimizu, and Y. Shinmen. "Production of arachidonic acid by *Mortierella elongata* 1S-5." *Agric Biol Chem*. 51:785–790.
130. Li, X., H. Xu, and Q. Wu. 2007. "Large-scale biodiesel production from microalgae *Chlorella protothecoides* through heterotrophic cultivation in bioreactors." *Biotechnol Bioeng*. 98:764–771.
131. Richmond, A. 2004. *Handbook of Microalgal Culture: Biotechnology and Applied Phycology*. Blackwell Science Ltd.: Oxford, UK.
132. Metting, F. B. 1996. "Biodiversity and application of microalgae." *J Ind Microbiol*. 17:477–489.
133. Spolaore, P. 2006. "Commercial applications of microalgae." *J Biosci Bioeng*. 1010:87–96.
134. De Swaaf, M. E., L. Sijtsma, and J. T. Pronk. 2003. "High-cell-density fed-batch cultivation of the docosahexaenoic acid producing marine alga *Crypthecodinium cohnii*." *Biotechnol Bioeng*. 811:666–672.
135. Miao, X., and Q. Wu. 2004. "High yield bio-oil production from fast pyrolysis by metabolic controlling of *Chlorella protothecoides*." 110(1):85–93.
136. Jones, D. 2007. "Extraction of Lipids from *Chlorella vulgaris* for use in Biodiesel." Undergraduate Research Thesis, Biosystems Engineering, Clemson University.
137. Wynn, J., et al. 2005. *Single Cell Oils*, Cohen, Z., and C. Ratledge (eds.), AOCS Press: Champaign, Ill.
138. Walker, T. H., H. D. Cochran, and G. J. Hulbert. 1999. "Supercritical carbon dioxide extraction of lipids from *Pythium irregulare*." *JAOCS*. 76(5):595–602.
139. Mann, J. E., and J. Myers. 1968. "On pigments, growth and photosynthesis of *Phaeodactylum tricornutum*." *J Phycol*. 4:349–355.
140. Molina Grima, E., et al. 2000. "Scale-up of tubular photobioreactors." *J Appl Phycol*. 12:355–368.
141. Drapcho, C. M., and D. E. Brune. 2000. "The partitioned aquaculture system: Impact of design and environmental parameters on algal productivity and photosynthetic oxygen production." *Aquacult Eng*. 21:151–168.
142. Weissman, J. C., and J. R. Benemann. 1978. "Biomass recycling and species control in continuous cultures." *Bioeng. Biotech*. 21:627–664.
143. Molina Grima, E., et al. 2003. "Recovery of microalgal biomass and metabolites: Process options and economics." *Biotechnol Adv*. 20:491–515.
144. Molina Grima, E., et al. 2001. "Tubular photobioreactors design for algal cultures." *J Biotechnol*. 92:113–131.
145. Molina Grima, E., et al. 1999. "Photobioreactors: Light regime, mass transfer, and scaleup." *J Biotechnol*. 70:231–247.
146. Pulz, O. 2001. "Photobioreactors: Production systems for phototrophic microorganisms." *Appl Microbiol Biotechnol*. 57:287–293.
147. Pearl, G. G. 2002. "Animal Fat Potential for Bioenergy Use, Bioenergy 2002." The Tenth Biennial Bioenergy Conference. Boise, ID, Sept. 22–26.
148. Canakci, M., and J. Van Gerpen. 2001. "A pilot plant to produce biodiesel from high free fatty acid feedstocks." *Trans ASAE*. 44:1429–1436.
149. Graham, S. A. 1989. "Cuphea: A new plant source of medium-chain fatty acids." *Crit Rev Food Sci Nutr*. 28:139–173.
150. Geller, D. P., J. W. Goodrum, and C. C. Campbell. 1999. "Rapid screening of biologically modified vegetable oils for fuel performance." *Trans ASAE*. 42:859–862.
151. Ratledge, C. 2004. "Fatty acid biosynthesis in microorganisms being used for single cell oil production." *Biochimie*. 86(11):807–815.
152. León-Bañares, R., et al. 2004. "Transgenic microalgae as green cell-factories." *Trends Biotechnol*. 22:45–52.
153. Eroshin, V. K., et al. 2000. "Arachidonic acid production by *Mortierella alpina* with growth-coupled lipid synthesis." *Process Biochem*. 35:1171–1175.
154. Quasney, M. E., et al. 2001. "Inhibition of proliferation and induction of apoptosis in SNU-1 human gastric cells by the plant sulfolipid sulfoquinovosyl diacylglycerol." *J Nutr Biochem*. 12(5):310–315.
155. Ben-Amotz, A., T. G. Tornabene, and W. H. Thomas. 1985. "Chemical profiles of selected species of microalgae with emphasis on lipids." *J Phycol*. 21:72–81.

156. Benning, C. 1998. "Biosynthesis and function of the sulfolipid sulfoquinovosyl diacylglycerol." *Ann Rev Plant Physiol.* 49:53–75.
157. Christie, W. W. 2005. "Mono- and digalactosyl diacylglycerols and related lipids from plants." www.lipid.co.uk.
158. Gustafson, K. R., et al. 1989. "AIDS-antiviral sulfolipids from cyanobacterial (blue-green algae)." *J Nat Cancer Inst.* 81:1254–1258.
159. Conwell, K. 2005. "Green algal production of sulfoquinovosyl diacylglycerol under various light intensities." MS Thesis, Clemson University, Clemson, SC.
160. Walker, T. H., C. M. Drapcho, and F. Chen. 2006. "Bioprocessing technology for production of nutraceutical compounds." In: *Engineering Technologies and Processing of Functional Foods: Processing Technologies.* John Shi (ed). CRC Press: Boca Raton, FL.
161. Archer, S. D., K. A. McDonald, and A. P. Jackman. 1997. "Effect of light irradiance on the production of sulfolipids from Anabaena 7120 in a fed-batch photobioreactor." *Appl Biochem Biotechnol.* 67:139–152.
162. Norman, H. A., et al. "Semi-preparative isolation of plant sulfoquinovosyl diacylglycerols by solid phase extraction and HPLC procedures." *J Lipid Res.* 37:1372–1376.
163. "From bioplastics to biodiesel." 2007. Green Energy News. 11(52) http://www.green-energy-news.com/arch/nrgs2007/20070040.html.
164. Kusdiana, D., and S. Saka. 2004. "Two-step preparation for catalyst-free biodiesel fuel production: Hydrolysis and methyl esterification." *Appl Biochem Biotechnol.* 113–116, 781–791.
165. Kusdiana, D., and S. Saka. 2001. "Methyl esterification of free fatty acids of rapeseed oil as treated in supercritical methanol." *Chem Eng Jpn.* 34:383–387.
166. Korus, R., et al. 1992. "Transesterification process to manufacture ethyl ester of rape oil." In: *Proceedings of 1st Biomass Conference of the Americas.* pp. 815–826.
167. Haas, M. J., and S. Bloomer. 2000. "Simple, high-efficiency synthesis of fatty acid methyl esters from soapstock." *JAOCS.* 77:373–379.
168. Fröhlich, A., B. Rice, and G. Vicente. 2001. "The conversion of waste tallow into biodiesel grade methyl ester." In: *Proc of 1st World Conf on Biomass for Energy and Industry.* Sevilla, Spain, June 5–9, 2000, James and James Ltd., pp. 695–697.
169. Boocock, D. G. B. 2002. "Biodiesel fuel from waste fats and oils: A process for converting fatty acids and triglycerides." In: Proceedings of Kyoto University International Symposium on Post-Petrofuels in the 21st Century, Prospects in the Future of Biomass Energy. Montreal, Canada, September 3–4, 2002, pp. 171–177.
170. AOCS. 1991. "Official method Ca 14-56 testing for total, free, and combined glycerol using iodometric-periodic acid."
171. Lepper, H., and L. Friesenhagen. 1986. "Process for the production of fatty acid esters of short-chain aliphatic alcohols from fats and/or oils containing free fatty acids." US Patent 4608202.
172. Knothe, G. 2006. "Analyzing biodiesel: Standards and other methods." *J Am Oil Chem Soc.* 83:823–833.
173. Joshi, H. C., J. Toler, and T. H. Walker. "Optimization of biodiesel production from cottonseed oil." *JAOCS.* 85:357–363.
174. Grace Application Note. 2006. "A Reversed Phase HPLC Method Using Evaporative Light Scattering Detection (ELSD) for Monitoring the Reaction and Quality of Biodiesel Fuels Romulus Gaita." Grace Davison Discovery Sciences, The application notebook. September 2006, Industrial 51.
175. Thomas, E. W., R. E. Fuller, and K. Terauchi. 2007. "Fluoroelastomer compatibility with biodiesel fuels." *SAE Int.* 1:4061. http://www.dupontelastomers.com/literature/viton/20E5483C5825D7398525736700470EB1.pdf
176. Duffy, M., and D. Smith. 2002. "Estimated costs of crop production in Iowa-2002." Iowa Cooperative Extension Service Bulletin No. FM 1712.
177. Lydersen, B. K., N. A. D'Elia, and K. L. Nelson. 1994. *Bioprocess Engineering: Systems, Equipment and Facilities.* John Wiley & Sons: New York.

# CHAPTER 7
# Biological Production of Hydrogen

## 7.1 Introduction

Hydrogen gas has great potential as a sustainable, environmentally friendly alternative fuel[1,2,3] because it combusts to form only water and energy, as shown in Eq. (7.1):

$$H_2 + \tfrac{1}{2} O_2 \rightarrow H_2O \tag{7.1}$$

The energy density of hydrogen (143 kJ/g) is the highest among energy fuels. Hydrogen can be produced through a variety of physical, chemical, or biological methods. Currently, 95 percent of global hydrogen production is from fossil fuel sources, primarily via the thermocatalytic reformation of natural gas.[3] Biological production contributes a small portion to the overall total, but this portion is expected to grow as techniques and systems improve.

Biological production of hydrogen can be carried out by two main mechanisms—photobiological production and fermentation.[4,5] Photobiological hydrogen production has the advantage that it utilizes solar radiation, a free energy source, to drive the process but advanced reactor designs are required to achieve moderate solar radiation conversion efficiencies and $H_2$ production rates. Fermentation processes can utilize free carbon/energy sources in agricultural by-products or wastes, but not all fermentative microbes are able to hydrolyze the complex lignocellulosic materials contained in many feedstocks. Therefore, both photobiological and fermentation processes hold great promise as future contributors to global $H_2$ production, but technical challenges remain that need to be overcome.

### 7.1.1 Important Enzymes

Biological hydrogen production reactions carried out in photobiological and fermentation pathways are dependent on hydrogenase or nitrogenase enzymes that catalyze the reduction of protons (Eq. 7.2):

$$2H^+ + 2e^- \leftrightarrow H_2 \qquad (7.2)$$

In general, hydrogenase enzymes can catalyze the reaction in either direction in vitro. In vivo, they primarily catalyze either production or uptake (oxidation) of hydrogen depending on the needs of the host organism, although they may reverse roles under inhibitory $H_2$ concentrations. In fermentative bacteria, hydrogenase enzymes catalyze the reduction of protons to molecular hydrogen as means of accepting excess electrons produced from the oxidation of organic substrates, while for photosynthetic bacteria, hydrogenases catalyze the use of hydrogen as an electron donor.[6]

At present, several classes of enzymes have been identified: nitrogenases, Fe-hydrogenases, NiFe-hydrogenases (including NiFeSe-hydrogenases), and nonmetal hydrogenases.

In general, Fe-hydrogenases are found primarily in the Domains Bacteria and Eucarya, while the NiFe-hydrogenases are primarily in Archaea and Bacteria.[6] NiFe-hydrogenases are typically characterized as uptake hydrogenases, and thus exhibit a low rate of $H_2$ evolution in vitro.[7] Fe-hydrogenases catalyze $H_2$ evolution at high rates, which is consistent with their physiological role. The turnover rates for the production of $H_2$ are on the order of 0.6, 100, and 6000/s for nitrogenase, NiFe-hydrogenase, and Fe-hydrogenase respectively,[5] indicating in general the orders of magnitude greater efficiency of Fe-hydrogenases than nitrogenases in $H_2$ production. Cyanobacteria contain two types of hydrogenase enzymes, an uptake hydrogenase induced under nitrogen fixing conditions, and a reversible (bidirectional) hydrogenase able to both produce and oxidize hydrogen. Cyanobacteria *Aphanocapsa variabilis, Synechocystis* sp., and *Anacystis nidulans* contain bidirectional NiFe-hydrogenases.[6] For Fe- and NiFe-hydrogenases obtained from fermentative bacteria *Desulfovibrio* and *Clostridium*, the activity ratio of $H_2$ evolution/$H_2$ uptake in vitro is less than 1 (~ 0.25 to 0.3).[7] However, the NiFe-hydrogenases extracted from the hyperthermophilic fermentative archeon *Pyrococcus furiosis* catalyze $H_2$ evolution at rates comparable to Fe-hydrogenases, and with a high activity ratio of $H_2$ evolution/$H_2$ oxidation of ~11 at 80°C,[7] indicating the high affinity of these hydrogenases for $H_2$ generation. Hydrogenase from *P. furiosis* is one of few hydrogenases that utilizes NADPH rather than NADH as the electron donor—a mesophilic form of this hydrogenase has not been identified.[8] Several structural genes encoding for Fe-hydrogenases in the fermentative bacteria *Thermotoga maritima* have been identified.[6] Hydrogenase activities can vary greatly even for closely related species. For example, hydrogenase activity was found to be 42 times higher for *Thermotoga neapolitana* than for *T. maritima*.[9]

### 7.1.2 Abiotic $H_2$ Production

The production of $H_2$ from organic compounds utilizing enzymes alone to catalyze the reaction without live cultures has not been widely studied. Woodward et al.[8] reported yields of 11.6 mol $H_2$ and 6 mol $CO_2$ per mol of glucose-6-phosphate (G6P) using two enzymes from the oxidative branch of the pentose phosphate cycle, G6P-dehydrogenase and 6-phosphogluconate dehydrogenase, plus hydrogenases obtained from *P. furiosis*. This yield represents 97 percent of the theoretical maximum yield of 12 mol $H_2$/mol glucose, indicating the reaction proceeds to near completion as a result of the rapid oxidation of NADPH by hydrogenase. This data suggests that higher yields of $H_2$ production may be possible from enzymatic reactors than bioreactors. However, although the rates of hydrogenase activities vary widely, there is no evidence to suggest that the quantity of hydrogenases in microbial cultures is the limiting factor.[5] Thus, addition of hydrogenase enzymes may not increase the efficiency of $H_2$ production in microbial reactors.

## 7.2 Photobiological $H_2$ Production

Photobiological hydrogen production has great potential because the energy available in solar radiation is immense. Yearly average solar irradiation can be as high as 5 kWh/m$^2$-day or 6.6 GJ/year ($10^9$ J/year).[5] At a high solar radiation conversion efficiency of 10 percent for light-driven $H_2$ production and a price $15/GJ for $H_2$ energy approximately $10/m$^2$-year would be generated from $H_2$ production.[5] At a more conservative conversion efficiency of 3 percent and price of $5/GJ, only $1/m$^2$-year would be generated.[5] The main limiting factor for all light driven processes is the low conversion efficiency of high-intensity solar energy. Under full sunlight conditions, in oxygenic (oxygen generating) photosynthesis, the rate of light energy capture by photosynthetic pigments is 10 times greater than the rate of transfer of electrons from Photosystem II to I. This means that ~90 percent of solar radiation captured is not utilized in photosynthesis; rather, it is released primarily as heat. To avoid light saturation, advanced reactor designs may use rapid mixing to take benefit of the "flashing light effect"—that is, the rapid exposure of algal cells to light to maximize light capture. However, water velocities generated by mechanical mixing should not exceed 25 to 35 cm/s to avoid excessive energy required for mixing.[5] Another strategy for avoiding light saturation is the use of light attenuation devices that transfer solar radiation to deeper depths of the algal culture. In these designs, large concentrating mirrors collect light energy and then channel the energy through optical fibers to the depths of the reactor. Although first demonstrated in the 1970s and currently under investigation, these designs have not yet proven technically or economically practical.[5]

Therefore, the goal of much photosynthetic $H_2$ production research is to increase solar energy conversion efficiencies beyond the 3 percent currently achievable with outdoor microalgal cultures at high solar intensity. The photo-dependent hydrogen processes include direct biophotolysis, indirect biophotolysis, and photofermentation.[4,5]

### 7.2.1 Direct Biophotolysis

In oxygenic photosynthesis, two photosystems are used (see Chap. 2). Light energy that strikes Photosystem II (PSII) is used to split water molecules [Eq. (7.3)] generating oxygen, protons, and electrons.

$$2H_2O \xrightarrow{\text{light energy}} O_2 + 4H^+ + 4e^- \qquad (7.3)$$

The electrons are accepted by PSII, reducing it to a potential of approximately −0.8 V. The electrons then flow through a series of electron carrier molecules to Photosystem I (PSI). Acceptance of the electrons at PSI reduces it to a potential that is sufficient to reduce ferrodoxin, which in turn reduces $NADP^+$ to NADPH. In the typical process for photoautotrophic growth, the NADPH is then used to reduce inorganic carbon for synthesis of new cell mass. This reaction is ubiquitous in algae and cyanobacteria, so its potential for energy production is great if it could be fully exploited.

In direct biophotolysis, an alternative route for the electrons is utilized. The electrons transferred to ferrodoxin in PSI are transferred to protons instead of $NADP^+$, reducing the protons to hydrogen gas through the activity of Fe-hydrogenase enzyme. The Fe-hydrogenase enzymes needed to catalyze this reaction are found in the chloroplasts of green algae such as *Scenedesmus obliquus*, *Chlamydomonas reinhardtii*, and *Chlorella fusca*[6] and in many cyanobacteria. The key problem with direct biophotolysis is that the Fe-hydrogenases that catalyze the reduction of $H^+$ are extremely sensitive to inhibition by oxygen. Thus, direct biophotolysis requires photobioreactor designs that allow for production, capture, and separation of $H_2$ and $O_2$.

Hydrogen production by direct biophotolysis has been demonstrated. A photon energy conversion efficiency of 22 percent of visible light energy into $H_2$ was demonstrated in vivo under low light conditions and low oxygen partial pressures with the green microalga *Chlamydomonas reinhardtii*.[5] Visible light, also known as the *photosynthetically active region*, encompasses the wavelength range of 400 to 700 nm and 50 percent of total solar radiation energy. Therefore, the equivalent solar energy conversion efficiency for these results translates to a total energy conversion of 11 percent.

Despite innovative photobioreactors designed to maximize $H_2$ production and attempts at molecular engineering of hydrogenases that are tolerant of oxygen, obstacles to efficient hydrogen production by direct biophotolysis remain.[5]

## 7.2.2 Indirect Biophotolysis

The process of indirect biophotolysis is accomplished by separating $H_2$ and $O_2$ production both temporally and spatially. In the first step, light energy is used to produce oxygen and stored carbohydrate [Eq. (7.4)], typically starch in green algae and glycogen in cyanobacteria.[10] Imposing nitrogen limitation triggers accumulation of large amounts of stored carbohydrate in both green algae and cyanobacteria, and results in the loss of oxygen evolving capacity. Then, in the second step, stored carbohydrates are converted to $H_2$ and $CO_2$ in a light-driven process under oxygen-depleted conditions [Eq. (7.5)].

$$6CO_2 + 12H_2O \xrightarrow{\text{light energy}} C_6H_{12}O_6 + 6O_2 \quad (7.4)$$

$$C_6H_{12}O_6 + 6H_2O \xrightarrow{\text{light energy}} 12H_2 + 6CO_2 \quad (7.5)$$

Cyanobacteria, such as *Aphanocapsa montana*, *Anabaena variabilis*, and *Spirulina* produce hydrogen through indirect biophotolysis.[4,10]

Alternatively, the stored carbohydrate could be fermented to 4 mol of hydrogen gas per mol of stored glucose in a dark anaerobic fermentation process. One proposed strategy would utilize a four-step process:[5,10] (1) photosynthetic production of algal biomass with high stored carbohydrate content and a potential 10 percent solar energy conversion in a low-cost, open raceway system; (2) concentration of algal biomass; (3) anaerobic, dark fermentation of stored carbohydrate in concentrated algal biomass to hydrogen gas and acetate (4 mol $H_2$ and 2 mol acetate produced per mol glucose fermented); and (4) final phase utilizing closed photobioreactors for indirect biophotolysis to use light-driven conversion of remaining stored carbohydrate and acetate produced by fermentation to hydrogen gas.

## 7.2.3 Photofermentation

Photosynthetic bacteria such as *Rhodopseudomonas capsulate* or *R. sphaeroides*, and other purple non-sulfur bacteria have the capacity to produce $H_2$ through the action of nitrogenase enzyme in a process called photofermentation.[11] In this process, organic substrates are oxidized under anaerobic conditions using light energy when nitrogen limitation exists. Excess $e^-$ are transferred to ferrodoxin and then to protons, producing $H_2$ gas as catalyzed by nitrogenase. The $H_2$ producing activity of nitrogenase is stimulated by light, but is strongly inhibited by oxygen, ammonium and nitrogen gas, and requires substantial input of energy.

$$2H^+ + 2e^- + 4ATP \xrightarrow{\text{light energy}} H_2 + 4ADP + 4P_i \quad (7.6)$$

Hydrogenase enzymes, which play a critical role in $H_2$ production in direct biophotolysis, primarily act to consume $H_2$ in photofermentation. Mutated cells with reduced production of hydrogenase result in increased $H_2$ production.[11] Organic substrates such as lactate or malate result in greater $H_2$ production rates than sugars such as glucose or sucrose. $H_2$ production rates of 0.016 to 0.035 L/L-hr (0.7 to 1.6 mmol/L-hr) have been reported for *R. sphaeroides* using malate or lactate substrates under indoor lighting and batch conditions.[11] Although wastewaters from dairy, tofu, or sugar processing can be used as substrates, lower rates of $H_2$ production were achieved.[11] Reported solar conversion efficiencies range from 1 to 5 percent on average.[11] Therefore, conditions required for optimum $H_2$ production by photofermentation include maximal nitrogenase activity, minimal hydrogenase activity, use of favorable substrates such as lactate or malate, high carbon to nitrogen ratio in media and light distribution throughout the culture depth.[11]

### 7.2.4  Photobiological $H_2$ Production Potential

Because of the low solar energy conversion of all photobiological $H_2$ production, Hallenbeck and Benneman[5] conclude that continued basic research and development are needed before photobiological hydrogen production can be fully utilized, and that conversion of organic substrates to $H_2$ through dark fermentation processes is more efficient and has greater potential for expanding biohydrogen production. Biohydrogen production by fermentation is the focus of the majority of this chapter.

## 7.3  Hydrogen Production by Fermentation

### 7.3.1  Overview

Hydrogen gas is produced under anaerobic conditions by certain chemoorganotrophic microorganisms that use organic substrates as their carbon and energy source and hydrogen ion ($H^+$) as electron acceptor. The biological production of hydrogen by fermentation is generally associated with the presence of an iron-sulfur protein called ferredoxin, an electron carrier of low redox potential. Hydrogen can be produced by mesophilic (25 to 40°C), thermophilic (40 to 65°C), extreme thermophilic (65 to 80°C), and hyperthermophilic (>80°C) microorganisms of the Archaea and Bacteria Domains. In the hyperthermophilic group, most of the $H_2$-producers are archaea, but one group, the Thermotogales, are bacteria. In addition to $H_2$, these fermentations result in the formation of a variety of organic acids such as acetate, butyrate, lactate, or propionate, along with lesser amounts of alcohols such as ethanol. The amounts and rates of formation of each of these products depends on the organism and substrate used, as well as culture conditions such as dissolved $H_2$ concentration, pH, and culture temperature.

## 7.3.2 Energetics

In theory, the microbial fermentation of glucose ($C_6H_{12}O_6$) to hydrogen could occur as;[12,13]

$$C_6H_{12}O_6 + 6H_2O \rightarrow 6CO_2 + 12H_2 \quad \Delta G^{\circ\prime} = -25.9 \text{ kJ/reaction} \quad (7.7)$$

In this reaction, 99 percent of the energy originally present in glucose is contained in the 12 mol $H_2$ produced. This is calculated by comparing the energy released from glucose combustion vs. hydrogen combustion:[12,13]

$$C_6H_{12}O_6 + 6O_2 \rightarrow 6CO_2 + 6H_2O \quad \Delta G^{\circ\prime} = -2872 \text{ kJ/reaction} \quad (7.8)$$

$$12H_2 + 6O_2 \rightarrow 12H_2O \quad \Delta G^{\circ\prime} = -2846 \text{ kJ/reaction} \quad (7.9)$$

The standard free energy of formation ($\Delta G^{\circ\prime}$) for Eq. (7.7) is believed to be too small to allow for microbial growth; therefore, this pathway is not known to occur in microbial systems. Under physiological conditions, Thauer et al.[13] estimated that 42 to 50 kJ/reaction are required for ATP synthesis at 100 percent efficiency and equilibrium conditions, and approximately 63 kJ/reaction under nonequilibrium conditions. By considering only microbial pathways that are thermodynamically favorable, Thauer et al.[13] predicted that a maximum of 4 mol $H_2$/mol glucose could be produced by microbial fermentation [Eq. (7.10)]:

$$C_6H_{12}O_6 + 2H_2O \rightarrow 2CH_3COO^- + 2H^+ + 2CO_2 + 4H_2$$
$$\Delta G^{\circ\prime} = -216 \text{ kJ/reaction} \quad (7.10)$$

The $\Delta G^{\circ\prime}$ value for this reaction (also reported as $-207$ kJ/reaction[14] for standard conditions of 25°C, 1 M solutes and 101 kPa total gas pressure) is sufficient for ATP production and microbial growth, and has been found to occur in natural and bioreactor systems. But, the $\Delta G$ values reported for actual culture conditions are more exergonic, and therefore more favorable to $H_2$ formation. For example, Schroder et al.[15] found that $\Delta G = -290$ kJ/mol glucose for culture conditions of 85°C, 10 mM glucose, 3 mM acetate, and 1 percent $H_2$ (gas phase), while Zinder[14] reported $\Delta G = -319$ kJ/reaction at conditions of 10 µM glucose and 0.01 kPa for partial pressure of $H_2$.

In addition, other researchers cite a critical value of only 19 kJ/reaction for $\Delta G^{\circ\prime}$ as the minimum required for microbial growth.[16] Therefore, the maximum ATP production by fermentation of glucose could potentially be greater than 4 mol/mol glucose.

### 7.3.3 Thermotogales

Hydrogen production by the hyperthermophilic bacteria of the order Thermotogales will be the focus of the fermentation section. Thermotogales consists of one family, Thermotogaceae, and six genera, including *Thermotoga*. The genus is named because it has a bag-shaped outer structure called "toga" surrounding its cell.[17] There are nine species of *Thermotoga* (*T. maritima T. neapolitana, T. thermarum, T. elfii, T. subterranea, T. hypogea, T. petrophila, T. naphthophila,* and *T. lettingae*), all of which are obligate anaerobic bacteria that ferment glucose to acetate, $CO_2$, and $H_2$. The first two described members, *T. maritima* and *T. neapolitana*, are the most closely related based on 16S rRNA gene sequence analysis. *T. maritima* was originally isolated from a geothermally heated, shallow marine sediment at Vulcano, Italy[17] while *T. neapolitana* was obtained from a submarine hot spring near Lucrino, the Bay of Naples, Italy.[18,19] Other species of *Thermotoga* have been isolated from freshwater environments. All *Thermotoga* spp. have temperature optima of 65°C or greater, and four of the most closely related (*T. neapolitana, T. maritima, T. petrophila,* and *T. naphthophila*) have optima at or near 80°C.[20]

### 7.3.4 Biochemical Pathway for Fermentative $H_2$ Production by *Thermotoga*

In the hyperthermophilic bacterium *T. maritima*, fermentation of glucose to pyruvate was determined to occur primarily via the Embden-Meyerhoff pathway (EMP)[15] with the transfer of electrons from ferredoxin to H+ catalyzed by hydrogenase enzyme (Fig. 7.1). Additionally, approximately 15 percent of their energy production derives from the Entner-Duodoroff (ED) pathway.[21] This contrasts with the hyperthermophilic archaeon that ferment glucose to acetate, $CO_2$, and $H_2$ primarily via a modified ED pathway.[15] Four mol of ATP/mol glucose were formed by *T. maritima*,[15] which is 2 ATP higher than many other fermentation processes.

The electrons removed as the organic substrate is oxidized are transferred to hydrogen ion. This use of hydrogen ion (H+) as an inorganic electron acceptor means that organic substrates do not serve as internal electron acceptors and thus reduced organic compounds are not formed. The oxidation states of carbon in glucose, acetate ($CH_3COO^-$), and carbon dioxide are 0, 0, and +4, respectively. One-third of the carbon contained in glucose (2 mol) becomes oxidized to $2CO_2$, with the removed 8e− transferred to 8H+, forming 4 mol of $H_2$. The other 4 mol of C contained in glucose are converted to acetate, with no gain or loss of electrons. Thus, $H_2$ production via the acetate pathway represents an atypical fermentation reaction. In most fermentation processes, inorganic electron acceptors are not used and instead, a metabolic organic intermediate compound becomes reduced in order to absorb electrons and regenerate NAD+. For

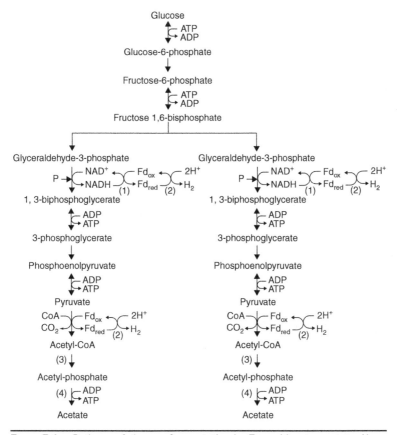

**FIGURE 7.1.** Pathway of glucose fermentation by *T. maritima* to acetate, $H_2$, and $CO_2$. Reactions mediated by (1) NADH: ferredoxin oxidoreductase and (2) hydrogenase; (3) phosphate acetyltransferase; and (4) acetate kinase. Adapted from Schroder et al.[15]

example, in ethanol fermentation, acetaldehyde is reduced to ethanol by NADH to reform $NAD^+$. Because of the differences between the hydrogen fermentation pathway and other fermentations, the term anaerobic oxidation[22] has been used to describe the hydrogen fermentation pathway.

### 7.3.5 Hydrogen Production by Other Bacteria

Thermophilic bacteria such as *Clostridium thermolacticum*,[23] *C. thermocellum*,[24] *Caldicellulosiruptor saccharolyticus*,[25] *Acetothermus paucivorans*,[26] *Acetomicrobium flavidum*,[27] and *Thermoanaerobacterium thermosaccharylyticum*[23,28] have been reported to produce $H_2$ (Table 7.1) Many other mesophilic organisms ferment glucose to produce $H_2$ and other reduced products such as ethanol, propionate, butanol, and lactate

| Organism | Temp (°C) | Peak H$_2$ produced (mmol H$_2$ gas/L-hr) | $Y_{P/S}$ (mol H$_2$/ mol hexose) | Substrate | Reactor | Ref |
|---|---|---|---|---|---|---|
| Mixed mesophilic culture | 26 | 1.8 | 0.92 | 4 g/L glucose | Batch | 29 |
| Thermotoga maritima | 80 | 1.3[a] | 4.3 | 2.2 g/L glucose | Batch | 15 |
| Thermotoga elfii | 65 | 1.2 | 3.3 | (a) 9 g/L glucose | Batch | 30 |
| Thermotoga elfii | 65 | 1.3 | 2.8 | (b) Miscanthus hydrolysate[c] | | |
| Caldicellulosiruptor saccharolyticus | 70 | 4.1[a] | NR | 10 g/L sucrose | Batch | 25 |
| Caldicellulosiruptor saccharolyticus | 70 | 8.4 | 3.0[b] | 10 g/L sucrose | Batch | 31 |
| Thermotoga elfii | 65 | 2.7 | 3.3 | 10 g/L glucose[d] | | |
| Thermotoga elfii | 65 | 4.5 | 2.8 | 10 g/L glucose[e] | | |
| Thermotoga neapolitana | 77 | 4.5 | 2.5 | 5 g/L glucose | Batch | 32 |
| Clostridium thermolacticum DSM 2910 | 58 | 3.1[a,f] | 3.2[a,f] | 20 g/L lactose | Batch | 23 |
| Thermoanaerobacterium thermosaccharylyticum | 58 | NR | 3.7[d] | 20 g/L milk permeate | Batch | 23 |

[a]Approximate value—calculated from data.
[b]equivalent $Y_{H2/sucrose}$ = 5.9 mol/mol.
[c]contained ~10 g/L glucose + 3.3 g/L xylose.
[d]media with tryptone.
[e]media without tryptone.
[f]equivalent $Y_{H2/lactose}$ = 7.4 mol/mol.
NR—not reported.

TABLE 7.1  A Comparison of Hydrogen Production Rates for Various Microorganisms

depending on conditions.[33,34,35] *Clostridium pastueurianum* is a classic high $H_2$ and volatile fatty acid producer when $H_2$ accumulates in gas phase. For example, *C. pasteurianum* ferments 1 mol glucose to 2.6 mol $H_2$, 0.6 mol acetate, 2 mol $CO_2$, 0.7 mol butyrate, and 3.3 mol ATP when the partial pressure of $H_2$ gas (p$H_2$) is high (101 kPa). Under low hydrogen partial pressures, *C. pasteurianum* ferments glucose to 4 mol $H_2$, 2 mol acetate, 2 mol $CO_2$, and 4 mol ATP.[15] Many mixed cultures of heterotrophic bacteria are also able to ferment glucose to acetate and $H_2$. Lower $H_2$ molar ratios were found using mixed mesophiles obtained from soil sources grown on glucose and sucrose substrates.[29]

Still other microbes are capable of producing acetate and $H_2$ from propionate, butyrate, or ethanol [Eq. (7.7)] when cocultured in syntrophic conditions with $H_2$ consumers, so that the $H_2$ partial pressure is very low.[36] Syntrophy is the synergistic relationship established between microbial groups that produce a product and those that consume that product. Syntrophic $H_2$ production is thermodynamically unfavorable, except when the $H_2$ partial pressure is maintained at very low levels. *Syntrophobacter* and *Syntrophomonas* are genera of bacteria that produce $H_2$ only under syntrophic conditions.

Another common pathway results in the formation of butyrate as the main product, with the theoretical yield of 2 mol $H_2$/mol glucose.

$$C_6H_{12}O_6 + 2H_2O \rightarrow CH_3CH_2CH_2COO^- + 2HCO_3^- + 3H^+ + 2H_2 \qquad (7.11)$$

Again, the $\Delta G^{c'}$ for standard conditions (−135 kJ/reaction) is less favorable for $H_2$ production than the reported $\Delta G$ value (−284 kJ/reaction) under nonstandard conditions (10 µM glucose and $H_2$ partial pressure of 0.01 kPa).[14] Clearly, it is essential to maintain low $H_2$ concentration to maximize $H_2$ production.

If ethanol and acetic acid are the end-products, 2 mol $H_2$ are produced from fermentation of 1 mol of glucose.[15,35] If propionic acid is the end-product of dark fermentation, no hydrogen is produced.[34,37]

## 7.3.6 Coproduct Formation

The conversion of glucose to methane in anaerobic digestion results in 85 to 97 percent recovery of energy.[13] The 4 mol of $H_2$ produced in the acetate pathway contain only 33 percent of the energy contained in glucose. However, in the acetate pathway, 2 mol acetate ($CH_3COO^-$) are formed per mol of glucose. Acetate is an economically valuable and recoverable product. Current industrial production of acetate is primarily achieved by petrochemical routes.[38] An important industrial use of acetate is for acetate deicers. These noncorrosive and environmentally friendly deicer compounds include calcium magnesium acetate for road deicer and potassium acetate and sodium acetate as airport runway deicers. Much of the current research into biological acetate production has focused on conversion of glucose to acetate[39,40] or lactose to acetate.[23,41,42] Biological $H_2$ production from agricultural feedstocks with recovery of the coproduct acetate from the fermentation broth may represent an economical route of acetate production and coincides with the biorefinery concept for utilization of all recoverable products.

In addition, the acetate formed could be used as substrate for electrical energy generation in microbial fuel cells, or as a substrate for methane formation. Oh et al.[43] investigated the use of acetate-rich fermentative media as substrate for electrical energy generation in microbial fuel cells. (See Chap. 9 for discussion of microbial fuel cells.)

### 7.3.7 Batch Fermentation

Hydrogen gas is produced as $H^+$ acting as an electron acceptor is reduced. A typical batch growth curve for a culture of *T. neapolitana* at 77°C (Fig. 7.2) reveals an increase in cell mass with concomitant increase in $H_2$ concentration[32] that is characteristic of growth-associated products. Hydrogen concentrations accumulated to ~27 mM (mmol $H_2$ gas in headspace/L media), or partial pressure (at 25°C) of ~38 kPa. Although the media was buffered, the pH declined to suboptimum levels. Both pH decline and $H_2$ accumulation are inhibitory to cell growth, and both must be maintained at non-inhibitory levels in order to fully utilize substrate and optimize $H_2$ production. A pH decline from 7.3 to 5.9 had been shown to inhibit cell growth and divert fermentation pathways away from acetate/$H_2$ production and toward lactate production in *Clostridium thermo-lacticum* in batch culture without pH control.[23]

This data for *T. neapolitana* growth indicates an observed biomass yield ($Y_{X/S}$) of 0.248 g biomass/g glucose.[32] The observed hydrogen

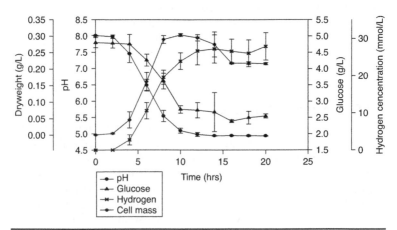

**FIGURE 7.2** Substrate, biomass, and hydrogen concentrations in batch *T. neapolitana*[32] fermentation at 77°C.

product yield ($Y_{P/S}$) from substrate is 0.0286 g $H_2$/g glucose or 2.55 mol $H_2$/mol glucose, indicating that $H_2$ production was inhibited and other reduced products such as butyrate or lactate were likely formed. The growth-associated product formation constant was calculated as 0.115 g $H_2$/g dry cell mass. The maximum specific growth rate ($\hat{\mu}$) with glucose as carbon source was 0.94 $hr^{-1}$ at 77°C, with half-saturation constant ($K_s$) of 0.57 g/L.

### 7.3.8 Hydrogen Inhibition

Inhibition of cell growth and $H_2$ production due to accumulation of dissolved $H_2$ in liquid is well established in a variety of $H_2$-evolving fermentation pathways due to inhibition of $H_2$-synthesizing hydrogenases. Hydrogen gas must be produced to continue to regenerate oxidized forms of NADH and ferrodoxin. The $H_2$ synthesis pathways are sensitive to high p$H_2$, and end-product inhibition will occur, shifting pathways away from $H_2$ production to more reduced substrates such as lactate, ethanol, acetone, or butanol.[44]

The p$H_2$ that is reported to inhibit $H_2$ production by fermentative organisms varies widely with species and culture conditions. The term "inhibition" is ambiguous, since for some it may indicate shifting of metabolic pathway to more reduced products, while for others it indicates complete cessation of growth. For example, Huber et al.[17] noted that p$H_2$ levels of 240 kPa inhibited growth of *T. maritima* in batch fermentations, while Schroder et al.[15] reported $H_2$ concentration of 6.9 mM at the end of batch fermentation of glucose by *T. maritima*. van Niel et al.[25] calculated the corresponding $H_2$ partial pressure (~2 kPa) for the Schroder data and concluded the $H_2$ gas partial pressure was inhibitory. However, other factors such as low pH or sulfur compounds in media (see Sec. 7.3.9) could have suppressed $H_2$ formation in the original fermentation. van Niel et al.[25] determined that p$H_2$ of 10 to 20 kPa caused a metabolic shift toward production of lactate for *C. saccharolyticus* in batch fermentation at 70°C, with growth and $H_2$ synthesis completely inhibited at a p$H_2$ of 56 kPa for low cell density cultures. For cultures with higher cell density, a microbial inhibition model[45] [Eq. (7.12)] was used to model $H_2$ inhibition in *C. saccharolyticus* cultures:[25]

$$\mu = \frac{\hat{\mu} S}{K_S + S}(1 - P/P_{crit})^n \qquad (7.12)$$

The authors found that the value of $P_{crit}$ for $H_2$ was 27.7 mM for lag phase, 25.1 mM for early exponential phase and 17.3 mM for late exponential phase[25], indicating inhibition by $H_2$ became more pronounced as growth proceeded.

### 7.3.9 Role of Sulfur—Sulfidogenesis

Hyperthermophilic $H_2$-producing archaea, such as *Pyrococcus furiosus*, and bacteria, such as *T. maritima* or *T. neapolitana* have developed strategies for growth under $H_2$ inhibition conditions. Sulfur compounds such as elemental sulfur, polysulfides, and cystine can be used by some hyperthermophiles as alternative electron acceptors, forming hydrogen sulfide ($H_2S$) in a process called sulfidogenesis.[7,9] Elemental sulfur ($S^0$), with valence of zero, often exists as polysulfides in aqueous systems. The sulfur in $H_2S$ has a valence of $-2$; therefore, $2e^-$ are accepted per mol of $S^0$ reduced. Polysulfides are linear chains of typically 4 or 5 sulfur atoms in which the internal sulfur atoms have a zero valence,[9] while the two outer sulfur atoms are at a valence of $-1$. In water, cystine is formed as two cysteine molecules react with oxygen. Notably, sulfate, a common electron acceptor for facultative organisms, is not used by the hyperthermophilic $H_2$ producer *T. neapolitana* as alternative electron acceptor.[9,18]

In the $H_2$ producing *P. furiosis* and *T. neapolitana*, two cytoplasmic polysulfide reducing enzymes have been purified. The first, a bifunctional flavoprotein called sulfide dehydrogenase, oxidizes NADH or NADPH with electrons going to reduce $S^0$ or polysulfides to produce hydrogen sulfide.[9]

$$H^+ + \text{NADH (or NADPH)} + S^0 \xrightarrow{\text{sulfide dehydrogenase}} \text{NAD}^+ \text{(or NADP}^+) + H_2S$$

(7.13)

Under inhibitory $H_2$ conditions, this "facilitated fermentation" mechanism[9] serves to divert electrons away from $H_2$ production to a sulfur reduction pathway, forming hydrogen sulfide. Childers[9] found the specific activity of sulfide dehydrogenase in *T. neapolitana* was up to 48 times greater than the activity of sulfhydrogenase, indicating that facilitated fermentation may be the primary mechanism of sulfidogenesis for this organism. Polysulfide (Km 0.15 mM) is preferred over $S^0$ as electron acceptor, and the enzyme is more active with NADH (Km 0.03 mM) than NADPH (Km 0.41 mM) in *T. neapolitana* in contrast to *P. furiosis* where the enzyme is more active with NADPH. This same enzyme can also catalyze the acceptance of electrons from reduced ferrodoxin and donate to NAD$^+$.

The second enzyme, a hydrogenase called sulfhydrogenase, catalyzes the reduction of elemental polysulfide using $H_2$ as electron donor.

$$H_2 + S^0 \xrightarrow{\text{sulfhydrogenase}} H_2S \qquad (7.14)$$

Thus, when $H_2$ concentrations become inhibitory to the growth of the microorganism, this enzyme catalyzes the oxidation of $H_2$ with the electrons accepted by sulfur compounds present in the media, forming hydrogen sulfide in a "detoxification" mechanism.[7] This reaction may be the primary sulfidogenesis reaction for archaea. As a hydrogenase, this dual function enzyme also catalyzes the reduction of $H^+$ from NADPH to form $H_2$. Sulfhydrogenase activity was found to be similar for *T. neapolitana* and *T. maritima*.[9]

| | $V_{gas}/V_{liquid}$ ratio | Mol formed/mol glucose consumed | | | | | $10^8$ Cell | $Y_B$ (g cells/g glucose)* |
|---|---|---|---|---|---|---|---|---|
| | | Acetate | $H_2$ | $CO_2$ | Lactate | $H_2S$ | | |
| No $S^0$ in media | Large | 2 | 4 | 2 | 0.02 | – | 1.4 | 0.25 |
| | Small | 1 | 2 | 1 | 0.1 | – | 0.65 | NR† |
| $S^0$ in media | Large | 2 | 0.3 | 2 | ND‡ | NR | 1.4 | 0.23 |
| | Small | 2 | 0.5 | 2.1 | 0.02 | 3.6 | 1.5 | 0.24 |

*Calculated from reported data.
†Not reported.
‡Not detected.
Source: Adapted from Schroder et al.[15]

**TABLE 7.2** Decrease in $H_2$ Production Due to Use of Sulfur as Electron Acceptor

Due to these effects, sulfur compounds increased the growth rate and biomass concentration and decreased the rate of $H_2$ evolution achieved by *T. neopolitana* in batch culture when $H_2$ was at inhibitory levels.[9] Unlike most archaea, the sulfide reducing bacteria are not dependent on $S^0$ for growth in the absence of $H_2$ inhibition. Huber et al.[17] found that growth of *T. maritima* was completely inhibited by $H_2$ when grown in MMS media in absence of $S^0$ at $H_2$ partial pressure of 240 kPa (300 kPa total pressure measured at culture temperature). However, when $S^0$ was added to the media, $H_2$ production was reduced by approximately 40 percent, but growth continued even at $pH_2$ of 240 kPa (300 kPa total pressure).

Schroder et al.[15] found that $H_2$ accumulation in gas phase inhibited $H_2$ production and cell growth of *T. maritima* in batch fermentations at 80°C when no $S^0$ was in the media (Table 7.2). However, when $H_2$ was not at inhibitory levels due to large gas/liquid volume ratio (11.5/1), addition of $S^0$ did not stimulate growth but did reduce $H_2$ production as $e^-$ were diverted from $H^+$ as $e^-$ acceptor to $S^0$. When $H_2$ was inhibitory, $S^0$ in the media both stimulated growth and diverted $e^-$ to $S^0$.

For biological hydrogen fermentations, optimizing the rate of $H_2$ production for energy use is the goal of the reactor. Further, $H_2S$ gas content in the biogas will interfere with efficient hydrogen fuel cell operation. Therefore, control of headspace $H_2$ to optimize $H_2$ production and avoid $H_2S$ gas production is highly desirable.

Finally, use of Fe(III) as electron acceptor by *T. maritima* producing Fe(II) with $H_2$ as electron donor has also been documented when $H_2$ levels became inhibitory.[46] Other strategies used by these organisms for remediating inhibitory $H_2$ conditions may be found. These alternative routes diminish the rate of $H_2$ produced, so $H_2$ inhibitory conditions should be avoided in the reactor.

### 7.3.10 Use of Other Carbon Sources Obtained from Agricultural Residues

Thermotoga spp. can utilize various carbon sources.[17,18,19,20,47] *T. maritima* can utilize glucose, ribose, xylose, galactose, sucrose, maltose, starch, glycogen, yeast extract, and whole cell extracts of bacteria (such as *Lactobacillus*) and archae (such as *Methanobacterium*, etc.) for growth.[17] *T. neapolitana* was found to utilize carbon sources such as glucose, ribose, xylose, sucrose, maltose, lactose, galactose, starch, and glycogen, but not volatile acids or alcohols such as acetate, formate, pyruvate, propionate, ethanol, methanol, glycerol, glutamate, glycine, or casamino acids,[19] although corresponding $H_2$ production rates were not reported.

Hydrogen production rates for *T. neapolitana* using various carbon (Table 7.3) and nitrogen sources (Table 7.4) in batch culture at 77°C indicated that *T. neapolitana* was able to utilize several carbon sources effectively.[32] The highest $H_2$ production occurred with glucose, sucrose, and xylan. Cellobiose, a disaccharide subunit of cellulose, was a good carbon source for the bacterium. However, low $H_2$ production was achieved with cellulose, indicating hydrolysis pretreatments would be required for cellulose to serve as feedstock for $H_2$ production by *T. neapolitana* fermentation. Inhibition due to low pH or $H_2$ gas accumulation in headspace was not avoided in these trials.

| Carbon source* | $H_2$ in gas phase† (%) | Absolute total pressure† (kPa) | $H_2$ produced‡ (mmol $H_2$ gas/L) |
|---|---|---|---|
| Glucose | 28.9 | 142 | 30.4$^a$ |
| Sucrose | 26.4 | 141 | 27.4$^{ab}$ |
| Xylan | 26.4 | 139 | 27.0$^{ab}$ |
| Rice flour | 25.7 | 137 | 25.9$^{bc}$ |
| Cellobiose | 24.8 | 128 | 23.5$^{bc}$ |
| Starch | 24.4 | 120 | 22.3$^c$ |
| Corn starch | 23.4 | 129 | 22.3$^c$ |
| Xylose | 22.7 | 130 | 21.7$^c$ |
| Beet pulp | 16.0 | 108 | 11.9$^d$ |
| Cellulose | 3.03 | 108 | 2.41$^e$ |

*Media contained per liter: 5.0 g carbon source listed, plus 2.0 g of yeast extract, 2.0 g of trypticase, 1.0 g of $NH_4Cl$, 0.3 g of $K_2HPO_4$, 0.3 g of $KH_2PO_4$, 0.2 g of $MgCl_2 \cdot 2H_2O$, 0.1 g $CaCl_2 \cdot 2H_2O$, 10.0 g of NaCl, 0.1 g of KCl, 1.0 g of cysteine HCl, 0.121 g of trizma base, and 10.0 mL of vitamin solution and 10.0 mL of trace element solution as specified in DSMZ media 141,. 30 hr incubation time.
†Measured at 25°C.
‡Values with same letter are not significantly different (LSD, $\alpha = 0.05$).

**TABLE 7.3** $H_2$ Production from Batch *T. neapolitana* Fermentation Using Various Organic Carbon Sources[32]

| Primary nitrogen source* | H₂ in gas phase† (%) | Absolute total pressure† (kPa) | H₂ produced‡ (mmol H₂ gas/L) |
|---|---|---|---|
| Yeast | 30.0 | 128 | 28.2ᵃ |
| Soybean meal | 26.0 | 119 | 22.8ᵇ |
| Canola meal | 22.5 | 126 | 21.0ᵇ |
| Linseed meal | 21.0 | 113 | 17.5ᶜ |
| Fish meal | 18.2 | 104 | 14.0ᵈ |
| Cottonseed meal | 3.68 | 110 | 2.95ᵉ |

*Media contained primary nitrogen source as listed (2 g/L), plus 2 g/L trypticase, 5 g/L glucose and other components as described in Table 7.3.
†Measured at 25°C.
‡Means not sharing common letter differed significantly (LSD test, $\alpha = 0.05$).

**TABLE 7.4** H₂ Production from Batch *T. neapolitana* Fermentation Using Various Organic Nitrogen Sources[32]

A 20-hr incubation time was sufficient at 77°C for batch culture using glucose, xylose, xylan, cellulose, beet pulp pellet, or cellobiose as carbon source, but 36 hr were required when sucrose, rice flour, starch and corn starch were used as carbon source.[32]

Alternative nitrogen sources were investigated by combining various primary N compounds (2 g/L) with trypticase (2 g/L) as secondary N source.[32] H₂ production was greatest when yeast extract was used as the primary N source, although soybean and canola meal were also effectively utilized (Table 7.4).

When both yeast extract and trypticase were substituted with 4 g/L of alternative N sources (Table 7.5), H₂ production was highest

| Nitrogen source* | H₂ in gas phase† (%) | Absolute total pressure† (kPa) | H₂ produced‡ (mmol H₂/L) |
|---|---|---|---|
| Soybean meal | 23.0 | 119 | 20.28ᵇ |
| Canola meal | 20.0 | 110 | 16.28ᶜ |
| Linseed meal | 8.29 | 109 | 6.59ᵈ |
| Fish meal | 0.00 | 120 | < 0.01ᵈ |
| Cottonseed meal | 0.00 | 120 | < 0.01ᵈ |

*Media contained 4 g/L of listed nitrogen source, 5 g/L glucose and other components as described in Table 7.3.
†Measured at 25°C.
‡Means not sharing common letter differed significantly (LSD test, $\alpha = 0.05$).

**TABLE 7.5** H₂ Production from Batch *T. neapolitana* Fermentation Using Various Single Organic Nitrogen Sources[32]

with soybean meal and canola meal. These rates of $H_2$ production are approximately 29 percent and 43 percent lower with soybean meal and canola meal, respectively, than with the combined yeast extract and trypticase, but could represent a substantial savings in media requirements for economical $H_2$ production.

Cull peaches are an example of an easily-fermentable agricultural waste that can be utilized for $H_2$ production. In South Carolina in a typical year, more than 200 million pounds of peaches are harvested for the fresh market, with approximately 20 million pounds discarded due to bruising and other imperfections. Peaches contain 4.6 to 9.6 percent sugar (wet weight basis), primarily as glucose and fructose. The sugars present in cull peaches can be used by *T. neapolitana* to produce hydrogen gas. Fermentation data on peach media containing 5 g/L (dry weight) depitted, blended whole peaches and other components as described in Table 7.3 by *T. neapolitana* resulted in 30 percent $H_2$ gas accumulation as compared to 29.3 percent for glucose media.[48] These results suggest that fruit and vegetable proces-sing wastes could be explored for hydrogen production.[49]

Hemicellulosic hydrolysates from agricultural residues such as sugarcane bagasse, rice straw, and wheat straw are plentiful and contain significant quantities of xylose and glucose in their hemicellulose and cellulose fractions. These sugars are fermentable to various biofuels, but most research has been focused on ethanol production. One ton of milled sugar yields 180 to 280 kg of sugarcane bagasse, with the typical bagasse containing 19 to 24 percent hemicellulose, 32 to 48 percent cellulose, and 23 to 32 percent lignin.[50] This composition represents typically 71 percent total reducing sugars, 25 percent xylitol and 41 percent glucose on a dry weight basis.[51] A wide range of sugar concentration is present in bagasse hydrolysate due to varying hydrolysis conditions of temperature, time, and acid concentration. A typical sugarcane bagasse hydrolysate may contain 17 to 105 g/L xylose and 7 to 30 g/L glucose, while a typical rice straw hydrolysate may contain 16 to 79 g/L xylose and 4 to 23 g/L glucose.[52]

Organisms such as *Clostridium* can effectively ferment cellulosic materials to $H_2$. Using delignified wood as feedstock, *C. thermocellum* 27405 produced 2.5 mmol $H_2$/L for low substrate (0.1 g/L) media and 6 mmol $H_2$/L for high substrate (4.5 g/L) media. An average yield of 1.6 mol $H_2$/mol glucose equivalent was obtained for delignified wood.[24] Acetate and ethanol were produced at ratio of 4:3, as well as lactate and formate as fermentation end-products.

Limited research has been conducted on pretreatment/hydrolysis of feedstocks for $H_2$ production by hyperthermophiles. De Vrije et al.[30] compared mechanical pretreatment (milling and extrusion), chemical pretreatment (12 N NaOH @ 70°C for 4 hr) and enzymatic hydrolysis (cellulose and beta-glucosidase) for *Miscanthus* biomass

## Biological Production of Hydrogen

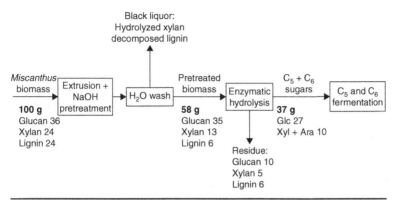

**FIGURE 7.3** Process for Treatment and Fermentation of *Miscanthus* Biomass to $H_2$. (*Adapted from de Vrije et al.*[30])

used as feedstock for *T. elfii* fermentation to $H_2$. They found the optimum process, consisting of 1 step extrusion-NaOH pretreatment, resulted in 77 percent delignification, with 95 percent cellulose and 56 percent hemicellulose yields (Fig. 7.3). After enzymatic hydrolysis, 69 percent of cellulose and 38 percent of hemicellulose was converted to glucose, xylose, and arabinose. Fermentation of the *Miscanthus* hydrolysate resulted in $H_2$ production rates equivalent to rates achieved with glucose media (approximately 1.3 and 1.2 mmol/L-hr, respectively). Utilization of both glucose and xylose occurred simultaneously and at approximately equal rates, with no clear preference by the organism for either substrate. An advantage of the extrusion pretreatment process is the moderate operation temperature used, preventing the formation of inhibitory compounds formed during other pretreatment processes.

### 7.3.11 Process and Culture Parameters

#### 7.3.11.1 Reactor Environment

Fermentative organisms that produce $H_2$ gas are obligate anaerobes, and hydrogen production by fermentative organisms is an anaerobic process. Thus, strictly anaerobic conditions must be initiated and maintained in the reactor vessel during production. Although some researchers have reported that low concentrations of oxygen are tolerated and even beneficial to growth and hydrogen production by *T. neapolitana*,[47,53] most have documented that the $H_2$ producers, including *Thermotoga* spp., only tolerate low levels of oxygen during storage[17] and that oxygen is strongly inhibitory to growth and oxygen production at culture temperatures (70 to 90°C).[17,19,32]

Anaerobic conditions can be initiated, maintained, and monitored in the reactor vessels by (1) flushing media with nitrogen gas,

(2) heating or boiling of the media to remove dissolved oxygen, (3) adding chemical agents such as sodium sulfite or cysteine hydrochloride to consume residual oxygen in the liquid, (4) adding resazurin to act as visible redox indicator, and (5) maintaining positive pressure in headspace to prevent air contamination.[54]

### 7.3.11.2 Hydrogen Gas Solubility and Recovery from Headspace

Hydrogen gas has a low solubility in water which facilitates its transfer to the gas phase and its recovery from the headspace of the bioreactor. The solubility of hydrogen is much less than ammonia, and about $1/10$ the solubility of methane or oxygen. As with most gases, the concentration of dissolved $H_2$ decreases with increasing salt content, and decreases with increasing temperature, although this effect is less pronounced than with oxygen or methane. The low solubility of hydrogen gas provides a competitive advantage for marine hyperthermophiles, as the elevated salinity and temperature of the culture environment reduce the equilibrium dissolved $H_2$ concentration, and therefore decrease the impact of inhibitory $H_2$ levels in solution.

Accumulated $H_2$ gas must be removed from the reactor headspace to avoid end-product inhibition of $H_2$-synthesizing hydrogenases. The main strategies for avoiding $H_2$ end-product inhibition are to flush gases out of headspace with $N_2$ or other inert gas or evacuate accumulated $H_2$ gas. $H_2$ partial pressures of <10 kPa may be needed to avoid $H_2$ gas inhibition.[25]

### 7.3.11.3 Reactor Temperature

Operating the reactor at optimum temperature for the microbial species cultured will maximize $H_2$ production rates. Optimum temperatures for *Thermotoga* spp. are shown in Table 7.6. Thermophilic fermentations have several advantages: (1) many industrial organic wastewaters, such as those arising from food processing facilities, are discharged at elevated temperatures; (2) elevated culture temperatures depress growth of contaminant bacteria,[55] and lessen or negate the need for sterilization of media and equipment; (3) thermophilic fermentation pathways tend to have fewer end-products and greater $H_2$ production rate; and (4) large scale fermentations generate excess heat that requires cooling for mesophilic cultures. Hyperthermophilic fermentations can make use of the biological heat generation to raise temperature of media to culture temperature.

### 7.3.11.4 Media

A variety of media have been used for culture of *Thermotoga* spp. (see summary in Huber and Hannig[20]). A typical media used by Yu[32] modified from van Ooteghem et al.[47] contained per liter: 5.0 g carbon source (glucose, starch, or xylose, etc.), 2.0 g of yeast extract, 2.0 g of trypticase, 1.0 g of $NH_4Cl$, 0.3 g of $K_2HPO_4$, 0.3 g of $KH_2PO_4$, 0.2 g of $MgCl_2 \cdot 2H_2O$, 0.1 g $CaCl_2 \cdot 2H_2O$, 10.0 g of NaCl, 0.1g of KCl, 1.0 g of cysteine HCl, 0.121 g of trizma base, and 10.0 mL of vitamin solution and 10.0 mL of trace element solution as described in DSMZ medium 141.

| Species | Temperature optimum (°C) | Temperature range (°C) | pH optimum | pH range | NaCl optimum (%) | NaCl range (%) |
|---|---|---|---|---|---|---|
| T. maritima | 77 | 55–90 | 6.5 | 5.5–9.0 | 2.7 | 0.25–6.0 |
| T. neapolitana | 77 | 55–90 | 7.0 | 5.5–9.0 | 2.0 | 0.25–6.0 |
| T. elfii | 66 | 50–72 | 7.5 | 5.5–8.7 | 1.0 | 0–2.4 |
| T. subterranea | 70 | 50–75 | 7.0 | 6.0–8.5 | 1.2 | 0–2.4 |
| T. petrophila | 80 | 47–88 | 7.0 | 5.2–9.0 | 1.0 | 0–5.5 |
| T. naphthophila | 80 | 48–86 | 7.0 | 5.4–9.0 | 1.0 | 0.1–6.0 |

Source: Adapted from Huber and Hannig.[20]

TABLE 7.6  Summary of Required Growth Conditions for *Thermotoga* spp.

### 7.3.11.5 Yeast Extract
Several researchers[15,19] found that a low concentration (~0.5 g/L) of yeast extract was required for growth for *T. maritima*. van Niel et al.[31] found that growth of *T. elfii* was also completely dependent on yeast extract. In the presence of 0.5 percent yeast extract as sole carbon source, Schroder et al.[15] found cell densities of 1 to 2 × 10$^7$ cells/mL were achieved on yeast extract alone, as compared to cell densities of 1.4 × 10$^8$ cells/mL that were achieved with ~2 g/L glucose concentration with yeast extract. Yu[32] found that yeast extract in combination with trypticase (2 g/L each) yielded highest production of $H_2$ for *T. neapolitana*, although alternative N sources such as soybean meal or canola meal alone supported growth with 30 to 50 percent reduction in $H_2$ production rate (see Table 7.5).

### 7.3.11.6 Salt Content
Most *Thermotoga* spp. require a minimum salinity level[20] (Table 7.6). Providing the lowest level while maintaining high $H_2$ production rates is desirable to decrease media cost. *T. neapolitana* growth occurs over a wide range of NaCl concentrations with peak growth at 2 percent (20 g/L).

### 7.3.11.7 pH Optimum
The pH optimum for *T. neapolitana* is 7.0, with range of 5.0 to 9.0, and pH control is critical to maintain cell growth and $H_2$ production (Table 7.6).

### 7.3.11.8 Cysteine Hydrochloride
Addition of cysteine hydrochloride ($C_3H_7NO_2S \cdot HCl$) at concentration of 0.5 to 1 g/L is common to provide reducing conditions in media and consume residual oxygen.[9,30,33,36,47,53] In water, 2 mol of cysteine react with 1 mol of $O_2$ to form 1 mol of cystine and 2 mol of water. Cysteine HCl contains <20 percent sulfur by weight, so this addition does not represent a substantial addition of sulfur. However, cystine is used preferentially by *Thermotoga* as electron acceptor, forming hydrogen sulfide instead of hydrogen gas. Therefore, minimum cysteine-HCl addition should be used to prevent $H_2S$ production or alternative methods for providing reducing conditions employed.

### 7.3.11.9 Storage and Preservation
Under anaerobic conditions, *Thermotoga* cultures can be stored for several months at 4°C, or for long-term preservation at –140°C in liquid nitrogen in presence of 5 percent dimethylsulfoxide.[20] *Thermotoga maritima* and *T. neapolitana* can be obtained from DSMZ— the German Resource Centre for Biological Material.

## 7.4 Hydrogen Detection, Quantification, and Reporting
Comparing and interpreting the results from reported fermentation processes is complicated by various reporting units and measurement

techniques used. Values of $H_2$ partial pressure and total pressure will be a function of the culture temperature, culture and headspace volume, concentration of substrate, and presence of elemental sulfur compounds in media. From a $H_2$ gas recovery point of view, the total gas pressure and $H_2$ concentration in the gas mixture are important values. From an engineering point of view for sizing fermentation reactors, the $H_2$ production rate per liter of reactor volume is needed. When reporting $H_2$ inhibition, the $H_2$ concentration in the aqueous phase is the most critical. Information is given below on how to measure, calculate, and report these various $H_2$ concentrations.

### 7.4.1 Hydrogen Detection

The $H_2$ concentration in the gas phase is commonly measured with gas chromatography (GC) with thermal conductivity detector (TCD), using argon or nitrogen as the carrier gas. Typical GC operating conditions include temperature of 100°C for the TCD and pressure of 151 kPa (22 psi) for the carrier gas. Silica columns (at 25°C) or microcapillary columns may be used for separation.

Headspace sampling for GC analysis can be completed as a static or dynamic process. In static headspace analysis, a sample of headspace gas is collected manually with a syringe and injected into GC for gas detection. The headspace may be at an elevated pressure due to fermentation gases and water vapor. Two types of gas sampling syringes may be used—a standard gas syringe or a gas-tight (pressure-lock) syringe. In both types of syringes, the sample of collected gas will equilibrate to the elevated pressure of the reactor headspace while the syringe needle is inserted into the headspace. In a standard syringe, after the syringe has been removed from the reactor, the gas in the syringe will expand and escape through the needle, and the sample in the syringe will equilibrate to atmospheric pressure.[56] If a gas-tight syringe is used, the elevated pressure in the syringe is maintained after it is removed. Thus, the means of sampling of the headspace gas will result in a different mass of gas injected into the GC for the same volume of gas injected, and therefore needs to be noted.

Once the sample is analyzed by GC, the percentage of $H_2$ in the sample is calculated by comparing the $H_2$ peak area on the chromatograph to those obtained for $H_2$ standards. Common means of preparing standards including standard addition, where known volumes of gas are added to samples, and external standards, where a series of gas solutions are prepared and analyzed. External standards should contain all gases that would be present in gas samples so are difficult to accurately prepare. Assuming the gas mixture in the headspace behaves in an ideal manner, the percentage of $H_2$ represents the mol fraction (mol $H_2$/mol gas mixture), the volume fraction (volume $H_2$/volume gas mixture), and the pressure ratio (partial pressure of $H_2$/total pressure of gas mixture).

### 7.4.2 Total Gas Pressure

The total gas pressure in the reactor can be measured with a manometer. The pressure reading of the manometer is typically displayed as gauge pressure, and should be corrected to absolute pressure by adding the equivalent of 1 atmosphere pressure to the gauge pressure measurement. Conversion for pressure measurements are 1 atm = 101.3 kPa = 760 mm Hg = 14.7 psi.

The total gases in the headspace will consist of (1) the inert gas used to flush headspace of reactor (i.e., $N_2$ or Ar); (2) the gases produced by fermentation ($H_2$, $CO_2$, CO, and/or $H_2S$); and (3) water vapor due to evaporation of water to gas phase. The partial pressure of each gas will increase with increasing temperature according to the ideal gas law. In addition, the water vapor pressure will increase with increasing temperature as calculated with Antoine's Equation [Eq. (7.15)]. Therefore, ideal gas law cannot be applied to the total pressure value for the headspace of a bioreactor, due to the impact of water vapor.

### 7.4.3 Water Vapor Pressure

To calculate the water vapor pressure for a given temperature, the Antoine equation[57] can be used:

$$\log_{10} p^* = A - \frac{B}{T+C} \qquad (7.15)$$

where $p^*$ = vapor pressure, mm Hg
$A$, $B$, and $C$ are constants
$T$ = temperature, °C.

The values of $A$, $B$, and $C$ for water vapor are given in Table 7.7.

The water vapor pressure calculated using Eq. (7.15) is low at ambient temperatures but increases exponentially at higher temperatures (Fig. 7.4).

### 7.4.4 Hydrogen Partial Pressure

The $H_2$ partial pressure for a given temperature is calculated as:

$$pH_2 = \%H_2 \cdot P_T \qquad (7.16)$$

| Temperature range (°C) | A | B | C |
| --- | --- | --- | --- |
| 0–60 | 8.10765 | 1750.286 | 235.000 |
| 60–150 | 7.96681 | 1668.210 | 228.000 |

*Values from Lange's Handbook of Chemistry,[58] as reported in Felder and Rousseau.[57]

TABLE 7.7 Values for Constants in Antoine Equation for Water Vapor*

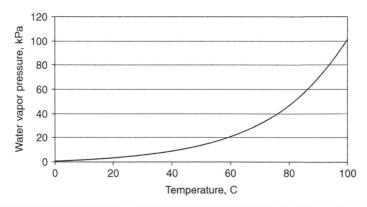

**FIGURE 7.4** Calculated water vapor pressure as function of temperature.

where $pH_2$ = $H_2$ partial pressure, kPa; and $P_T$ = total absolute pressure, kPa.

Because of the increase in water vapor pressure with increasing temperature, the percentage of $H_2$ in the fermentation gas will decrease as the temperature increases. Therefore, it is imperative that measurements of the total pressure of the fermentation gas and the $H_2$ detection with GC analysis are made at the same temperature, or erroneous $H_2$ partial pressure values will be calculated.

### 7.4.5 Hydrogen Gas Concentration

The concentration of $H_2$ gas in the headspace can be calculated through the ideal gas law relationship for gas mixtures using the gas partial pressure. First, the mol of $H_2$ gas in the headspace is calculated as:

$$PV = nRT \quad \text{or} \quad n_{H_2} = \frac{pH_2 \cdot V}{RT} \qquad (7.17)$$

where  $V$ = total volume of gas headspace, L
 $n_{H_2}$ = mol $H_2$ in gas mixture, mol
 $R$ = universal gas constant, 8.314 L-kPa/mol-K
 $T$ = gas temperature (at measurement), K

If the total pressure was measured at a temperature other than the culture temperature (i.e., if reactor was cooled prior to measurement), the $pH_2$ value at the actual culture temperature can be calculated as:

$$\frac{pH_{2,T1}}{T_1} = \frac{pH_{2,T2}}{T_2} \qquad (7.18)$$

The $pH_2$ value at the culture temperature is needed to determine the dissolved $H_2$ concentration in the media.

### 7.4.6 Hydrogen Concentration Expressed as mol $H_2$/L Media

To report the mol of $H_2$ produced per liter of culture volume, divide the mol of $H_2$ in the gas headspace by the liquid volume.

### 7.4.7 Hydrogen Production Rate

For design purposes, it is meaningful to report the rate of $H_2$ production expressed as mmol $H_2$/L-hr. For batch reactors, this is calculated as the increase in $H_2$ concentration in headspace over a defined period of time. For a continuous flow reactor, the $H_2$ production rate is calculated as the steady state concentration of $H_2$ gas divided by the hydraulic retention time.

### 7.4.8 Dissolved $H_2$ Concentration in Liquid

The equilibrium dissolved $H_2$ concentration can be calculated using Henry's law that relates the equilibrium concentration of dissolved gas in a liquid to the partial pressure of the gas in contact with the liquid. Henry's law is routinely presented in two forms, the first form is given below:

$$C_A = k_H \cdot p_A \qquad (7.19)$$

where $C_A$ = concentration of dissolved gas in liquid, M
$k_H$ = Henry's law constant at given temperature, M/kPa
$p_A$ = pressure of gas A in contact with liquid, kPa

The second form of Henry's law is as follows:[58]

$$p_A = H \cdot x_A \qquad (7.20)$$

where $H$ = Henry's constant at given temperature, kPa; and $x_A$ = mol fraction of dissolved gas A in solution, mol A/mol solution.

Conversion of $k_H$ value from $H$ value is as follows:[58]

$$k_H = \frac{1}{H} \cdot \frac{X \text{ mol solution}}{L} \qquad (7.21)$$

For dilute aqueous solutions, $X = 55.6$ mol $H_2O$/ L may be used.

In general, the solubility of gases decreases with increasing temperature and increasing salinity, and the value of the Henry's constant reflects these changes. Henry's constant, $k_H$, at a given temperature can be calculated from the $k_H$ value at a baseline temperature of 25°C as follows:[59]

$$k_H = k_H^\ominus \cdot \exp\left(\frac{-\Delta_{soln} H}{R} \cdot \left(\frac{1}{T} - \frac{1}{T^\ominus}\right)\right) \qquad (7.22)$$

| Temperature (°C) | Henry's constant, $k_H$ (M/kPa) | |
|---|---|---|
| | Calculated* | Tabulated† |
| 0 | 8.98E-06 | 9.47E-06 |
| 20 | 7.92E-06 | 8.03E-06 |
| 40 | 7.11E-06 | 7.31E-06 |
| 60 | 6.46E-06 | 7.17E-06 |
| 80 | 5.93E-06 | 7.26E-06 |
| 100 | 5.49E-06 | 7.36E-06 |

*Calculated using Eq. (7.23).[59]
†Converted tabulated values from International Critical Tables.[60]

**TABLE 7.8** Compilation of Henry's Constant Values

where  $k^\ominus_H$ = Henry's constant at 298.15 K (25°C), M/kPa
$\Delta_{soln}H$ = enthalpy of solution
$T$ = temperature, K
$T^\ominus$ = 298.15 K

The temperature dependence of $k_H$, $\left[\frac{d \ln k_H}{d(1/T)}\right]$ is equal to $\left[\frac{-\Delta_{soln}H}{R}\right]$. Therefore, Eq. (7.22) can be rewritten as:

$$k_H = k^\ominus_H \times \exp\left(\frac{d \ln k_H}{d(1/T)} \times \left(\frac{1}{T} - \frac{1}{T^\ominus}\right)\right) \qquad (7.23)$$

As reported in Sander,[59] the standard value for Henry's constant for $H_2$ gas, $k^\ominus_H$, is equal to 7.8 × 10⁻⁴ M/atm, and a typical value for $\left[\frac{d \ln k_H}{d(1/T)}\right]$ for $H_2$ gas is 500.

Utilizing this relationship, a table of Henry's constants for hydrogen as a function of temperature can be calculated (Table 7.8). These values reflect the decrease in $k_H$ as temperature increases.

However, tabulated values[60] for a temperature range of 0 to 100°C differ from the calculated values (Table 7.8 and Fig. 7.5). Because the tabulated values listed in the International Critical Tables[60] explicitly extend to 100°C, these are recommended for use at high temperatures.

**Example 7.1** The gas in the headspace of a 1.0-L bioreactor (0.8 L liquid volume and 0.2 L gas volume) operated as a batch reactor is sampled. The $H_2$ concentration of gas was determined at 70°C with GC to be 0 percent at the beginning of exponential phase and 25.5 percent after 1.5 hr. The gauge pressure in the headspace of the reactor at 1.5 hr is 175 kPa at the culture temperature of 70°C. Calculate the $H_2$ concentration (as mmol $H_2$ gas/L media), the rate of $H_2$ production (as mmol $H_2$ gas/L-hr), the equilibrium concentration of dissolved $H_2$ in the liquid (mmol dissolved $H_2$/L), and the water vapor pressure at culture temperature.

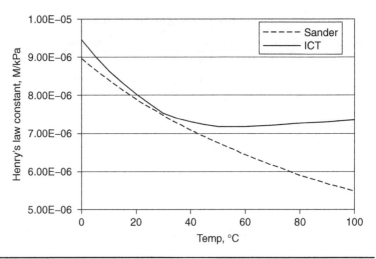

FIGURE 7.5  Comparison of Henry's law constant, $k_H$, values.

**Solution**

(1) Absolute total pressure $(P_T) = 175$ kPa $+ 101.3$ kPa $= 276.3$ kPa
(2) The $H_2$ gas partial pressure:

$$pH_2 = \%H_2 \cdot P_T = 0.255 \times 276.3 \text{ kPa} = 70.46 \text{ kPa}$$

(3) The mol of $H_2$ in gas phase:

$$n_{H_2} = \frac{PV}{RT} = \frac{pH_2 \cdot V_g}{RT} = \frac{(70.46 \text{ kPa}) \cdot (0.2 \text{ L})}{(8.314 \text{ L} \cdot \text{kPa/mol} \cdot \text{K}) \cdot (343 \text{ K})} = 4.94 \times 10^{-3} \text{ mol}$$

(4) The $H_2$ concentration (mmol $H_2$ gas/L media):

$$H_2 = \frac{n_{H_2}}{V_l} = (4.94 \text{ mmol}/0.8 \text{ L}) = 6.18 \text{ mmol } H_2/\text{L media}$$

(5) The rate of $H_2$ production:

$$r_{H_2} = \frac{H_2 \text{ concentration}}{\Delta t} = 6.18 \text{ mmol } H_2/\text{L media}/1.5 \text{ hr}$$

$$= 4.12 \text{ mmol H2/L-hr}$$

(6) Dissolved $H_2$ concentration calculated using Henry's law:
Henry's constant, $k_H$, at 70°C $= 7.21 \times 10^{-6}$ M/kPa (from Table 7.8)
$C_{H_2}$ in liquid $= k_H \cdot pH_2 = 7.21 \times 10^{-6}$ M/kPa·70.46 kPa $= 0.000508$ M $= 0.508$ mM
(7) Water vapor pressure @ 70°C calculated using Antoine equation:

$$\log_{10} p_{H_2O}* = 7.96681 - \frac{1668.21}{70°\text{C} + 228} = 2.369$$

The water vapor pressure $= p_{H_2O}* = 233.8$ mm Hg $= 31.2$ kPa.

## 7.5 Fermentation Bioreactor Sizing for PEM Fuel Cell Use

The ultimate goal of biohydrogen production is for use in hydrogen fuel cells for electrical energy generation. In the anode of the fuel cell, $H_2$ is oxidized as follows:

$$H_2 \rightarrow 2H^+ + 2e^- \tag{7.24}$$

Thus, 2e⁻ are generated for every molecule of $H_2$ reacted. The rate of hydrogen fuel needed by a proton exchange membrane (PEM) fuel cell to produce a desired current generated can be calculated as:[44]

$$R_{H_2} = \frac{I}{zF} \tag{7.25}$$

where $R_{H_2}$ = rate of $H_2$ fuel used, mol/s
$I$ = current, A (C/s)
$z$ = mol of electrons transferred per mol of fuel
$F$ = Faraday's constant, 96,485 Coulombs (C)/mol of electrons

For $H_2$ gas, $z = 2$.

The output power is given by

$$P_C = V_C In \tag{7.26}$$

where $P_C$ = output power of fuel cell, W
$V_C$ = voltage of a stack of cells, V
$n$ = number of cells

The fuel cell voltage, $V_C$, is selected to achieve desired efficiency at nominal power, and is typically in range of 0.6 to 0.8 V.[61]

Rearranging Eq. (7.26) yields:

$$I = \frac{P_C}{V_C n} \tag{7.27}$$

Substituting this equation into Eq. (7.25) yields

$$R_{H_2} = \frac{P_C}{V_C nzF} \tag{7.28}$$

To convert this fuel usage to a $H_2$ mass flow rate, multiply $R_{H_2}$ by the MW of hydrogen (2.02 g/mol). For a PEM fuel cell, the $H_2$ fuel usage can be calculated with Eq. (7.28), using an average $V_C$ value of 0.779 V.[44] For example, the $H_2$ fuel usage required for a single 1-kW PEM fuel cell ($P_C$ = 1000 W = 1000 J/s) is calculated as follows:

$$R_{H_2} = \frac{1000 \text{ J/s}}{(0.779 \text{ V}) \cdot (1) \cdot \frac{2 \text{ mol e}^-}{\text{mol H}_2} \cdot \frac{96,485 \text{ C}}{\text{mol e}^-}} = 6.65 \times 10^{-3} \text{ mol H}_2/\text{s}$$

$$= 48.4 \text{ g H}_2/\text{hr}$$

PEM fuel cells operate best at high flow rates of hydrogen. In practical terms, the limit to the hydrogen flow rate is 1.1 times the stoichiometric hydrogen required flow rate, therefore, 90 percent efficiency for $H_2$ utilization. Therefore, for a 1-kW PEM fuel cell, a practical $H_2$ gas flow rate would be 1.1 × 48.4 g $H_2$/hr = 53.2 g $H_2$/hr. For oxygen flow rate to the cathode, typically 2 times the stoichiometric utilization rate is needed.[61]

PEM fuel cells typically operate at a temperature of 60 to 80°C and require hydrogen gas saturated with water vapor prior to use.[61] Therefore, $H_2$ gas production by hyperthermophilic fermentation may not need heating or humidification prior to use. Carbon dioxide present in the gas mixture, and nitrogen gas used for flushing of the headspace will constitute diluents of the hydrogen gas. Hydrogen sulfide at concentrations above 50 ppm will interfere with efficient fuel cell performance, so it should be removed prior to fuel cell use.[62]

A 1-kW PEM fuel cell would generate 8760 kWh over the course of 1 year (1 kW × 365 days/year × 24 hr/day). According to the US DOE Energy Information Administration[63] the average yearly residential consumption in the United States in 2003 was 10,872 kWh (906 kWh/month), while the yearly average for South Carolina was 14,148 kWh. Therefore, a PEM fuel cell of 1.25 kW size would be needed to provide the total energy needs for an average home in the United States, while a 1.75-kW PEM fuel cell would be needed for the average home in South Carolina.

By considering the maximum rate of biohydrogen production (4.5 mmol $H_2$/L-hr or 9.09 mg $H_2$/L-hr) obtained for batch *T. neapolitana* cultures grown on 5 g/L glucose media,[32] the corresponding bioreactor volumes can be calculated for various PEM fuel cell sizes as follows (Table 7.9):

$$V = \frac{R_{H_2} \times \text{efficiency factor}}{H_2 \text{ production rate}} = \frac{48.4 \text{ g/hr} \times 1.1}{9.09 \times 10^{-3} \text{gH}_2/\text{L} \cdot \text{hr}} = 5850 \text{ L}$$

If greater $H_2$ production rates are achieved, the bioreactor volume needed would decrease. Simulations of batch and continuous flow stirred tank bioreactors (CSTRs) for fermentative $H_2$ production

| PEM fuel cell size (kW) | Stoichiometric $H_2$ flow rate required (g/hr) | Practical $H_2$ flow rate required (g/hr) | Fermentative bioreactor volume (L) |
|---|---|---|---|
| 1.0 | 48.4 | 53.2 | 5,850 |
| 2.5 | 121 | 133 | 14,600 |
| 5.0 | 242 | 266 | 29,300 |

TABLE 7.9 Calculated Sizes of Bioreactor Needed for Various PEM Fuel Cell Sizes

(Chap. 3) indicated that maximum $H_2$ production rates were predicted for a CSTR at a hydraulic retention of 1.6 hr. With 20 g/L glucose media and $H_2$ gas removal to prevent end-product inhibition, the predicted maximum $H_2$ production rates were 170 mmol/L-hr. At this high rate of $H_2$ production, the calculated reactor volume needed to fuel a 1.0-kW PEM fuel cell is 154 L. Optimization of bioreactor operation for $H_2$ production is needed to enable economical energy generation with a PEM fuel cell.

## Acknowledgment

The authors would like to thank Biosystems Engineering graduate student Dr. Xioahui Yu for his contributions to this chapter.

## References

1. Das, D. and T.N. Veziroglu. 2001. Hydrogen production by biological processes: a survey of literature. *International Journal of Hydrogen Energy.* 26(1):13–28.
2. Lee, S. 1996. *Alternative Fuels.* Taylor and Francis: Washington, DC.
3. Sperling, D. and J.S. Cannon, (eds.). 2004. *The Hydrogen Energy Transition: Moving Toward the Post Petroleum Age in Transportation.* p. 80.
4. Benneman, J.R. 2000. Hydrogen Production by Algae. *Journal of Applied Phycology.* 12:291–300.
5. Hallenbeck, P.C., and J.R. Benemann. 2002. Biological hydrogen production; fundamentals and limiting processes. *International Journal of Hydrogen Energy.* 27:1185–1193.
6. Vignais, P.M., B. Billoud, and J. Meyer. 2001. Classification and phylogeny of hydrogenases. FEMS Microbiology Reviews. 25(4):455–501.
7. Adams, M.W.W. 1990. "The metabolism of hydrogen by extremely thermophilic, sulfur-dependent bacteria." *Fems Microbiology Reviews.* 75:219–237.
8. Woodward, J. et al. 2000. Enzymatic production of biohydrogen. *Nature.* 405:1014–1015
9. Childers, S.E. 1997. Sulfur Reduction in the Hyperthermophilic Bacterium *Thermotoga neapolitana*. PhD Dissertation, University of Connecticut, CT.
10. Benneman, J.R. 2001. Biohydrogen Production: Final Summary Report 1996–2000. Report to the US Department of Energy Hydrogen Program.
11. Koku, H., et al. 2002. Aspects of the metabolism of hydrogen production by Rhodobacter sphaeroides. *International Journal of Hydrogen Energy.* 27:1315–1329.
12. Thauer, R. 1976. Limitation of microbial $H_2$-formation via fermentation. In: H.G. Schlegel, Barnea, J. (eds.) *Microbial Energy Conversion.* Erich Goltze KG: Gottingen, pp. 201–204.
13. Thauer, R., K. Jungermann, and K., Decker. 1977. Energy converservation in chemotrophic anaerobic bacteria. *Bacteriological Reviews.* 41:100.
14. Zinder, S. 1984. Microbiology of anaerobic conversion of organic wastes to methane: Recent developments. *Am. Soc. Microbiol. News.* 50:294–298.
15. Schroder, C., M. Selig, and P. Schonheit. 1994. Glucose fermentation to acetate, $CO_2$ and $H_2$ in the anaerobic hyperthermophilic eubacterium *Thermotoga maritima*—Involvement of the Embden-Meyerhof pathway. *Archives of Microbiology.* 161:460–470.
16. Fennell, D.E., and J.M. Gossett. 1998. Modeling the production of and competition for hydrogen in a Dechlorinating culture. *Environmental Science Technology.* 32:2450–2460.

17. Huber, R., et al. 1986. Thermotoga-Maritima Sp-nov represents a new genus of unique extremely thermophilic eubacteria growing up to 90-degrees-C. *Archives of Microbiology.* 144:324–333.
18. Belkin, S., C.O. Wirsen, and H.W. Jannasch. 1986. A new sulfur-reducing, extremely thermophilic eubacterium from a submarine thermal vent. *Applied and Environmental Microbiology.* 51:1180–1185.
19. Jannasch, H.W., et al. 1988. Thermotoga neapolitana sp. nov. of the extremely thermophilic, eubacterial genus *Thermotoga*. *Arch Microbiol.* 150:103–104.
20. Huber, R., and M. Hannig. 2006. Thermotogales. In: M. Dworkin (ed.). *Prokaryotes.* Springer: New York, p. 899.
21. Selig, M. et al. 1997. Comparative analysis of Embden-Meyerhof and Entner-Doudoroff glycolytic pathways in hyperthermophilic archaea and the bacterium *Thermotoga*. *Archives of Microbiology.* 167(4):217–232.
22. Grady, C.P.L., G.T. Daigger, and H.C. Lim. *Biological Wastewater Treatment*, 2nd (ed.). Marcel Dekker, Inc: New York, NY. 1999.
23. Talabardon, M., J.P. Schwitzguebel, and P. Peringer. 2000. Anaerobic thermophilic fermentation for acetic acid production from milk permeate. *Journal of Biotechnology.* 76:83–92.
24. Levin, D. B., R. Islam, N. Cicek, and R. Sparling, 2006. Hydrogen production by Clostridium thermocellum 27405 from cellulosic biomass substrates. *International Journal of Hydrogen Energy.* 31:1496–1503.
25. van Niel, E.W.J., P.A.M Claassen, and A.J.M. Stams. 2003. Substrate and product inhibition of hydrogen production by the extreme thermophile, *Caldicellulosiruptor saccharolyticus. Biotechnology and Bioengineering.* 81:255–262.
26. Dietrich, G., N. Weiss, and J. Winter. 1988. *Acetothermus paucivorans*, gen. nov., sp. nov., a strictly anaerobic, thermophilic bacterium from sewage sludge, fermenting hexoses to acetate, $CO_2$ and $H_2$. Syst. *Applied Microbiology.* 10:174–179.
27. Soutschek, E., et al. 1985. *Acetomicrobium flavidum*, gen. nov., sp. nov., a thermophilic, anaerobic bacterium from sewage sludge, forming acetate, $CO_2$ and $H_2$ from glucose. Syst. *Applied Microbiology.* 5:377–390.
28. Ueno, et al. 2001. Characterization of a microorganism isolated from the effluent of hydrogen fermentation by microflora. *Bioscience and Bioengineering.* 92(4):397–400.
29. Logan, B.E., et al. 2002. Biological hydrogen production measured in batch anaerobic respirometers. *Environmental Science & Technology.* 36:2530–2535.
30. de Vrije, T., et al. 2002. Pretreatment of *Miscanthus* for hydrogen production by *Thermotoga elfii. International Journal of Hydrogen Energy.* 27:1381–1390.
31. van Niel, E.W.J., et al. 2002. Distinctive properties of high hydrogen producing extreme thermophiles, *Caldicellulosiruptor saccharolyticus* and *Thermotoga elfii. International Journal of Hydrogen Energy.* 27:1391–1398.
32. Yu, X. 2007. Biohydrogen production by the hyperthermophilic bacterium *Thermotoga neapolitana*. PhD Dissertation, Clemson University, Clemson SC.
33. George, H.A., et al. 1983. Acetone, isopropanol, and butanol production by *Clostridium beijerinckii* (syn. *Clostridium butylicum*) and *Clostridium aurantibutyricum*. *Applied and Environ. Microbiology.* 45(3):1160–1163.
34. Hawkes, F.R., et al. 2002. Sustainable fermentative hydrogen production: challenges for process optimization. *International Journal of Hydrogen Energy.* 27:1339–1347.
35. Hwang, M. H., et al. 2004. Anaerobic bio-hydrogen production from ethanol fermentation: the role of pH. *Journal of Biotechnology.* 111(3):297-309.
36. Lee, M. J. and S. H. Zinder 1988. Hydrogen partial pressures in a thermophilic acetate-oxidizing methanogenic coculture. *Applied and Environmental Microbiology.* 54(6):1457–1461.
37. Ren, N.Q., et al. 2006. Biohydrogen production from molasses by anaerobic fermentation with a pilot-scale bioreactor system. *International Journal of Hydrogen Energy.* 31:2147–2157
38. Ghose, T.K., and A. Bhadra. 1985. Acetic acid. In: *Comprehensive Biotechnology, The Practice of Biotechnology: Current Commodity Products*. M. Moo-Young (ed.) London: Pergamon Press. 3:701–729

39. Maryknowski, C.W., et al. 1985. Fermentation as an advantageous route for the production of an acetate salt for roadway deicing. *Industrial Engineering Chemical Research Development.* 24:457–465.
40. Parekh, S. R., and M. Cheryan. 1991. Production of acetate by mutant strains of *Clostridium thermoaceticum. Applied Microbiology Biotechnology.* 36:384–387
41. Collet, C., et al. 2005. Acetate production from lactose by *Clostridium thermolacticum* and hydrogen-scavenging microorganisms in continuous culture—Effect of hydrogen partial pressure. *Journal of Biotechnology.* 118(3):328–338.
42. Huang, Y., and S-T. Yang. 1998. "Acetate production from whey lactose using co-immobilized cells of homolactic and homoacetic bacteria in a fibrous-bed bioreactor." *Biotechnology and Bioengineering.* 60(4): 498–507.
43. Oh, S., B. Min, and B.E. Logan. 2004. Cathode performance as a factor in electricity generation in microbial fuel cells. *Environmental Science and Technology.* 38:4900–4904.
44. Levin, D.B., L. Pitt, and M. Love. 2004. Biohydrogen production: prospects and limitations to practical application. *International Journal of Hydrogen Energy.* 29:173–185.
45. Han, K., and O. Levenspiel. 1988. Extended Monod kinetics for substrate, product, and cell inhibition. *Biotechnology and Bioengineering.* 32:430–437.
46. Vargas, M., et al. 1998. Microbiological evidence for Fe(III) reduction on early Earth. *Nature.* 395:65–67.
47. Van Ooteghem, S.A., S.K. Beer, and P.C. Yue. 2002. Hydrogen production by the thermophilic bacterium *Thermotoga neapolitana. Appl Biochem Biotechnol.* 98–100, 177–189.
48. Yu, X., and C. Drapcho. 2008. Biohydrogen Production using Various Organic and Nitrogen Sources. In preparation.
49. Gavala, H.N., et al. 2006. Thermophilic anaerobic fermentation of olive pulp for hydrogen and methane production: modeling of the anaerobic digestion process. *Water Science & Technology.* 53(8):271–279.
50. Scurlock, J. *Bioenergy feedstock characteristics.* Oak Ridge National Laboratory Publication. http://bioenergy.ornl.gov/papers/misc/biochar_factsheet.html
51. Pessoa, A., I.M, Mancilha, and S. Sato. 1996. Cultivation of *Candida tropicalis* in sugar cane hemicellulosic hydrolyzate for microbial protein production. *J. Biotechnol.* 51:83.
52. Parajo, J.C., H. Dominguez, and J.M. Dominguez. 1998. Biotechnological production of xylitol. Part-3: operation in culture media made from lignocellulose hydrolsates. *Bioresource Technol.* 66:25.
53. Van Ooteghem, S.A., et al. 2004. $H_2$ production and carbon utilization by *Thermotoga neapolitana* under anaerobic and microaerobic growth conditions. *Biotechnology Letters.* 26:1223–1232.
54. Valentine D.L, W.S. Reeburgh, and D.C. Blanton. 2000. A culture apparatus for maintaining $H_2$ at sub-nanomolar concentrations. *Journal of Microbiological Methods.* 39:243–251.
55. van Groenestijn, et al. 2002. Energy aspects of biological hydrogen production in high rate bioreactors operated in the thermophilic temperature range. *International Journal of Hydrogen Energy.* 27:1141–1147.
56. Kolb, B. and L.S. Ettre. 2006. *Static Headspace-Gas Chromatography: Theory and Practice.* Wiley-INterscience, New York, NY.
57. Felder, R.M. and R. W. Rousseau. 2000. Elementary *Principles of Chemical Processes. John Wiley & Sons,* Inc.: New York.
58. Speight, J. Lange's Handbook of Chemistry. 16th ed. McGraw-Hill Professional Publishing: New York, NY.
59. Sander, R. 1999. Compilation of Henry's Law Constants for Inorganic and Organic Species of Potential Importance in Environmental Chemistry, version 3. http://www.henrys-law.org
60. Washburn, E.W (ed). 2003. *International Critical Tables of Numerical Data, Physics, Chemistry and Technology (1st Electronic Edition).* Knovel Publishers.

61. Barbir, F. 2003. System Design for Stationary Power Generation. In: Vielstich, V., Lamm, A. and Gasteiger, H.A. (ed.). *Handbook of Fuel Cells*. John Wiley & Sons: Hoboken, NJ.
62. Landsman, D.A. and F. J. Luczak. 2003. Catalyst sudies and coating technologies. In: Vielstich, V., Lamm, A. and Gasteiger, H.A. (ed.). *Handbook of Fuel Cells*. John Wiley & Sons: Hoboken, NJ.
63. US DOE Energy Information Administration. 2001. A Look at Residential Energy Consumption in 2001. http://www.eia.doe.gov/emeu/recs/

# CHAPTER 8
# Microbial Fuel Cells

## 8.1 Overview

One of the most exciting technologies for biological production of energy is the microbial fuel cell (MFC). An MFC is a specialized biological reactor where the electrons processed during microbial metabolic activity are intercepted to provide useful electrical power. Under normal aerobic growth conditions, chemotrophic microbes oxidize chemical compounds, transfer the electrons to intermediate electron carrier molecules, then to an electron transport system and finally to the terminal electron acceptor, oxygen. (See Chap. 2 for a review of biochemical pathways). In an MFC, the oxidation of the electron donor compound is physically separated from the terminal electron acceptor. The microbes are grown in the anode chamber where the electron donor compound is oxidized, with the electrons transferred to the anode instead of oxygen or other external electron acceptor (Fig. 8.1). The electrons pass from the anode through a circuit that includes an external resistance load, then to a cathode and finally to the terminal electron acceptor contained in the cathode chamber. Typically, the anode compartment is separated from the cathode compartment by a proton exchange membrane (PEM) or cation exchange membrane (CEM). Protons pass from the anode compartment through the membrane to the cathode compartment, and finally to oxygen, that then combines with hydrogen to form water. By diverting electron flow from microbial respiration to the electrodes, MFCs convert chemical energy to electrical energy. Although environmental designs incorporating use of anodes inserted into anaerobic sediments with cathodes placed in aerobic layers have been investigated,[1] this chapter will focus on the use of constructed MFCs.

## 8.2 Biochemical Basis

In the typical process of aerobic respiration, the oxidation of an organic compound such as glucose would result in the following half-reaction:

$$C_6H_{12}O_6 + 6H_2O \rightarrow 6CO_2 + 24H^+ + 24e^- + 4ATP \qquad (8.1)$$

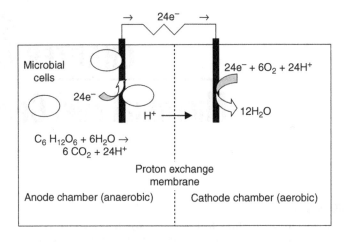
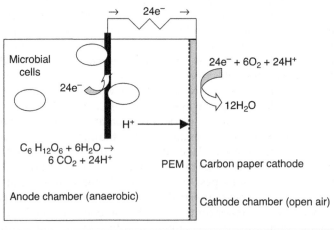

**Figure 8.1** (a) Diagram of two-chamber microbial fuel cell with aqueous cathode and anode chambers with solid graphite electrodes. (b) Diagram of single-chamber microbial fuel cell with aqueous anode chamber and air cathode chamber.

This summary reaction represents the biochemical pathways of glycolysis (Embden-Meyerhoff pathway) and the tricarboxylic acid (TCA) cycle used by many chemoorganotrophic microorganisms. The theoretical coulombic yield can be calculated using Faraday's constant, 96,485 C/mol of electrons. For glucose, since 24 mol $e^-$ are removed per mol of glucose oxidized to $CO_2$, the theoretical yield is $2.3 \times 10^6$ C/mol glucose.

The half-reaction for the reduction of oxygen by aerobic microbes through the electron transport system is given as:

$$24H^+ + 24e^- + 6O_2 \rightarrow 12H_2O + 34ATP \quad (8.2)$$

The proton motive force that develops through the transfer of electrons in the electron transport system is used by the cell to generate adenosine triphosphate (ATP). This step produces the bulk of the ATP formed from aerobic oxidation of an organic compound.

The overall equation for the aerobic oxidation of an organic compound, not considering cell growth, is therefore:

$$C_6H_{12}O_6 + 6O_2 \rightarrow 6CO_2 + 6H_2O + 38ATP \qquad (8.3)$$

In an MFC, these processes are physically separated. The growth of the organism occurs in the anode, with the final use of oxygen as terminal electron acceptor in the cathode. The protons produced by the oxidation of glucose pass through the exchange membrane to the cathode chamber. Since the electron transport system of the microbe is being circumvented, the proton motive force that drives ATP generation does not develop. Therefore, the ATP yield for microbial growth in the anode is low, possibly as low as 4 mol ATP/mol glucose if the transfer of electrons from the microbe to the anode is 100% efficient and no oxygen diffuses into anode from cathode. The low ATP yield results in low biomass yield typically found in anaerobic growth. Measurement of glucose consumption and metabolite production by microbes growing under aerobic, anaerobic, and MFC conditions confirms that the metabolic pathway of bacterial growth in an MFC lies between aerobic and anaerobic conditions.[2]

## 8.3 Past Work Summary

The application of microbial growth with electron capture for energy production in a designed structure called "biochemical fuel cell" was proposed as early as 1966.[3] The critical elements of an MFC (anode, cathode, and redox mediator compounds) were discussed. Early work in the development of MFCs by Bennetto and collaborators.[2,4,5,6,7,8] focused on the use of heterotrophic, organotrophic bacterial cultures to oxidize organic substrate (Table 8.1). They reported high conversion of lactose, sucrose, and glucose to electricity by *Escherichia coli*[6,7] and *Proteus vulgaris*[2,5,6] in two-chamber MFCs using thionine as anodic redox mediator and ferricyanide as electron acceptor in cathode chamber. Application of MFC technology for waste treatment was suggested by early researchers[9] and was demonstrated for food processing wastewater in 1983.[7] The range of organic compounds that can be metabolized by chemotrophic microorganisms (and their enzymes) is extremely large so almost any waste stream may be utilized in an MFC. MFC technology for waste treatment has been advanced by Logan and collaborators, using domestic,[10,11] animal,[12] food processing[13] and synthetic biprocessing[11,14,15] wastewaters using mixed bacterial cultures isolated from wastes or sediments. This group has also developed single-chamber MFC technology, replacing the aqueous

| | | | Anode | Cathode | Mediator | | Cathode |
|---|---|---|---|---|---|---|---|
| Ref | Year | Organism or inoculum | Energy source | Acceptor | Anode | Cathode | Type |
| 7 | 1983 | Escherichia coli | Lactose | Ferricyanide | Yes [1] | | Aqueous |
| 6 | 1984 | A. eutrophus, B. subtilis E. coli, P. vulgaris | glucose, succinate | Ferricyanide | Yes [1,8] | | Aqueous |
| 2 | 1985 | Proteus vulgaris | Glucose | Ferricyanide | Yes [1] | | Aqueous |
| 5 | 1985 | Proteus vulgaris | Sucrose | Ferricyanide | Yes [1] | | Aqueous |
| 16 | 1997 | Synechococcus sp. | Illumination | Ferricyanide | Yes [2] | --- | Aqueous |
| 17 | 2001 | Synechococcus sp. PCC7942 | Illumination | Bilirubin oxidase | Yes [3] | Yes [5] | Aqueous |
| 18 | 2003 | Blue-green aglae Anabaena | Illumination Glucose | Ferricyanide | Yes [4] | --- | Aqueous |
| 19 | 2003 | Rhodoferax ferrireducens | Glucose | Iron (III) | --- | --- | Aqueous |
| 20 | 2003 | Anaerobic sludge | Glucose | Oxygen | --- | Yes [6] | Aqueous |
| 21 | 2004 | mixed: A. faecalis; E. gallinarum, P. aeroginosa | Glucose | Ferricyanide | --- | --- | Aqueous |
| 22 | 2004 | Domestic Wastewater | Glucose (G) Wastewater (W) | Oxygen Oxygen | --- --- | --- --- | Air Air |
| 11 | 2004 | Primary Clarifier | Acetate Starch Glucose Butyrate | Oxygen | --- | --- | Air |
| 15 | 2004 | Anaerobic digestor | Acetate (20mM) | Oxygen Ferricyanide | --- | --- | Aqueous |
| 23 | 2005 | Anaerobic sludge | Artificial WW | Ferricyanide | --- | --- | Aqueous |
| 14 | 2005 | Wastewater | Acetate (800mg/L) Butyrate (1000mg/L) | | --- | --- | Air |
| 24 | 2005 | Gammaproteobacteria marine sediment | 385 mg/lLcysteine: 770 mg/L cysteine: | Oxygen | Yes [7] | --- | Aqueous |
| 12 | 2005 | Swine WW | Swine WW | Oxygen | --- | --- | Aqueous Air |
| 25 | 2006 | Anaerobic, aerobic sludge Proteobacteria majority | Na-acetate | Ferricyanide | --- | --- | Aqueous |
| 26 | 2006 | Primary Clarifier | 500 mg/L glucose | Oxygen | --- | --- | Air |
| 27 | 2006 | Shewanella oneidensis | Lactate | Ferrocyanide | --- | --- | Aqueous |
| 28 | 2007 | Secondary Clarifier Anaerobic Sludge | Ethanol | Oxygen | --- | --- | Aqueous Air |
| 29 | 2007 | Primary Clarifier | Cube: 1g/L acetate Bottle: 1g/L glucose | Oxygen | --- | --- | Air |
| 30 | 2007 | Domestic Wastewater | Glucose | Oxygen | --- | --- | Air |

[a] thionine.
[b] 2-hydroxy-1,4-napthoquinone (HNQ).
[c] 2,6-dimethyl-1,4-benzoquinone (DMBQ).
[d] 0.01 M methylene blue.
[e] 2,2'-azinobis(3-ethylbenzothiazolin-6-sulfonate) (ABTS$^{2-}$).
[f] potassium hexacyanoferrate.
[g] cysteine.
[h] variety of compounds compared.
Source: Adapted from Little et al., 2007.

**TABLE 8.1** Summary of Representative MFC Literature

# Microbial Fuel Cells

| Electrodes | | Membrane | Reactor | Treatments | Power | Density |
|---|---|---|---|---|---|---|
| Anode | Cathode | | | | | |
| vitreous carbon | Pt foil | Cation EM | Batch | | 18 | mW/L |
| vitreous carbon | Pt foil | Cation EM | Batch | *P. vulgaris* with thionine | 4.48 | mW/m$^2$ |
| vitreous carbon | Pt foil | Cation EM | Batch | | 11 (45) | mW/m$^2$ (mW/L) |
| Carbon | Carbon | Ion-exchange | Batch | | 82 | mW/L |
| Carbon Cloth | Carbon Cloth | Ion-exchange | Batch | | 18 (35) | mW/m$^2$ (mW/L) |
| Carbon Felt 2.25 cm$^2$ | Carbon Felt 2.25 cm$^2$ | KCl- salt bridge | Batch | | 0.4 | mW/m$^2$ |
| Cr/Au | Cr/Au | Nafion 117 PEM | Batch | | 0.061 | mW/L |
| Graphite Rod | Graphite Rod | Nafion 117 PEM | | | 961 | mW/m$^2$ |
| Graphite | Graphite | Ultrex PEM | Fed-Batch | | 3,600 | mW/m$^2$ |
| Graphite | Graphite | Ultrex PEM | Fed-Batch | | 4,310 | mW/m$^2$ |
| Carbon Paper | Carbon Paper 0.5 mg/cm2 Pt | Nafion 117-Bonded | Batch | G-PEM: G-no PEM: | 262 (6.6) 494 (13) | mW/m$^2$ (mW/L) mW/m$^2$ (mW/L) |
| Carbon Paper | Carbon Paper 0.5 mg/cm2 Pt | Nafion 117-Bonded | Batch | WW-PEM WW-no PEM | 28 (0.7) 146 (3.7) | mW/m$^2$ (mW/L) mW/m$^2$ (mW/L) |
| Carbon Paper 100 cm$^2$ | Carbon Paper 0.5 mg/cm$^2$ catalyst containing 10% Pt | Nafion 117-Bonded | Continuous 30 C | Acetate Starch Glucose Butyrate | 286 242 212 220 | mW/m$^2$ mW/m$^2$ mW/m$^2$ mW/m$^2$ |
| Carbon Paper 11.25 cm$^2$ | Carbon Paper 0.5 mg-Pt/cm2 | Nafion 117 PEM | Batch 30 C | Oxygen Ferricyanide | 86 151 | mW/m$^2$ mW/m$^2$ |
| vitreous carbon | reticulated carbon | PEM | Continuous | | 170 | mW/m$^2$ |
| Carbon Paper | Carbon Paper 0.35 mg/cm$^2$ Pt | None | Batch | Acetate: Butyrate: | 506 (13) 305 (7.6) | mW/m$^2$ (mW/L) mW/m$^2$ (mW/L) |
| Carbon paper (two) | Carbon Paper 0.35 mg/cm$^2$, 10% Pt | Nafion 117 PEM | Batch | 385 mg/Ll 770 mg/l | 19 39 | mW/m$^2$ mW/m$^2$ |
| Carbon Paper 11.25 cm$^2$ | Carbon Paper 0.35 mg/cm$^2$, 10% Pt | Nafion 117 PEM | Fed Batch | Aqueous Air | 45 261 | mW/m$^2$ mW/m$^2$ |
| Graphite Granuals | Graphite Granuals | Cation Exchange CM17000 | Continuous | | 258 | mW/L |
| Graphite Rod | Graphite Rod | | | | | |
| Carbon Cloth | Carbon Cloth | Carbon/PTFE | Batch | Air | 1540 (51) | mW/m$^2$ (mW/L) |
| Vitreous carbon | Vitreous carbon | Nafion 117 PEM | Continuous | Aqueous | 24 | mW/m$^2$ |
| Graphite Felt | Graphite Felt | | | | 10 | mW/m$^2$ |
| Carbon Paper | Carbon Paper 0.35 mg/cm$^2$ | Nafion | Fed Batch | Aqueous Air | 38 (0.27) 514 (26) | mW/m$^2$ (mW/L) mW/m$^2$ (mW/L) |
| Fiber Brush | Carbon Cloth wet proofed | Carbon/PTFE | Batch | Cube: Bottle: | 2400 (73) 1430 (2.3) | mW/m$^2$ (mW/L) mW/m$^2$ (mW/L) |
| Carbon Paper | Carbon Paper | trafiltration membrane | | Carbon Paper | 8.8 | mW/L |
| | Graphite Brush | polyester carrier | | Graphite Brush | 17.7 | mW/L |

**TABLE 8.1** *(Continued)*

cathode chamber with an open-air chamber.[10,14,22,29] Some of the highest power densities produced to date have been reported by Rabaey et al.[20,21,31] using a two-chamber MFC with enriched bacterial cultures obtained from wastewater and added redox mediator (potassium hexacyanoferrate)[20] or ferricyanide as electron acceptor in the cathode.[21]

Early researchers proposed use of photosynthetic cultures in fuel cells in both anode compartment for capture of electrons from photosynthetic pathways,[9] and in the cathode for production of oxygen[3] but less work in this area has been reported. The potential for harnessing the electron flow from photosynthesis is great, due to the vast energy available in solar radiation that could potentially be harvested, but interception of electrons from photosynthetic pathways appears more difficult. Investigations into the use of cyanobacteria, such as *Synechococcus* sp. and *Anabaena,* cultured in the anode chamber to harvest electrons from photosynthetic pathways have met with some success.[16,17,18]

## 8.4 Fuel Cell Design

The primary objective in the design of any fuel cell is to maximize the energy output while minimizing the fuel cell volume and cost.[32] In MFC development for wastewater treatment, additional objectives are to maximize the oxidation of organic substrate or to produce an effluent of specified discharge quality. Strategies to achieve this include maximizing the electrode surface area to volume ratio to maximize mass transport, maximizing electron transport between cells and electrodes through natural or added electron mediators, and use of continuous reactor operation as opposed to batch mode to maximize power output with minimal anode volume.

### 8.4.1 Anode Compartment

The anode compartment consists of the anode electrode, substrate, and microbial culture. Oxygen and alternative electron acceptors such as sulfate or nitrate should not be present in anode to avoid aerobic or anaerobic respiration from occurring in anode chamber. Metabolism of the substrate occurs in the anode compartment, with transfer of the electrons to the anode rather than to a final electron acceptor. The power output of an MFC increases with increasing substrate concentration in the anode, and can be modeled as a Monod relationship (see Sec. 8.6.1). Substrates tested in MFC designs include glucose and other prepared media and a variety of wastewaters. The anode material should be electrically conductive, biocompatible, and chemically stable. Solid graphite rods or plates, carbon paper, and carbon cloth are often used due to high electrical conductivity, low cost, and availability. Graphite may be embedded with various compounds such as platinum or manganese to increase efficiency of electron transfer between the microbe and anode. Increasing surface area with use of crushed graphite, graphite felt, and reticulated vitreous carbon can achieve high power outputs due to the high surface area available for microbial growth and electron transfer but reducing clogging by biofilm growth and maintaining high porosity is essential.[33]

Power output of an MFC is typically normalized on the basis of anode surface area ($m^2$) for designs using solid anode or carbon paper. For designs with crushed graphite or graphite brush anodes that may take up considerable volume of anode chamber, power density can be calculated on basis of total anode volume, total anode chamber volume, or net anode volume ($m^3$).

### 8.4.2 Microbial Cultures

Fuel cells in which enzymes or catalysts are used in place of living cells are considered enzymatic or "bio" fuel cells, but not MFCs.[33] Organotrophic microorganisms used in an MFC must be able to carry out respiration, and inoculum obtained from wastewater is commonly used.[34] Early work utilized cultures of *E. coli*[6,7] or *P. vulgaris*[2,5,6] *and later S. cerevisiae*. More recently, pure cultures of *Shewanella oneidensis*,[10,37] *S. putrefaciens*,[36] or *Geobacter*[10,37] have been used due to specific advantages relating to electron transfer. *Geobacter* spp. are advantageous when anodes embedded with iron are used, since the attached growth of *Geobacter* cultures facilitates electron transfer directly to Fe (III), reducing to Fe (II).[37] *Shewanella* is advantageous because it produces external electron mediators (quinones) that facilitate electron transfer. *S. putrefaciens* can be found in many aquatic environments, both natural and wastewater.[38] Further, *Shewanella* and other bacteria have been found to produce "nanowires"—electrically conductive protein structures[39] that extend from the outer membrane in response to electron acceptor-limited conditions. These pilli-like appendages enhance electron transport between the environment and cell. In *S. oneidensis* strain MR-1, electrochemically active appendages ranging from 50 to 150 nm in diameter and tens of microns in length[39] were found. These appendages were also found in the cyanobacterium *Synechocystis* when grown under $CO_2$ limited conditions and in the fermentative thermophile *Pelotomaculum thermopropionicum*.[39] These results indicate that the formation of electrically conductive nanowires may represent a common bacterial strategy for improved electron transfer under electron acceptor-limited conditions. Mixed cultures of heterotrophic bacteria obtained from wastewater or sediments have been used successfully in MFCs and show a great diversity of genera and species.[10,24,33] Mixed culture systems have been shown to achieve higher power densities than pure cultures in many circumstances.[21,33] Bacteria present in wastewater and sediments that have been shown to produce electricity in an MFC include *Alcaligenes faecalis*,[21] *Brevibacillus agri*,[25] *Enterococcus gallinarum*,[21] *Geobacter sulfurreducens*,[37,40] *Geobacter metallireducens*,[10,37] *Proteobacteria* spp.,[25] *Pseudomonas aeruginosa*,[21] and *S. putrefaciens*.[36]

### 8.4.3 Redox Mediators

Redox mediators provide a means of transferring electrons from within the cell to the electrode and can be either exogenous or naturally produced by the organism. Efficient redox mediators should (1) have oxidized and reduced states that easily enter and exit the cell membrane; (2) have a redox potential favorable to provide rapid electron transfer without significant loss of potential; and (3) be nontoxic, chemically stable, soluble, and nonabsorbing to cell walls or electrode surfaces.[41]

Addition of exogenous redox mediators in the anode compartment may significantly increase both the rate of electron transfer and the proportion of available electrons transferred.[7] Earlier researchers considered their use essential for efficient MFC operation.[2,5,7,42] Redox mediators proposed or used in earlier MFC research included thionine,[2,5,7] (Fig. 8.2) resorufin,[9] or methylene blue.[8] These compounds have the advantages that they are easily oxidized and reduced, and change colors when reduced thus allowing a visual indication of rate of reduction.[9] However, many are not chemically stable and must be replenished. More recent investigations have investigated the use of cysteine,[24,40] anthraquinone-2,6-disulfonate (AQDS)[37] for organotrophic cultures, or quinone mediators such as 2-hydroxy-1,4-napthoquinone (HNQ)[16] or 2,6-dimethyl-1,4-benzoquinone (DMBQ)[17] for photosynthetic MFCs using cyanobacterial cultures (Table 8.1) Fewer studies have investigated the impact of added redox mediators in the cathode

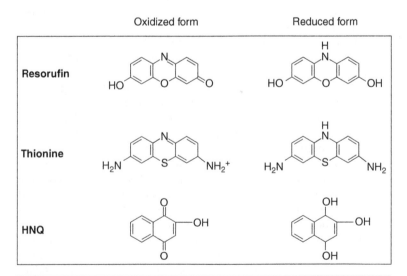

FIGURE 8.2 Redox mediators. (*Adapted from Bennetto, 1987.*)

chamber, although 2,2'-azinobis (3-ethylbenzothiazoline-6-sulfonate) (ABTS$^{2-}$)[17] and potassium hexacyanoferrate[20] have been reported.

Many recent investigations have utilized MFCs without added redox mediators.[10,14,19,22] Extracellular electron transfer to the anode in mediatorless MFCs can occur by several mechanisms: (1) by direct electron transfer from attached bacterial growth; (2), by natural mediator compounds excreted by bacteria;[43] or (3) by electroactive nanowires produced by certain bacteria acclimated to low-electron acceptor environments.[39] Many types of pure and enriched cultures have been shown to be electrochemically active, not requiring addition of external redox mediators.[10,14,21,24] Some studies have shown that the predominant mode of power generation for mixed cultures of heterotrophic bacteria obtained from wastewater is from direct electron transfer to the anode by attached bacterial growth,[10,14] while other studies have attributed the majority of power generation to production of naturally produced mediators.[20,21] Bacteria such as *S. putrefaciens*[43] or enriched cultures obtained from wastewater containing primarily *Alcaligenes faecalis*, *Enterococcus gallinarum*, and *Pseudomonas aeruginosa*[21] have been shown to excrete quinones which are redox mediators. Redox potential data obtained from single-chamber MFC without added redox mediators with inoculum obtained from wastewater suggest that electrons are diverted from respiratory enzymes in between NADH and cytochrome c.[14] Application of MFC technology for large-scale wastewater treatment or for continuous flow reactors would likely rely on the use of naturally electrochemically active species to avoid the high cost of redox mediator addition.

### 8.4.4 Cathode Compartment

The cathode chamber consists of the cathode and the final electron acceptor. The cathode may be a solid graphite rod, graphite granules, vitreous carbon, or carbon paper. For cathodes using oxygen as electron acceptor, Mn or Pt catalysts are often embedded on the surface to enhance electron reduction of oxygen. Pt applications as low as 0.1 mg Pt/cm$^2$ have been shown to be effective.[26]

Both two-chamber and single-chamber MFC designs are being researched and developed. The cathode compartment in two-chamber designs is filled with aqueous solution containing dissolved oxygen or other electron acceptor compound such as potassium ferricyanide, $K_3$(Fe(CN)$_6$). Oxygen has a greater redox potential (+0.8 V at neutral pH), is relatively easy to replenish with mechanical or bubble aeration, and produces no chemical wastes. Ferricyanide used in many studies[2,5,6,7,16,21] in an aqueous cathode has a lesser redox potential (+0.356 V) than oxygen but has been shown to increase power density by a factor of 1.5 to 1.8 over an oxygenated cathode,[33] due to enhanced electron transport.[22] However, ferricyanide is not easily reoxidized with oxygen and must be chemically regenerated; therefore, its use would not be feasible in large or continuous flow systems.

In a single-chamber design, the cathode chamber is simply an open air chamber[22] with carbon cloth or carbon paper serving as the cathode. Single-chamber designs have smaller total reactor volume since there is no cathode chamber, and do not need aeration to provide oxygen. In these designs, wet-proofed carbon cloth is often hot-pressed directly to the PEM so that direct electron transfer can occur.[14,22] When the PEM is omitted from these designs, the carbon cathodes are wet-proofed by applying polytetrafluoroethylene to carbon cloth[26] or coating an ultrafiltration hydrophilic membrane with graphite paint[30] to prevent water from leaking out of anode. Greater power densities have been achieved in single-chamber designs than two-chamber,[22] under lab-scale condition.

### 8.4.5 Exchange Membrane

A PEM or CEM is often used to physically separate the anode and cathode chambers in MFC designs, allowing the passage of H⁺ from the anode chamber to the cathode while restricting the diffusion of oxygen into the anode. Protons produced from the oxidation of substrate in the anode compartment decrease the pH of the solution if not transferred or neutralized. Diffusion of oxygen into the anode results in substrate oxidation without the transfer of electrons to the anode.

Due to their prior use in hydrogen fuel cells, Nafion (DuPont) membranes are commonly used in MFC research as PEMs. Nafion membranes consist of a hydrophobic fluorocarbon backbone with attached hydrophilic sulfonate groups ($-SO_3$).[44] Most membranes also allow for a high passage of other cations such as sodium, potassium, ammonium, calcium, and magnesium that are typically found in MFC substrates.

To quantify the diffusion of oxygen across a PEM, the one-dimensional steady state form of Fick's Law can be used:[22]

$$N_{O_2} = D_{AB} A \frac{\Delta C}{\Delta x} \qquad (8.4)$$

where $N_{O_2}$ = oxygen flux, mol/cm²

$D_{AB}$ = binary diffusion coefficient, m²/s

$\frac{\Delta C}{\Delta x}$ = concentration gradient across membrane, mol/cm³-cm

For MFC environments, the oxygen concentrations at the membrane surfaces can be estimated as saturation on the cathode side (~2.6 × 10⁻⁷ mol $O_2$/cm³)[24] and zero on the anode side. For Nafion 117 PEM, a diffusion coefficient of oxygen ($D_{AB}$) through PEM was determined to be 4.4 × 10⁻⁶ cm²/s[22] and membrane thickness of 190 μm.[45]

## 8.4.6 Power Density as Function of Circuit Resistance

Power production in an MFC is a function of circuit resistance, as each MFC design has a characteristic internal resistance that affects the optimum resistance load to apply to the MFC. The optimum resistance for maximum power output of the cell can be determined by varying the resistance and measuring the voltage output[33] and constructing a polarization curve of voltage vs. current. If the resulting curve is linear, it indicates an MFC with high internal resistance. The slope of the linear regression of the polarization plot is equal to the internal resistance of the fuel cell.[33]

A power curve is a plot of the resulting power (or power density) as a function of current, calculated from the polarization curve data. A symmetrical power curve will result in an MFC with high internal resistance, with peak power produced at the optimum resistance load. For MFC designs that display high internal resistance, the optimum external resistance load is equal to the internal resistance value.

Characteristic polarization and power curves for an MFC with high internal resistance are shown in Figs. 8.3 and 8.4. For this two-chamber MFC, an internal resistance of 657 Ω is indicated, with maximum power of approximately 0.2 mW.[46]

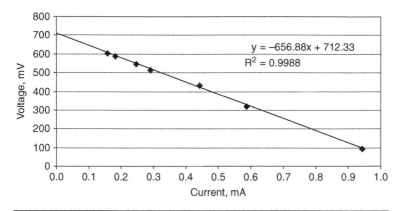

**FIGURE 8.3** Polarization curve for two-chamber MFC. (*Adopted from Little, 2008.*)

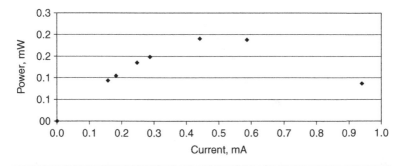

**FIGURE 8.4** Power curve for two-chamber MFC. (*Adopted from Little, 2008.*)

## 8.5 MFC Performance Methods

### 8.5.1 Substrate and Biomass Measurements

In MFC applications, the substrate concentration may be measured by analytical techniques such as GC or HPLC for specific compound determination. For mixtures of substrate found in municipal, animal, or food processing wastewaters, the organic substrate concentration is often represented as the chemical oxygen demand (COD). The COD represents the number of electrons contained in a compound, expressed as the mass of oxygen required to accept the electrons as the compound is completely oxidized. The power output from an MFC is directly related to the electrons captured from the substrate.

For solutions containing both soluble and particulate organic compounds, the soluble COD content is determined by filtering the sample prior to COD analysis. The particulate COD is determined by difference of the total COD and soluble COD values. If the growth media contains only soluble organic substrate such as glucose, suspended biomass will be the only particulate organic fraction present in the mixture. Therefore, the biomass concentration expressed as COD may be estimated as the difference between the total and soluble COD values.

Other means of determining suspended microbial growth in the anode chamber include optical density (OD) measurements at 680 nm with a digital spectrophotometer or dry weight measurements after filtration through 0.45 or 0.2 μm filters. Growth curves obtained during batch growth of *E. coli* using deproteinated milk serum substrate in a two-chamber MFC are shown in Fig. 8.5.[47]

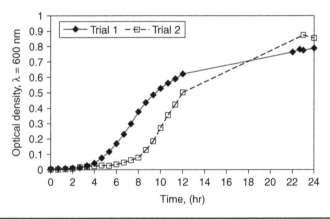

**FIGURE 8.5** Growth curves of *E. coli* in anode of MFC. (*Adopted from Pourciau et al., 1999.*)

## 8.5.2 Basic Power Calculations

The voltage across the external resistor or load in an MFC can be measured using a multimeter. Voltage measurements are converted to current values using Ohm's law:

$$V = IR \tag{8.5}$$

where $V$ = voltage, V
$I$ = current, A
$R$ = resistance, $\Omega$

The power output from an MFC is calculated as

$$P = IV \tag{8.6}$$

where $P$ = power, W.

Power density is used to relate power output to the anode surface area or anode chamber volume. Power density is calculated based on anode surface area as follows:

$$P_{DA} = IV/A_A \tag{8.7}$$

where $P_{DA}$ = power density on area basis, W/m²; and $A_A$ = anode surface area, m².

The power density may be calculated based on total or net anode chamber volume as follows:

$$P_{DV} = IV/\mathcal{V} \tag{8.8}$$

where $P_{DV}$ = power density on volume basis, W/m³; and $\mathcal{V}$ = anode chamber volume, L or m³.

The efficiency of an MFC is often expressed as the Coulombic efficiency.[33] The Coulombic efficiency ($E_C$) is calculated in two ways. In the first, $E_C$ is expressed as the ratio of coulombs transferred from the substrate to anode to the theoretical maximum coulombs that would be produced if all of the substrate was oxidized. The coulombs of energy produced ($C_P$) and transferred to the anode is calculated for batch reactors by integrating the area under a curve of current versus time. Commonly, this is calculated simply as:[33]

$$C_P = I \cdot t \tag{8.9}$$

where $t$ = time of stable voltage output, s.

For continuous flow reactors, the $C_P$ produced at steady state is calculated as:

$$C_P = I \cdot \tau \tag{8.10}$$

where $\tau$ = hydraulic retention time in the anode chamber, s.

The theoretical maximum coulombs ($C_{max}$) that may be transferred from a specific substrate is calculated as[33]

$$C_{max} = FbM\forall = FbS\forall/MW \qquad (8.11)$$

where $F$ = Faraday's constant, 96,485 C/mol of electrons
$b$ = mol of electrons available for removal per mol of substrate
$M$ = substrate concentration, mol/L
$S$ = substrate concentration, g/L
$MW$ = molecular weight of the substrate

The Coulombic efficiency ($E_C$) is then calculated as:

$$E_C = C_P/C_{max} \cdot 100\% \qquad (8.12)$$

For example, if a medium containing 5 g/L glucose is used in the MFC, $b_{glu}$ = 24 mol electrons/mol glucose, $MW_{glu}$ = 180 g/mol, and $S_{glu}$ = 5 g/L. For a 0.1 L anode chamber, the theoretical maximum coulombs that could be produced are 6566 C.

If an undefined media is used, such as food processing or municipal wastewater, the substrate concentration will often be expressed as COD, and the ratio of electrons transferred per mol of substrate and the molecular weight of the compounds in the substrate will often be unknown. By electron balance, 1 mol of electrons transferred is equivalent to 8 g COD (see Chap. 2). Therefore, a simplified formula can be used to calculate $C_{max}$ for all cases of undefined or defined media where COD is used to express the organic substrate concentration:

$$C_{max} = Ff\ S_{COD}\forall \qquad (8.13)$$

where $f$ = factor of 1 mol electrons/ 8 g COD; and $S_{COD}$ = substrate concentration, g COD/L.

For the second way to calculate $E_C$, the Coulombic efficiency is expressed as the ratio of the coulombs produced to the *recoverable* coulombs based on the amount of substrate removed from the system. In this calculation, $\Delta S$ or $\Delta S_{COD}$ is used in place of $S$ or $S_{COD}$ in the calculation of $C_{max}$ as shown below:

$$C_{max} = Fb\Delta S\ \forall/MW \qquad \text{(for defined media)} \qquad (8.14)$$

$$C_{max} = Ff\Delta S_{COD}\forall \qquad \text{(for undefined media)} \qquad (8.15)$$

An important factor for evaluating the performance of an MFC is the energy recovery.[33] The overall energy efficiency, $\varepsilon_C$, is the ratio of power produced over time interval $t$ to the heat of combustion of the organic substrate added in that time interval, calculated as:[33]

$$\varepsilon_C = \int_0^t \frac{VIdt}{\Delta Hm_{added}} \qquad (8.16)$$

where $\Delta H$ = heat of combustion, J/mol; and $m_{added}$ = mol of substrate added.

However, $\Delta H$ is not known for mixed substrates such as wastewaters. Reported energy efficiencies for MFC using pure substrates (2 to 50 percent)[20,22] are favorable in comparison to other energy processes (i.e., $\varepsilon_C \leq 40$ percent for thermal conversion of methane).[33]

### 8.5.3 Calculation Example

Liu and Logan[22] evaluated the performance of a single-chamber MFC design using an air cathode, operated in batch mode, with anode (and PEM) surface area of $7.065 \times 10^{-4}$ m², anode chamber volume of 0.028 L, 1000 Ω resistor, with and without a PEM, and 0.6 g/L glucose media. For the experiments with the PEM, the voltage output increased quickly to 0.35 V and slowly dropped to 0.25 V over 95 hr, for an average of 0.30 V. Without the PEM, the maximum voltage increased suddenly to 0.52 V but then decreased rapidly to <0.04 V after 15 hr. The basic power calculations for the glucose experiments with and without PEM, and oxygen diffusion rate with PEM are as follows:

#### 8.5.3.1 Calculate Peak Power Density for MFC with PEM

$I = V/R$. Peak $I = 0.35$ V/1000 Ω = $0.35 \times 10^{-3}$ A
Peak $P_{DA} = IV/A_A = 0.35 \times 10^{-3}$ A · 0.35 V/ $7.065 \times 10^{-4}$ m² = 173 mW/m²
Peak $P_{DV} = IV/\Psi = 0.35 \times 10^{-3}$ A · 0.35 V/0.028 L = 4.38 mW/L

#### 8.5.3.2 Calculate Coulombic Efficiency for MFC with PEM

Using average voltage output of 0.30 V over 95 hr;

$Cp = I \cdot t = 0.30 \times 10^{-3}$ A · 95 hr · 3600 s/hr = 102.6 C
$C_{max} = FbS\Psi/MW = 96,485$ C/mol e⁻ · 24 mol e⁻/mol glucose · 0.6 g/L glucose · 0.028 L/180 g glucose/mol = 216.1 C
Average Coulombic efficiency = 102.6 C/216.1 C = 47%

#### 8.5.3.3 Calculate Peak Power Density for MFC without PEM

$I = V/R$. Peak $I = 0.52$ V/1000 Ω = $0.52 \times 10^{-3}$ A
Peak $P_{DA} = IV/A_A = 0.52 \times 10^{-3}$ A · 0.52 V/ $7.065 \times 10^{-4}$ m² = 382 mW/m²
Peak $P_{DV} = IV/\Psi = 0.52 \times 10^{-3}$ A · 0.52 V/0.028 L = 9.66 mW/L

#### 8.5.3.4 Calculate Coulombic Efficiency for MFC without PEM

Using average voltage output of 0.28 V over 15 hr;

$Cp = I \cdot t = 0.28 \times 10^{-3}$ A $\cdot$ 15 hr $\cdot$ 3600 s/hr = 15.1 C
$C_{max} = FbS\mathcal{V}/MW$ = 216.1 C (same value as with PEM)
Average Coulombic efficiency = 15.1 C/216.1 C = 7%

To demonstrate the use of the COD-based calculations, calculate the power density and Coulombic efficiency for the MFC with PEM data using COD units. Using the relationship that 1 g/L glucose = 1.07 g/L COD, the initial substrate concentration (0.6 g/L glucose) is equivalent to 0.64 g/L COD.

#### 8.5.3.5 Calculate Coulombic Efficiency for MFC with PEM on COD basis

Using average voltage output of 0.30 V:

$Cp = I \cdot t = 0.30 \times 10^{-3}$ A $\cdot$ 95 hr $\cdot$ 3600 s/hr = 102.6 C
$C_{max} = Ff S_{COD} \mathcal{V}$ = 96,485 C/mol e$^-$ $\cdot$ 1 mol e$^-$/8g COD $\cdot$ 0.64 g/L COD $\cdot$ 0.028 L = 216.1 C
Average Coulombic efficiency = 102.6 C/216.1 C = 47%

#### 8.5.3.6 Calculate Oxygen Diffusion Rate Through PEM

Using average PEM thickness of 0.019 cm$^2$, saturation oxygen concentration (~2.6 $\times$ 10$^{-7}$ mol/cm$^3$) at the cathode surface of PEM and zero on anode side, with $D_{AB}$ = 4.4 $\times$ 10$^{-6}$ cm$^2$/s:[14]

$$N_{O_2} = D_{AB} A \frac{\Delta C}{\Delta x}$$

$$= 4.4 \times 10^{-6} \text{ cm}^2/\text{s} \cdot 7.065 \times 10^{-4} \text{ m}^2 \cdot 10{,}000 \text{ cm}^2/\text{m}^2 \cdot$$

$$\frac{(2.6 \times 10^{-7} - 0) \text{mol O}_2/\text{cm}^3}{0.019 \text{ cm}}$$

$$= 4.2 \times 10^{-10} \text{ mol O}_2/\text{s} = 0.05 \text{ mg/hr}$$

## 8.6 MFC Performance

### 8.6.1 Power Density as Function of Substrate

Power produced from an MFC is a function of substrate concentration. The response can be modeled using Monod (Michelis-Menton) relationship:[14]

$$P = P_{max} \cdot S/(K_S + S) \tag{8.17}$$

where  $P$ = power density, mW/m²
 $P_{max}$ = maximum power produced, mW/m²
 $S$ = substrate concentration, mg/L
 $K_S$ = half-saturation constant for substrate, mg/L

In general, maximum power density values are lower for difficult to degrade substrates, and half-saturation values are greater. For a single-chamber MFC, the $P_{max}$ and $K_S$ values were 661 mW/m² and 141 mg/L, respectively, for acetate[14] while values of $P_{max}$ and $K_S$ were 225 mW/m² and 1510 mg/L total COD, respectively, for treating swine wastewater (Table 8.2).[12]

By assuming several factors of an MFC, including bacterial growth as monolayer on anode, a high bacterial growth rate (0.35 hr⁻¹), and complete oxidation of all substrate, Liu and Logan[22] estimated that the highest possible power density achievable for a mediatorless MFC was on the order of 1000 to 2000 mW/m². Greater power could be achieved with additional layers of bacteria or use of mediators. The highest power densities achieved thus far are 3600 to 4310 mW/m² for two-chamber MFC systems with added redox mediator in the

| $P_{max}$ mW/m² | Ks, mg/L | Substrate | MFC design | Resistor, Ω | Ref |
|---|---|---|---|---|---|
| 226* | 79 | Glucose | Single chamber, with PEM | 1,000 | 22 |
| 453* | 103 | Glucose | Single chamber, without PEM | 1,000 | 22 |
| 661 | 141 | Acetate | Single chamber, without PEM | 218 | 14 |
| 349 | 93 | Butyrate | Single chamber, without PEM | 1,000 | 14 |
| 225 | 1,512 | Swine wastewater | Single chamber | 1,000 | 12 |
| 166 | 461 | Domestic wastewater | Single chamber, flat plate with PEM (trial 1) | 470 | 11 |
| 218 | 719 | Domestic wastewater | Single chamber, flat plate with PEM (trial 2) | 470 | 11 |

*Calculated from reported maximum voltage data.

**TABLE 8.2** Reported Monod Kinetic Parameter Values Obtained in Mediatorless MFCs

cathode[20] or ferricyanide as electron acceptor.[21] Even greater power density values were obtained for an MFC with a specialized anode (platinum carbon anode with electrocatalytic conductive polymer) and ferricyanide as electron acceptor,[48] therefore, the maximum power density achievable may be higher than expected.

### 8.6.2 Single-Chamber Versus Two-Chamber Designs

Single-chamber MFC designs have been shown to produce higher power densities than two-chamber designs, due to differences in internal resistances for two systems. Maximum power density using acetate as substrate was 43 mW/m$^2$ for a two-chamber MFC[15] versus 506 mW/m$^2$ for single-chamber design.[14] For treatment of swine wastewater, Min et al.[12] reported a maximum power density of 45 mW/m$^2$ for a two-chamber MFC, which is consistent with other reported values (40 and 43 mW/m$^2$)[15,36] for the same two-chamber MFC using acetate as substrate. Operating the MFC as a single-chamber system with swine wastewater, a maximum power density of 261 mW/m$^2$ ($R = 200\,\Omega$) was obtained. In similar experiments using the same single-chamber MFC, maximum power density ranged from 146 mW/m$^2$ for municipal wastewater to 494 mW/m$^2$ for glucose.[10,14]

These results highlight a notable difference between two- and single-chamber MFC designs. For a specific two-chamber MFC, maximum power generation appears to be limited by internal resistance of the design, and does not vary greatly with the particular type of substrate used. However, for single-chamber MFC designs, power generation is impacted by type of substrate used, with greater power density achieved with more easily degraded substrates, perhaps due to the bacterial community that develops.[12]

### 8.6.3 Single-Chamber Designs

#### 8.6.3.1 With and without PEM

The primary function of the PEM is to allow passage of protons from anode to cathode, while preventing the diffusion of oxygen from cathode to anode. Oxygen flux through the membrane will allow for oxidation of substrate in the anode using oxygen as electron acceptor, bypassing the anode and potentially reducing power production. In the studies conducted by Liu et al.[10] the COD removal attributable to oxygen transfer through a single-chamber MFC with PEM operated under continuous flow conditions ranged from 9 percent at 3 hr hydraulic retention time to 93 percent at 33 hr retention time, for a relatively dilute (220 mg/L COD) wastewater. Liu and Logan[22] compared a single-chamber MFC with and without a PEM and found that the oxygen diffusion rate from cathode to anode for a single-chamber MFC was greater when a PEM was not used (0.187 mg $O_2$/hr without PEM vs. 0.05 mg $O_2$/hr with PEM). However, increased diffusion into anode without PEM may not translate into reduced power density. Peak power density has been shown to increase when PEM was removed from a single-chamber MFC.[14,22] Liu and Logan[22] found that using wastewater (200 to 300 mg/L COD), peak power density for single-chamber MFC without PEM was 146 mW/$m^2$ as compared to 28 mW/$m^2$ with PEM. Using glucose media (600 mg/L), peak power density was 381 mW/$m^2$ without PEM and 173 mW/$m^2$ with PEM.[22] However, a stable voltage was achieved for a much longer time period with PEM (i.e., for glucose runs, 95 hr with PEM vs. 10 hr, without PEM), resulting in much lower Coulombic efficiency without PEM (40 to 55 percent with PEM vs. ~10 percent without PEM). The authors noted that for the MFC without PEM, a biofilm developed on the interior portion of the cathode surface. This biofilm likely consumed any oxygen diffusing into the anode, thus maintaining anaerobic conditions at the anode surface. Due to the oxygen diffusion into anode, increased substrate removal was achieved in the MFC without PEM (for wastewater trials 55 percent of COD was removed with PEM vs. 75 percent without PEM). The authors conclude that although lower coulombic efficiencies are obtained, the decreased cost and increased COD removal associated with single-chamber MFC operation without PEM are advantageous for systems used for wastewater treatment.[22] However, the increased power density without PEM was unexplained.

### 8.6.3.2 Mixing

For single-chamber designs, mixing has a negative impact on power density. Maximum power achieved was lower with mixing of the anode solution (118 vs. 128 mW/$m^2$ for mixed vs. non-mixed)[12] and lower with forced air flow in open cathode vs. passive air flow.[12,22] This effect may be due to increased oxygen transfer into anode with increased mixing of anode or increased air flow rate in open-air cathode, resulting in an increase in redox potential in anode.[12] This theory is supported by the result that COD removal was greater for mixed anode chamber than non-mixed and greater for forced air flow than passive air flow, while the coulombic efficiencies were lower for mixed.[12]

### 8.6.4 Wastewater Treatment Effectiveness

The effectiveness of MFC treatment of wastewater at lab scale has been well documented. COD removal from wastewater in single-chamber MFCs ranged from 40 to 80 percent for 3 to 33 hr hydraulic retention times, versus 20 to 50 percent COD removal for anaerobic treatment at same retention times.[10] Reported COD loading rates for

MFCs have approached 3 kg COD/m3-day, in the same range as that for activated sludge systems (0.5 to 2 kg COD/m3-day).[33] For a single-chamber MFC operated in continuous flow mode with municipal wastewater, the characteristic response of decreasing effluent substrate concentration with increasing hydraulic retention time was observed[11] with concomitant decrease in power density. COD removal rates ranged from 42 to 79 percent removal at hydraulic retention times of 1.1 and 4 hr, respectively.[11]

Biomass yield values reported for bacteria in MFCs are low, ranging from 0.07 to 0.22 g biomass COD/g $S_{COD}$,[33] which is an advantage for MFCs applied as wastewater treatment systems. Typical biomass yield values for waste treatment systems range from 0.05 to 0.4 g biomass COD/g $S_{COD}$ for anaerobic and aerobic wastewater environments, respectively. Further, biomass yield values are reported to be lower at greater external resistances,[49] due to low energy gain by the cells at high resistance. For systems characterized as high-internal resistance MFCs, optimum power density is obtained when external resistance is equal to the internal resistance. Potentially, MFC systems could be designed to minimize biomass yield by control of internal resistance if desired. Therefore, the use of MFC systems to treat wastewaters and produce power may provide added benefits including greater efficiency of substrate removal, lower production of sludge (cell mass) to be disposed of, and means for continuous monitoring of the degradation process. Further development of MFC technology is needed at pilot and industrial scale for wastewater treatment systems.

## 8.7 Fabrication Example

A lab-scale MFC that can easily be constructed by students allowing for investigation of growth and electrical power generation due to choice of substrates, microbial cultures, or anode/cathode selection is described below (Fig. 8.6).[47]

Many MFCs constructed for lab and bench-scale testing are constructed of materials to allow for ease of fabrication with routine radial arm saws and drills. Acrylic (Plexiglas) or polycarbonate (Lexan) are often used in 6 mm (1/4 in) or 3 mm (1/8 in) wall thicknesses. Acrylic and polycarbonate can be purchased as flat sheet for rectangular designs or as tubes for cylindrical designs[10,22] (U.S. Plastics Corporation). In this example, a two-chamber MFC was designed with 6 mm thick acrylic sheets and a Nafion NE-117 PEM (DuPont Corporation). Neoprene rubber gaskets, 3 mm thick, were placed on either side of the Nafion membrane and the two halves of the fuel cell were attached by stainless steel, threaded tie-rods, and 6 mm thick acrylic end-plates. Each chamber contained a packed bed electrode of mesh filled with 200 g screened, crushed, recycled graphite particles (Shuler Industries) with average diameter of 3.2 mm. Alternatively, activated carbon (Marineland Aquarium products) used as filter media for aquaria may be used. The

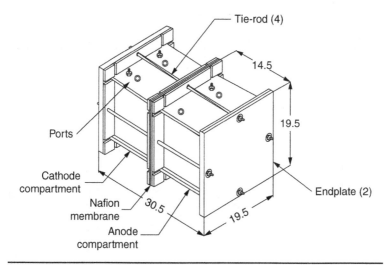

**Figure 8.6** A two-chamber microbial fuel cell utilizing a packed-bed electrode. (*Adopted from Pourciau et al., 1999.*)

*Note*: Dimensions in centimeter.

anode chamber had ports for sampling and insertion of pH probe, and cathode chamber has port for air flow and sampling. Resistors can be connected across the positive and negative terminals of the fuel cell using "bread boards" (All Electronic Corporation) so that variable resistance loads can be compared.

Inoculum from anaerobic sediments, wastewater sludge, or pure cultures may be used, depending on laboratory capabilities and skillset of students. For educational purposes, a non-objectionable substrate such as deproteinated milk serum (DPMS) may be used. DPMS is the permeate produced from cheese whey and milk ultrafiltration[50] that is comprised principally of lactose, water, low molecular weight proteins, and trace minerals (Table 8.3). The lactose concentration of DPMS is extremely high and exacts an enormous chemical oxygen demand on waste treatment systems (about 50,000 mg/L COD). Therefore, it represents a real wastewater disposal problem and relevant application of MFC systems for educational purposes. Deproteinated milk serum can be prepared by ultrafiltration of skim milk with a 10,000 NMWC cross flow continuous ultrafilter (A/G Technology, UFP-10-E-5A).

## 8.8 Future Directions

There are many areas of research and development that could increase power density, lower cost, and increase substrate utilization in MFCs. More advanced anode and cathode electrodes, specialized MFC designs,

| Component | Whole milk | Skim milk | Cheddar cheese whey | Lactic casein whey |
|---|---|---|---|---|
| Total solids | 5.60 ± 0.23 | 5.77 ± 0.19 | 6.41 ± 0.15 | 5.97 ± 0.45 |
| Lactose | 5.03 ± 0.20 | 5.06 ± 0.31 | 5.80 ± 0.23 | 4.55 ± 0.45 |
| Total nitrogen | 0.052 ± 0.02 | 0.060 ± 0.015 | 0.047 ± 0.009 | 0.062 ± 0.011 |
| Nonprotein nitrogen | 0.032 ± 0.008 | 0.023 ± 0.006 | 0.036 ± 0.005 | 0.042 ± 0.006 |
| Mineral (ash) | 0.46 ± 0.02 | 0.47 ± 0.02 | 0.54 ± 0.05 | 0.74 ± 0.05 |
| Lactate | — | — | — | 0.62 |
| Calcium | 0.03 | 0.02 | 0.05 | 0.14 |
| Sodium | 0.03 | 0.06 | 0.06 | 0.05 |
| Potassium | 0.12 | 0.16 | 0.18 | 0.17 |
| Magnesium | 0.01 | — | 0.01 | 0.07 |
| Phosphate | 0.11 | 0.09 | 0.12 | 0.26 |
| Chloride | 0.10 | 0.12 | 0.15 | 0.11 |

*Source*: Adapted from Hobman, 1984.

**TABLE 8.3** Percent Composition (wt/vol) for Ultrafiltration-Derived DPMS from Several Sources

and added mediators in the anode and cathode can boost power density and may have wide application for use as biological sensors or in sensor networks.[35] For municipal, animal, or bioprocessing wastewater treatment, mediatorless systems with simple anode/cathode designs using acclimated cultures from natural sources will likely be used due to their low cost. One MFC design approach that could be employed in an integrated waste treatment system is biological production of oxygen in the cathode of a two-chamber MFC. Oxygen used in an aqueous cathode chamber as external electron acceptor is typically generated with forced oxygen or air flow. However, biological oxygen production in the cathode through photosynthetic cyanobacterial growth could provide an economical means of oxygen generation, and potential increase in power density due to redox mediators excreted into solution by the cells.[46] Further, valuable products may be formed by the cyanobacteria in MFCs and recovered in addition to electrical power generation. For example, the potential for hydrogen production in the cathode by direct biophotolysis exists if photosynthetic cyanobacteria are used. Direct biophotolysis is normally inhibited by elevated oxygen concentrations under natural conditions, but the rapid consumption of oxygen at the cathode of an MFC should remove this inhibition, allowing for hydrogen recovery. Internal cyanobacterial storage products and other cellular constituents such as high-value polar lipids could be produced along with electron flow for MFC designs utilizing algae in the cathode. If the MFC is used for wastewater treatment, the overall scheme would include oxidation of organic substrate in the anode chamber by organotrophic bacteria, with treated effluent from the anode chamber pumped to the cathode chamber. Cyanobacterial growth would occur in the cathode chamber using solar energy and utilizing the nutrients contained in the treated wastewater. Thus, secondary and tertiary treatment (organic matter oxidation with nutrient removal) of the wastewater with hydrogen production could be achieved in single MFC.

# References

1. Lovely, D. 2006. "Microbial fuel cells: Novel microbial physiologies and engineering approaches." *Curr Opin Biotechnol.* 17:327–332.
2. Thurston, C., et al. 1985. "Glucose metabolism in a microbial fuel cell. Stoichiometry of product formation in a thionine-mediated *Proteus vulgaris* fuel cell and its relation to coulombic yields." *J Gen Microbiol.* 131:1393–1401.
3. Williams, K. R. 1966. *An Introduction to Fuell Cells.* Elsevier, Amsterdam, Chap. 10.
4. Bennetto, H. P. and J. L . Stirling. 1983. "Anodic reactions in microbial fuel cells." *Biotechnol Bioeng.* 25(2):559–568.
5. Bennetto, H., et al. 1985. "The sucrose fuel cell: Efficient biomass conversion using a microbial catalyst." *Biotechnol Lett.* 7(10):699–704.
6. Delaney G., et al. 1984. "Electron-transfer coupling in microbial fuel cells. 2. Performace of fuel cells containing selected microorganism-mediator-substrate combinations." *J Chem Technol Biotechnol.* 34B:13–27.
7. Roller, S., et al. 1983. "A bio-fuel cell for utilization of lactose wastes." *Biotech '83: Proceedings of the International Conference on the Commercial Application and Implications of Biotechnology,* pp. 655–663.
8. Roller, S., et al. 1984. "Electron-transfer coupling in microbial fuel cells: 1. Comparison of redox-mediator reduction rates and respiratory rates of bacteria." *J Chem Technol.* 34B:3–12.
9. Bennetto, H. P. 1987. "Microbes come to power." *New Sci.* 114:36–39.
10. Liu, H., R. Ramnarayanan, and B. E. Logan. 2004. Production of electricity during wastewater treatment using a single chamber microbial fuel cell. *Environ Sci Technol.* 38:2281–2285.
11. Min, B. and B. E. Logan. 2004. "Continuous electricity generation from domestic wastewater and organic substrate in a flat plate microbial fuel cell." *Environ Sci Technol.* 38(21):5809–5814.
12. Min, B., et al. 2005. "Electricity generation from swine wastewater using microbial fuel cell." *Water Research.* 39:4961–4968.
13. Oh, S. E. and B. E. Logan, 2005. "Hydrogen and electricity production from food processing wastewater using fermentation and microbial fuel cell technologies." *Water Resources.* 39(10):4673–4682.
14. Liu, H., S. Cheng, and B. E. Logan. 2005. "Production of electricity from acetate or butyrate using a single-chamber microbial fuel cell." *Environ Sci Technol.* 39(2):658–662.
15. Oh, S., B. Min, and B. E. Logan. 2004. "Cathode performance as a factor in electricity generation in microbial fuel cells." *Environ Sci Technol.* 38:4900–4904.
16. Yagishita, T., et al. 1997. "Effects of intensity of incident light and concentrations of *Synechococcus* sp. and 2-hydroxy-1,4-naphthoquinone on the current output of photosynthetic electrochemical cell." *Solar Energy.* 61(5):347–353.
17. Tsujimura, S., et al. 2001. "Photosynthetic bioelectrochemical cell utilizing cyanobacteria and water-generating oxidase." *Enzyme and Microb Technol.* 29: 225–231.

18. Lam, K. B., B. Chiao, and L. Lin. 2003. "A micro photosynthetic electrochemical cell." *Proceedings: 16th Annual International Conference on Micro Electro Mechanical Systems, 2003.* Kyoto. pp. 391–394.
19. Chaudhuri, S. and D. R. Lovely. 2003. "Electricity generation by direct oxidation of glucose in mediatorless microbial fuel cells." *Nat Biotechnol.* 21: 1229–1232.
20. Rabaey, K., et al. 2003. "Microbial fuel cell capable of converting glucose to electricity at high rate and efficiency." *Biotechnol Lett.* 25:1531–1535.
21. Rabaey, K, et al. 2004. "Biofuel cells select for microbial consortia that self-mediate electron transfer." *Appl Environ Microbiol.* 70(9):5373–5382.
22. Liu, H. and B. E. Logan. 2004. "Electricity generation using an air cathode single-chamber microbial fuel cell in the presence and absence of a proton exchange membrane." *Environ Sci Technol.* 38(14):4040–4046.
23. He, Z., S. D. Minteer, and L. T. Angenent. 2005. "Electricity generation from artificial wastewater using an upflow microbial fuel cell." *Environ Sci Technol.* 39:5262–5267.
24. Logan, B. E., et al. 2005. "Electricity generation from cysteine in a microbial fuel cell." *Water Research.* 39:942–952.
25. Aelterman, P., et al. 2006. "Continuous electricity generation at high voltages and currents using stacked microbial fuel cells." *Environ Sci Technol.* 40(10):3388–3394.
26. Cheng, S., H. Liu, and B. H. Logan. 2006. "Power densities using different cathode catalysts (Pt and CoTMPP) and polymer binders (Nafion and PTFE) in single chamber microbial fuel cells." *Environ Sci Technol.* 40:364–369.
27. Ringeisen, B.R., R. Ray, and B. Little. 2006. A miniature microbial fuel cell operating with an aerobic anode chamber. *J Power Sources.* 165(2):591–597.
28. Kim, J. R., et al. 2007. "Electricity generation and microbial community analysis of alcohol powered microbial fuel cells." *Bioresour Technol.* 98:2568–2577.
29. Logan, B., et al. 2007. "Graphite fiber brush anodes for increased power production in air-cathode microbial fuel cells." *Environ Sci Technol.* 41(9):3341–3346.
30. Zuo, Y., et al. 2007. "Tubular membrane cathodes for scalable power generation in microbial fuel cells." *Environ Sci Technol.* 41:3347–3353.
31. Rabaey, K., et al. 2005. "Tubular microbial fuel cells for efficient electricity generation." *Environ Sci Technol.* 39:8077–8082.
32. Eisenberg, M. 1962. Design and scale-up considerations for electrochemical fuel cells. In: *Advances in Electrochemistry and Electrochemical Engineering, Vol. 2 Electrochemical Engineering.* C. Tobias (ed.). Interscience: New York, pp. 235–291.
33. Logan, B. E., et al. 2006. "Microbial fuel cells: Methods and technology." *Environ Sci Technol.* 40(17):5181–5192.
34. Kim, J. R., B. Min, and B. E. Logan. 2005. "Evaluation of procedures to acclimate a microbial fuel cell for electricity production." *Appl Microbiol Biotechnol.* 68(1):23–30.
35. Ringuisen, B. R., et al. 2006. "High power density from a miniature microbial fuel cell using *Shewanella oneidensis* DSP10." *Environ Sci Technol.* 40(8):2629–2634.
36. Kim, H. J., et al. 2002. "A mediator-less microbial fuel cell using a metal reducing bacterium, *Shewanella putrefacians*." *Enzyme Microbiol Technol.* 30:145–152.
37. Bond, D. R. and D. R Lovely. 2003. "Electricity production by *Geobacter sulfurreducens* attached to electrodes." *Appl Environ Microbiol.* 30:1548–1555.
38. Khashe, S. and M. Janda. 1998. "Biochemical and pathogenic properties of *Shewanell alga* and *Shewanella putrefaciens*." *J Clinical Microbiol.* 783–787.
39. Gorby Y. A., et al. 2006. "Electrically conductive bacterial nonwires produced by *Shewanella oneidensis* strain MR-1 and other microorganisms." In: *Proceedings of the National Academy of Sciences of the United States of America.* 13(30):11358–11363.
40. Doong, R. A. and B. Schink. 2002. "Cysteine-mediated reductive dissolution of poorly crystalline iron(III) oxides by *Geobacter sulfurreducens*." *Environ Sci Technol.* 36:2939–2945.
41. Katz E., A. N. Shipway, and I. Willner. 2003. "Biochemical fuel cells." In: W. Vielstich, H. A. Gasteiger, and A. Lamm (eds.), *Handbook of Fuel*

*Cells-Fundamentals, Technology and Applications,* Vol. 1. John Wiley & Sons: New York, N, pp. 355–381.
42. Bennetto, H. P. 1990. "Electricity generation by microorganisms." *Biotechnol Education.* 1(4):163–168.
43. Newman, D. K. and R. Kolter. 2000. "A role for excreted quinones in extracellular electron transfer." *Nature.* 405(6782):94–97.
44. Slade, S., et al. 2002. "Ionic conductivity of an extruded Nafion 1100 EW series of Membranes." *J Electrochem Soc.* 149:12.
45. Basura, V. I., P. D. Beattie, and S. Holdcroft. 1992. "Solid-state electrochemical oxygen reduction at Pt/Nafion 117 and Pt/Bam3G 407 interfaces." *J Electroanalytical Chem.* 139:2530–2537.
46. Little, D. 2008. "Investigation of a photosynthetically aerated cathode in a microbial fuel cell treating biohydrogen fermentation waste, MS thesis. Clemson University, Clemson SC.
47. Pourciau, B., C. M. Drapcho, and M. Lima. 1999. "Design of a novel biological fuel cell for the utilization of deproteinated milk serum." *Proceedings: 4th Annual Meeting of the Institute for Biological Engineering,* June 18–29, Charlotte, NC. 2:B19–B29.
48. Schroder, U., J. Nieben, and F. Scholz. 2003. "A generation of microbial fuel cells with current outputs boosted by more than one order of magnitude." *Angew Chem Int Ed.* 42:2880–2883.
49. Freguia, S., et al. 2007. "Electron and carbon balances in microbial fuel cells reveal temporary bacterial storage behavior during electricity generation." *Environ Sci Technol.* 41(8):2915–2921.
50. Hobman, P. 1984. "Review of processes and products for utilization of lactose in deproteinated milk serum." *J Dairy Sci.* 67:2630–2653.

# CHAPTER 9
# Methane

With J. Michael Henson, Ph.D.
*Clemson University*
*Clemson, South Carolina*

## 9.1 Introduction

Anaerobic digestion of animal manures, municipal wastewater or sludge, and food processing wastewater has been practiced in various forms for many decades. Anaerobic digestion decreases the organic matter content in the waste through the biological conversion of organic carbon and produces a biogas containing primarily methane and carbon dioxide. Due to the development of sophisticated treatment technologies, thousands of anaerobic treatment processes are now installed and operating worldwide[1] in addition to the millions of small low-tech systems operating throughout less industrialized nations. Because of the expansive information available on the biological production of methane and anaerobic digestion process design,[2-4] this chapter serves as a brief overview of the microbial process and systems used for anaerobic digestion.

## 9.2 Microbiology of Methane Production

### 9.2.1 Methanogenic Environments

The microbiological production of methane occurs in engineered systems such as anaerobic digesters used for municipal and animal waste treatment, in the gastrointestinal tract of ruminants and nonruminant herbivores, and in many natural environments such as freshwater, estuarine, and marine sediments. In the late 1770s, the Italian physicist, Alessandro Volta, most well-known for developing an early battery, performed experiments with what he termed *combustible air*.[5] Volta collected gas bubbles from the disturbed sediment of a shallow lake and described the combustion of the gas that was later called methane. For his contribution toward the understanding of methanogenesis, the methanogenic archaean,

*Methanococcus voltae*, is named in his honor.[5] True anaerobic conditions, devoid of inorganic terminal electron acceptors such as oxygen, nitrate, ferrous iron, and sulfate, are required for methane production. Should any of these electron acceptors become available to the microbial community, the flow of electrons will be diverted to that electron acceptor, resulting in diminished or complete cessation of methanogenesis.[6] Under these circumstances, other reduced products will result, for example, nitrogen gas if nitrate is present or hydrogen sulfide if sulfate is present. The late Dr. Robert E. Hungate developed many of the techniques used in the study of anaerobic microbiology, including the conditions required to obtain the − 300 mV redox potential required for methanogenesis.

## 9.2.2 Methane Process Description

Methane is produced by a metabolically diverse community of bacteria and archaea that act as an integrated metabolic unit to produce methane and carbon dioxide through a series of sequential and concurrent reactions. The end-products of one group's metabolism are used as substrate by the next group. In general, the biological production of methane from complex organic compounds contained in biomass and waste sources involves four main phases: hydrolysis, fermentation (acidogenesis), acetogenesis, and methanogenesis (Fig. 9.1).

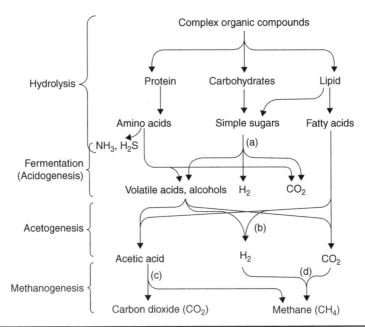

**FIGURE 9.1** Four stages of biological methane production. Numbered reactions involving biofuel production are: (1) hydrogen production by fermentative bacteria, (2) hydrogen production by syntrophic bacteria, (3) methane production by acetoclastic methanogens, and (4) methane production by hydrogen-oxidizing methanogens. (Adapted from Grady et al.[3])

### 9.2.2.1 Hydrolysis

Many of the potential biomass sources for methane production are high molecular weight, insoluble polymers such as polysaccharides, proteins, and fats that are too large to be transported across bacterial cell membranes. The initial conversion reactions may require several different types of enzymes. One unifying characteristic of these enzymes is that they are synthesized within the bacterial cells in small amounts and are secreted into the environment surrounding the bacterium until they contact the polymers. These enzymes catalyze hydrolysis reactions that cleave polymers and incorporate water, thus producing soluble monomers. Polysaccharides such as cellulose and hemicellulose are hydrolyzed to glucose and xylose by cellulase and hemicellulase enzymes. Proteins and lipids are hydrolyzed to their constituent amino acids and long-chain fatty acids by proteases and lipases, respectively. These compounds then enter the cell through active transport, and once an increase in the specific degradation products are sensed by the bacteria, genes that produce these enzymes are upregulated to increase the amount of these enzymes being secreted into the environment. In this way, the bacteria do not expend cellular energy producing these enzymes at high rates when not needed. This interconnectedness of the genetic regulation and microbial degradation process is fundamental to the ability to direct specific reactions. The rate of hydrolysis is a function of several factors, such as pH, substrate composition, and particle size.

### 9.2.2.2 Fermentation (acidogenesis)

The second phase of the overall process is fermentation that begins with the conversion of the sugar monomers to pyruvate ($C_3H_4O_3$), ATP, and the electron carrier molecule NADH by central metabolic pathways. The central metabolic pathways found within most bacteria are the Embden-Meyerhof pathway (glycolysis) and the pentose phosphate pathway (see Chap. 2 for Biochemical Pathways Review).

Next, these fermentative bacteria convert pyruvate and amino acids to a variety of short-chain organic acids—primarily acetate, propionate, butyrate, and succinate—and alcohols, $CO_2$, and $H_2$ through various fermentation pathways. The specific fermentation product denotes the fermentation pathway; for example, the pathway that yields ethanol is referred to as the ethanol fermentation pathway. During the fermentation reactions, NADH is oxidized to regenerate $NAD^+$, while organic intermediates of the fermentation pathways are reduced. Because the fermentation process results in the formation of various short-chain organic acids, this stage of the methane fermentation is also referred to as the *acid-forming stage* or *acidogenesis*.

### 9.2.2.3 Acetogenesis

The short-chain organic acids produced by fermentation and the fatty acids produced from the hydrolysis of lipids are fermented to acetic acid, $H_2$, and $CO_2$ by acetogenic bacteria. Syntrophic bacteria that oxidize organic acids to acetate, $H_2$, and $CO_2$ are reliant on the subsequent oxidation of $H_2$ by the next group, the methanogens, to lower the $H_2$ concentration and prevent end-product inhibition.[7] Syntrophic $H_2$ production is thermodynamically unfavorable, except when the $H_2$ partial pressure is extremely low. Inhibition or disruption of methanogenesis will cause an accumulation of $H_2$, which rapidly inhibits acetogenesis. *Homoacetogenesis* is the conversion of $H_2$ and $CO_2$ to acetic acid.[8] This route can be encouraged by collecting biogas containing $H_2$ and $CO_2$ at this stage if formation of acetate is desired.

### 9.2.2.4 Methanogenesis

In the final phase, methane is produced through two distinct routes by two different microbial groups. One route is by the action of the lithotrophic $H_2$-oxidizing methanogens that use $H_2$ as electron donor and reduce $CO_2$ to produce methane [Eq. (9.1)]. In the second route, the organotrophic acetoclastic methanogens ferment acetic acid to methane and carbon dioxide [Eq. (9.2)]. Approximately two-thirds of the methane produced derives from the acetoclastic methanogens. The complete conversion of a carbohydrate such as glucose to methane in anaerobic digestion results in 85 to 90 percent recovery of energy.[9]

$$4H_2 + CO_2 \rightarrow CH_4 + 2H_2O \qquad (9.1)$$

$$CH_3COOH \rightarrow CH_4 + CO_2 \qquad (9.2)$$

Methanogens occupy a unique position within the overall methane process in that they utilize acetate thereby removing a significant source of acidity, and they consume $H_2$ which allows for the continued growth of the syntrophic bacteria that convert propionate and butyrate to acetate. If the activity of the syntrophic bacteria is impeded, short-chain organic acids will accumulate, with a resulting decrease in pH that creates an environment unfavorable for the methanogenic archaea. At this point the overall fermentation essentially ceases.

### 9.2.3 Microbial Communities

The production of methane involves diverse groups of microbes responsible for each stage. The anaerobic bacteria that produce the hydrolytic enzymes needed for hydrolysis of complex organic compounds include species of *Bacteroides, Clostridium, Acetivibrio, and Fibrobacter*.[6] Some of the enzyme systems are very complex, especially those responsible for the degradation of cellulose. For example, cellulose-degrading anaerobic bacteria often produce cellulosome structures that contain multiple enzymes and may be associated with the surface of the bacteria or free within the immediate environment of the bacterium.

Many of the bacteria that carry out fermentation are the same as those that carried out the hydrolytic reactions, such as *Bacteroides* or *Clostridium*. Others, such as *Lactobacillus* are also involved in fermentation. Recently, four filamentous bacteria that belong to a new class, *Anaerolineae*, were isolated from anaerobic wastewater treatment sludges.[1] Two thermophilic species of the genus *Anaerolinea*, and one mesophilic species each of genera *Levilinea* and *Leptolinea* were identified. All strains ferment a range of carbohydrate substrates, suggesting that members of this new class play a key role in the primary degradation of carbohydrate and protein in methanogenic digestion processes.[1]

Syntrophic bacteria are involved in the conversion of propionate and butyrate to acetic acid in the acetogenesis phase. Species of *Syntrophomonas* including *S. wolfei, S. bryantii, and S. sapovorans* have been identified as butyrate-oxidizers that produce $H_2$ under syntrophic conditions in anaerobic digestion.[8] Members of the genus *Syntrophobacter* that oxidize propionate are also important syntrophic $H_2$ producers. Recently identified strains of the genus *Pelotomaculum* found in anaerobic waste treatment reactors include the propionate-oxidizing *P. thermopropionicum* and *P. schinkii*, and phthalate-degrading *P. terrephthalicum* and *P. isophthalicum*.[1] Members of genus *Pelotomaculum* are also key in propionate degradation in natural anaerobic environments such as wetlands.[1]

The terminal reactions of methanogenesis are carried out by a highly specialized group of strictly anaerobic archaea. While the methanogens resemble bacteria under the microscope in size and shape, they differ physiologically in several respects. For example, bacteria have cell walls composed of peptidoglycan, while the methanogenic archaea have cell walls of pseudomurein, protein, or polysaccharide. The polar lipids of the cytoplasmic membranes of bacteria and eukarya are phospholipids with fatty acids of various lengths containing one or more double bonds attached via ester linkages. Methanogens, on the other hand, have polar lipids in their cytoplasmic membranes attached via ether linkages to isoprene units of $C_{20}$ or $C_{40}$ chain lengths. These membrane structures may be either bilayers or monolayers.

The $H_2$-oxidizing methanogens include members of *Methanobrevibacter, Methanobacterium, Methanospirillum,* and *Methanogenium*.[3] As autotrophs, their biomass yield values are extremely low, but their growth rates are high enough so most of the $H_2$ produced is consumed. The acetoclastic methanogens include *Methanosaeta* spp., which are favored at low acetate concentrations, and *Methanosarcina* spp., which are favored when acetate concentrations are high.[3,10]

Overall, the biological production of methane is unique among the major biofuel production processes in that an integrated microbial community is involved rather than a pure culture of a single species. The advantage of this consortium is that members can perform the specialized functions of hydrolysis, fermentation, acetogenesis, or methanogenesis, that could not be performed by a single species. The presence of the hydrolytic bacteria allows for complex biomass sources containing cellulose to be utilized. In the case of many other biofuel processes, such as natural strains of *Saccharomyces cerevisaea* that ferment sugars to ethanol, the microbe that produces the biofuel cannot produce the enzymes needed to degrade cellulose or other complex organic compounds. The process design for ethanol production must include hydrolysis reactors that rely on added acid, base, or enzymes to convert biomass feedstock to soluble compounds, thus adding substantially to the cost of the bioprocess. The fact that the microbial community involved in methane production includes microbes capable of hydrolyzing cellulosic compounds make it highly versatile for biofuel production using waste biomass sources.

## 9.3 Biomass Sources for Methane Generation

The main categories of biomass usable for methane production include municipal and animal waste and wastewaters, municipal solid waste, food and vegetable processing waste, postharvest agricultural by-products, and dedicated energy crops. A complete review of methane generation from a variety of biomass sources is presented by Gunaseelan.[11] The composition of biogas produced from anaerobic digestion varies with biomass source. Biogas from digestion of biomass with high carbohydrate content will have lower methane and higher carbon dioxide content than biogas from biomass with greater lipid and protein content.[3,12] The theoretical maximum methane yield from carbohydrate, protein, and lipid is 0.37, 1.0, and 0.58 m$^3$ CH$_4$/kg organic dry matter.[12] Most animal manures contain all three components and so can effectively be digested for methane production. Because of the relatively low cost of anaerobic treatment, animal manures have been treated by anaerobic digestion in the United States for many years, but often without optimization for methane production or collection of biogas. Confined animal production facilities such as dairy farms, beef cattle feedlots, swine operations, and laying hen facilities produce large quantities of animal manures that are typically treated by anaerobic digestion. The dry matter content (or total solids) and organic dry matter (or volatile solids) of the manure affect the biogas yield (Table 9.1). Cosubstrates of corn or grass silage may be added to boost the methane content of biogas. These cosubstrates may also contain higher nutrient content (N and P) than the biomass source, which may be advantageous for land application of solids after digestion.[4]

| Feedstock | Dry matter (DM %) | Organic dry matter (ODM, % of DM) | Biogas yield (m³/ton ODM) | Biogas yield (m³/wet ton) |
|---|---|---|---|---|
| Cow manure | 7–15 | 65–85 | 200–400 | 25 |
| Pig manure | 3–13 | 65–85 | 350–550 | 27 |
| Chicken manure | 10–20 | 70–80 | 350–550 | 51 |
| Vegetable waste | 10–20 | 65–85 | 400–700 | 75 |
| Corn silage | 15–40 | 75–95 | 500–900 | 200 |
| Grass silage | 30–50 | 80–90 | 500–700 | 220 |
| Fat slurry | 8–50 | 70–90 | 600–1300 | 310 |

*Source*: Adapted from Planning and Installing Bioenergy Systems: A Guide for Installers, Architects and Engineers, 2005.

**TABLE 9.1** Manure and Cosubstrate Feedstock Characteristics and Biogas Yield

Municipal wastewater and bacterial sludge produced from aerobic wastewater treatment are often treated by anaerobic digestion. Advantages of anaerobic wastewater treatment over aerobic treatment include formation of valuable biogas, decreased sludge production due to low cell yield, and lower operating costs since aeration is not required. The methane generation from anaerobic processes is proportional to the decrease in chemical oxygen demand (COD) of the waste or sludge. The COD value of methane gas is 4 g COD/g $CH_4$.[3] At standard temperature and pressure (0°C and 1 atm pressure), this corresponds to 0.35 m³ $CH_4$ produced/kg COD removed, or 0.7 m³ $CH_4$ produced/kg VS destroyed for primary sludge.[3] A biochemical methane potential of 0.59 m³/kg VS for digestion of primary sludge has been reported.[13] Municipal solid wastes can also be digested anaerobically, with typical methane production rates ranging from 0.15 to 0.29 m³/kg VS.[11]

A significant source of biomass is found in various plants. Crops containing starch, such as corn, potatoes, rice, and wheat are excellent for fermentation to methane as are aquatic plants, including various kelp and algae. Several plants are being evaluated as to their suitability as possible biofuel crops. These plants include various poplars, willows, sorghum, sugarcane, pine trees, and switchgrass. Digestion of the cellulosic and hemicellulosic fractions of these plants may differ in rate and yield of methane production, since the hydrolysis of these materials will differ from the starchy plants. A potential biofuel crop is switchgrass (*Panicum virgatum*), a perennial grass found on the prairies of the Great

Plains. Under research conditions, this biofuel crop can produce significant quantities of biomass per acre, averaging 6 to 8 tons/acre with 11.5 tons/acre under optimal conditions.[14] Furthermore, this native prairie grass potentially may not only serve as a source of biofuel but it may also serve as significant carbon sink as well because of the extensive root system that it develops. For many plant biomass sources, sufficient nutrients may not be present so that addition of nitrogen or phosphorus boosts methane generation.[11]

Fruit and vegetable wastes (FVW) represent a substantial waste source that can be digested for methane production. The composition of various types of FVW indicates that a high percentage of the total solids are organic (volatile) solids with low cellulose content (Table 9.2).

The C:N:P ratio (or COD:N:P ratio) may be balanced for some types of FVW so that nutrient addition is not required for digestion, although for other wastes addition of N or P may be required. For example, the COD:N:P ratio was determined as 333:4:1 for cull peaches and 500:9:1 for honeydew, while a COD:N:P ratio of 300:5:1 is generally adequate for digestion.[16] This would indicate that N addition would boost methane generation for cull peach digestion while P addition would aid honeydew digestion. Due to the high organic solids content of most FVW, dilution is often required to obtain the desired organic loading rate and pH adjustment is often needed.[15,16]

**Example 9.1** Anaerobic digestion of manure. (Adapted from *Planning and Installing Bioenergy Systems: A Guide for Installers, Architects and Engineers*, 2005.)

Pig manure, 5000 m³, (1 ton/m³) and 1000 m³ of grass silage (0.8 ton/m³) are digested annually. The pig manure has a dry matter (DM) content of 10 percent, organic matter (OM) content of 80 percent of dry matter, and biogas yield of 450 m³/ton OM. The grass silage has a DM of 45 percent, OM of 90 percent, and biogas yield of 600 m³/ton ODM. Determine the biogas production per year.

**Solution**

$$\text{Rate}_{biogas, manure} = \frac{5000 \, m^3_{manure}}{yr} \times \frac{1 \, ton}{m^3} \times \frac{0.10 \, ton_{DM}}{ton_{manure}} \times \frac{0.80 \, ton_{DM}}{ton_{DM}}$$

$$\times \frac{450 \, m^3_{biogas}}{ton_{OM}} = 180,000 \, \frac{m^3_{biogas}}{yr}$$

$$\text{Rate}_{biogas, silage} = \frac{1,000 \, m^3_{silage}}{yr} \times \frac{0.8 \, ton}{m^3} \times \frac{0.45 \, ton_{DM}}{ton_{silage}} \times \frac{0.90 \, ton_{OM}}{ton_{DM}} \times \frac{600 \, m^3_{biogas}}{ton_{OM}}$$

$$= 194,000 \, \frac{m^3_{biogas}}{yr}$$

Total biogas production = 180,000 + 194,400 = 374,400 m³/yr.

| Waste | Total solids (TS), g/kg | Volatile solids (VS), g/kg | Volatile solids (% of TS) | Cellulose (g/kg) | Kjeldahl-N, g/kg | Total P, g/kg |
|---|---|---|---|---|---|---|
| Cull whole peach (depitted) | 81.2 | 69.7 | 85.9 | — | 1.38 | 0.357 |
| Honeydew | 69.0 | 62.5 | 90.6 | — | 1.58 | 0.166 |
| Potato peelings | 119.2 | 105.5 | 88.5 | 12.9 | — | — |
| Salad waste | 79.4 | 72.1 | 90.8 | 16.1 | — | — |
| Green peas and carrots | 179.4 | 171 | 95.3 | 16.1 | — | — |

*Source:* Adapted from Bouallagui, 2005, and Hills and Roberts, 1982.

**TABLE 9.2** Composition of Different Fruit and Vegetable Wastes

This calculation emphasizes the increased biogas production due to addition of a cosubstrate such as silage.

**Example 9.2** Anaerobic digestion of VW. (*Adapted from Hills and Roberts, 1982.*)

Cull fresh peaches are added to a digester at the rate of 8600 kg wet weight/day for 90 days. The peaches have a total solids (TS) content of 8.1 percent and a volatile solids (VS) content of 86 percent of TS, and a biogas generation rate of 0.81 m³ biogas/kg VS. The biogas contains 52 percent methane. Calculate the daily methane generation rate.

**Solution**

$$\text{Rate}_{CH_4} = \frac{8600 \text{ kg}_{peaches}}{\text{day}} \times 0.081 \frac{\text{kg}_{TS}}{\text{kg}_{peaches}} \times 0.86 \frac{\text{kg}_{VS}}{\text{kg}_{TS}} \times 0.81 \frac{\text{m}^3_{biogas}}{\text{kg}_{VS}}$$

$$\times 0.52 \frac{\text{m}^3_{CH_4}}{\text{m}^3_{biogas}} = 252.3 \frac{\text{m}^3_{CH_4}}{\text{day}}$$

## 9.4 Systems

Systems for anaerobic digestion for methane production range from low-rate, unmixed, unheated anaerobic lagoons used for small farm animal waste treatment to large, high-rate, process-controlled, heated, mixed digesters that may include both wet (low solids) and dry (high solids) fermentation systems. Wet fermentation systems are more common, although dry fermentation processes have been developed for municipal solid wastes, yard wastes, and dehydrated wastewater treatment sludges.[17] A typical anaerobic digestion system (Fig. 9.2) may

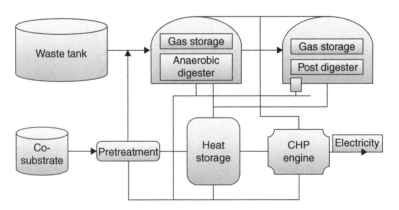

**Figure 9.2** Schematic of anaerobic digester system starting with manure waste and cosubstrate storage. The biogas engine functions as a combined heat and power (CHP) unit. (*Adapted from Planning and Installing Bioenergy Systems: A Guide for Installers, Architects and Engineers*, 2005.)

range in size from small (<100 m$^3$), to farm-scale (20 to 800 m$^3$), to industrial[4] (>800 m$^3$). The main components of most systems include an enclosed bioreactor, mixing system, heating system, and gas-liquid separation.[3] Also, some means of biomass storage may be needed, but methane generation will occur during storage, so for maximum energy recovery, manure should be moved to digester as soon as possible. Cosubstrates are typically stored in separate tank, since they may require pretreatment prior to digester, such as chopping or grinding. Mixing of the bioreactor can be accomplished with mechanical mixing, or recirculation of digestion liquid or biogas.

## 9.4.1 Reactor Conditions

The conditions required for effective anaerobic digestion are influenced by many factors, including system design. In general, the main operating conditions are:

### Environment
Anaerobic conditions must be maintained for methane production. A redox potential of −300 mV is required for growth of methanogens.[6]

### Moisture Content
For wet fermentation systems, the total solids content of the biomass slurry is typically less than 12 percent, with 2 to 10 percent most common for wastewater treatment.[2] For dry fermentation systems, the total solids content may exceed 20 percent.[2]

### Temperature
Anaerobic digestion can occur under psychrophilic (10 to 20°C), mesophilic (20 to 40°C), and thermophilic (40 to 60°C) conditions, although most systems operate in the mesophilic or thermophilic range. Maximum methane generation rates are achieved at a temperature of 35 to 37°C for mesophilic systems and ~55°C for thermophilic systems.[2] Within a temperature range, an increase in temperature increases methane generation rate. For example, reported rates of methane generation for digestion of crop residues were 310 and 366 mL/L/day for 25°C and 37°C, respectively, as compared to 108 mL/L/day in the reactor initially stabilized at 18°C.[18]

### Retention Time
For continuous flow digester systems, the necessary retention time decreases with increasing temperature as reaction rates increase with increasing temperature. Typical retention times are 40 to 100 days for psychrophilic, 25 to 40 days for mesophilic, and 15 to 25 days for thermophilic digestion. A greater hydraulic retention time for a given influent flow rate will result in larger required reactor volume.

### Loading Rate

The loading rate of waste to the digester is calculated as the concentration of total solids (TS) or volatile solids (VS) in the influent flow (mass/volume) divided by the hydraulic retention time. Loading rates range from 0.5 to 5 kg TS/$m^3$/day, with typical values of 1 to 3 kg TS/$m^3$/day for batch or single stage continuous flow systems.[4] Calculations of methane generation rates for anaerobic digestion typically report volume of methane generated per unit of VS or TS removed.

### pH

A neutral pH (range of 6.7 to 7.4) is required for methanogenesis.[2,15] Balanced growth of the fermentative, acetogenic, and methanogenic microbes will maintain pH in the proper range, but perturbations of the process (i.e., sudden changes in loading rate, temperature, or feed constituents) may upset the microbial balance. The formation of acids during the acidogenesis phase requires that the subsequent reactions proceed to consume these acids. Once the pH of the digester falls below 6.7, the interactions between these different groups of microbes become unbalanced and digestion begins to fail.

### 9.4.2 Process Design

Anaerobic digestion can be accomplished with batch or continuous flow processes.[3] In batch systems (Fig. 9.3), digesters are filled with biomass and the reactions proceed through all degradation steps sequentially. Batch systems are simpler and easier to operate than continuous flow systems, which make them attractive in low-tech applications.

Continuous stirred tank reactor (CSTR) systems (Fig. 9.4), are often used for anaerobic digestion. The loading rate of solids to an anaerobic digester is controlled by the kinetics of microbial activity,[2] with the slowest phase of the overall digestion process acting as the rate-limiting step. For biomass sources containing high amounts of cellulosic materials, hydrolysis is typically the rate-limiting step.[2]

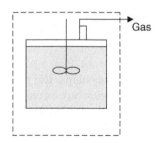

FIGURE 9.3   Batch system.

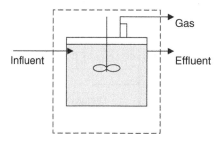

**FIGURE 9.4** Simple CSTR (no cell recycle).

However, for fruit and vegetable waste (FVW) with a high sugar and low cellulose content, methanogenesis is most often the rate-limiting phase. In FVW, the sugars are rapidly acidified to volatile acids and tend to inhibit methanogenesis when the feedstock is not adequately buffered. The acidogenic and methanogenic organisms differ with respect to pH optima, growth rates, and nutrient uptake kinetics. Because of the multistep nature of anaerobic digestion, two-stage reactor systems have been shown to be advantageous for continuous flow systems. The hydraulic retention times in the two reactors are set independently to optimize the rate of hydrolysis/fermentation in the first reactor and methanogenesis in the second reactor, thus maximizing the rate and amount of methane production. Sequencing batch reactors can also be used to separate the hydrolysis/fermentation and methanogenesis phases similar to two-stage continuous systems. However, for biomass sources with high lipid content, a two-stage system is not preferable, since hydrolysis and fermentation of lipids may not occur without the presence of the methanogens.[2]

Because of the slow growth rate of microbes under anaerobic conditions, especially the methanogens, long cell retention times are typically required in batch and single and two-stage continuous flow systems. In systems that do not employ a means of biomass recycle or retention, the hydraulic retention time is equal to the cell retention time. A long cell retention time can be accomplished in any continuous flow system by using a long hydraulic retention time; however, increasing the hydraulic retention time increases the reactor volume for a given flow rate. Alternatively, continuous flow systems can be designed to include a means of biomass (or cell) recycle, so that a long cell retention time can be maintained while using a short hydraulic retention time and relatively small digester volume. CSTR systems can employ biomass recycle as external settling basins with return of concentrated sludge (Fig. 9.5), or an internal retention of biomass through use of screens or filters. Retained biomass designs such as the upflow anaerobic sludge blanket (UASB) and upflow anaerobic fixed-film (attached growth) filter designs can be utilized to maintain a long cell retention time with short hydraulic retention time within

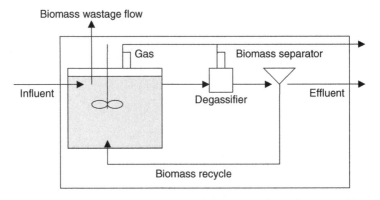

**FIGURE 9.5** CSTR with external biomass recycle.

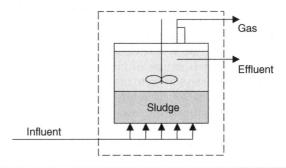

**FIGURE 9.6** Upflow anaerobic sludge bed (UASB) system. (Adapted from Reith, 2003.[2])

the reactor (Figs. 9.6 and 9.7). Low loading rate systems (i.e., relatively long hydraulic retention times) are usually applied to manures or slurries, while high rate systems (short hydraulic retention times with high cell retention times) are usually applied to more dilute wastewaters. A comparison of methane generation for various

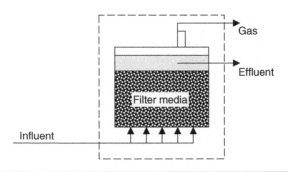

**FIGURE 9.7** Upflow anaerobic filter. (Adapted from Reith, 2003.[2])

| Process | Loading rate, kg VS/m³/day | Hydraulic retention time, days | Methane yield, m³/kg VS |
|---|---|---|---|
| Batch | 1.06 | 47 | 0.16 |
| Batch | 0.9 | 32 | 0.26 |
| CSTR, one-stage | 1.6 | 20 | 0.47 |
| CSTR, one-stage | 3.6 | 23 | 0.37 |
| Two-stage (batch + anaerobic filter) | 4.4 | 7 + 10 | 0.34 |
| Two-stage (CSTR + anaerobic filter) | 5.65 | 2 + 2.3 | 0.42 |

Source: Adapted from Bouallagui et al., 2005.

**TABLE 9.3** Methane Generation for Several Fruit and Vegetable Waste Digestion Processes

digester designs for FVW (Table 9.3) indicates the greater loading rate and shorter hydraulic retention time possible with two-stage, retained biomass systems.

Approximately 90 percent of the full-scale anaerobic digestion plants in use in Europe (as of 2005) rely on one-stage continuous flow reactor systems despite the reported advantages of two-stage systems, including the combined production of hydrogen and methane in two-stage systems.[17] The finding may lie in the greater simplicity and lower cost of the one-stage systems.

## 9.5 Biogas Composition and Use

Methane produced in anaerobic digesters is easily recovered in the gas phase due to its low solubility in water. The methane content of the biogas (Table 9.4) varies, depending on the characteristics of the

| Component | Volume percent[17] (% of dry gas) | Volume percent[2] (% of dry gas) |
|---|---|---|
| Methane ($CH_4$) | 50–80 | 55–75 |
| Carbon dioxide ($CO_2$) | 50–20 | 45–25 |
| Hydrogen ($H_2$) | <1 | — |
| Ammonia ($NH_3$) | <1 | 0–0.05 |
| Hydrogen sulfide ($H_2S$) | <1 | 0–1.5 |

**TABLE 9.4** Typical Biogas Composition

biomass feedstock. As mentioned earlier, biogas from manures or biomass with high carbohydrate content will have lower methane content and higher carbon dioxide content than biogas from biomass with greater protein and lipid content.[3,4] In addition to methane, carbon dioxide, and water vapor, biogas may contain hydrogen sulfide and ammonia in varying concentrations, depending on feedstock that may require removal prior to biogas use.[19] For example, biogas should contain less than 1000 ppm (0.1 percent) hydrogen sulfide for use in a boiler to prevent corrosion and less than 100 ppm (0.01 percent) for use in an internal combustion engine.[2]

Biogas can be used for heat production in a boiler, for combined heat and (electrical) power (CHP), or for upgrading to fuel gas quality. Most boilers convert biogas to heat with a 75 percent efficiency[12] and do not require a high gas quality. Removal of water vapor is advisable and will result in substantial removal of $H_2S$.[2] Other applications include combined heat and power (CHP) engine systems. The conventional CHP engine is a piston engine driving an electrical generator.[4] The biogas can be converted in typical CHP systems to electricity (35 percent) and heat (50 percent) with 15 percent energy loss.[2] Removal of $H_2S$ and $NH_3$ may be required as per manufacturer's specifications. Certain high temperature fuel cells (solid oxide and molten carbonate fuel cells) that operate at temperatures above 600°C can utilize methane directly, although conversion of methane to hydrogen gas via reformation prior to use in low-temperature proton exchange membrane (PEM) fuel cells is more common.[4]

## References

1. Nahiro, T., and Y. Sekiguchi. 2007. "Microbial communities in anaerobic digestion processes for waste and wastewater treatment: A microbiological update." *Curr OpinBiotechnol.* 18:273–278.
2. Reith, J. H., R. H. Wijffels, and H. Barren (eds). 2003. *Bio-methane and Bio-hydrogen: Status and Perspectives of Biological Methane and Hydrogen Production.* Dutch Biological Hydrogen Foundation: The Netherlands.
3. Grady, C. P. L., G. T. Daigger, and H. C. Lim. 1999. *Biological Wastewater Treatment*, 2d ed. Marcel Dekker, Inc: New York, NY.
4. *Planning and Installing Bioenergy Systems: A Guide for Installers, Architects and Engineers.* 2005. James & James Science Publishers Ltd.: UK.
5. Wolfe, R. S. 1993. "An historical overview of methanogenesis." In: Ferry, J. G. (ed.). *Methanogenesis.* Chapman & Hall, Inc: New York, NY, pp. 1–32.
6. Hungate, R. E. 1985. "Anaerobic biotransformations of organic matter." In: E. R. Leadbetter, and J. S. Poindexter (eds.). *Bacteria in Nature.* Plenum Press: New York, NY. 1:39–95.
7. Sobieraj, M., and D. R. Boone. 2006. "Syntrophomonadaceae." In: Dworkin, M., et al. (eds). *The Prokaryotes.* Springer. 4:1041–1049.
8. Nie, Y. Q., et al. 2007. "Enhancement of acetate production by a novel coupled syntrophic acetogenesis with homoacetogenesis process." *Process Biochem.* 42:599–605.
9. Thauer, R. 1976. "Limitation of microbial $H_2$-formation via fermentation." In: Schlegel, H. G., and J. Barnea (eds.). *Microbial Energy Conversion.* Erich Goltze KG: Gottingen, pp. 201–204.

10. Galand, P. E., et al. 2005. "Pathways for methanogenesis and diversity of methanogenic archae in three boreal peatland ecosystems." *Appl Environ Microbiol.* 71(4):2195–2198.
11. Gunaseelan, V. N. 1997. "Anaerobic digestion of biomass for methane production: A review." *Biomass Bioenerg.* 13(12):83–114.
12. European Energy Manager: Biogas Preparation Material. http://www.energy-manager.eu/getResource/10018/biogas.pdf.
13. Chynoweth, D. P., et al. 1993. "Biochemical methane potential of biomass and waste feedstocks." *Biomass Bioenerg.* 5:95–111.
14. http://bioenergy.ornl.gov/papers/misc/switgrs.html, and http://bioenergy.ornl.gov/papers/misc/switchgrass-profile.html.
15. Bouallagui, H., et al. 2005. "Bioreactor performance in anaerobic digestion of fruit and vegetable wastes." *Process Biochem.* 40:989–995.
16. Hills, D. J., and D. W. Roberts. 1982. "Conversion of Tomato, Peach and Honeydew Solid Waste into Methane Gas." *Transactions of ASAE.* 25(3):820-826.
17. Nishio, N., and Y. Nakashimada. 2007. "Recent developments of anaerobic digestion processes for energy recovery from wastes." *J Biosci Bioeng.* 103(2):105–112.
18. Bohn, I., L. Bjornsson, and B. Mattiasson. 2007. "Effect of temperature decrease on the microbial population of a mesophilic anaerobic bioreactor." *Environ Technol.* 28(8):943–952.
19. Zacari, S. M. 2003. "Removal of Hydrogen Sulfide from Biogas Using Cow-Manure Compost." MS Thesis, Cornell University.

# APPENDIX
# Conversion Factors and Constants

| Measure | Metric | US |
|---|---|---|
| Length | | |
| 1 m | 100 cm | 3.048 ft |
| 1 cm | 0.01 m | 0.3937 in |
| 1 in | 2.54 cm | 0.0833 ft |
| 1 mi | 1.609 km | 5,280 ft |
| Area | | |
| 1 $m^2$ | 10,000 $cm^2$ | 10.764 $ft^2$ |
| 1 hectare (ha) | 10,000 $m^2$ | 2.471 acres |
| 1 acre | 0.40469 ha | 43,560 $ft^2$ |
| Volume | | |
| 1 L | 1,000 $cm^3$ (cc) | 0.2642 gal |
| 1 L | 1,000 mL | |
| 1 cubic meter ($m^3$) | 1,000 L | 264.2 gal |
| 1 cubic meter ($m^3$) | | 35.31 $ft^3$ |
| 1 barrel (bbl) | 159 L | 42 gal |
| 1 gal | 3.785 L | |
| Weight | | |
| 1 g | | |
| 1 kg | 1,000 g | 2.205 lb |
| 1 ton (English) | | 2000 lb |
| 1 ton (metric) | | 1.1 ton (English) |
| Density | | |
| 1 g/mL | 1.0 kg/L | 8.33 lb/gal |
| Flow rate | | |
| 1 gal/min | 3.785 L/min | 8.021 $ft^3$/h |
| Heat transfer coefficient | | |
| 1 kW/$m^2$ K | 86.042 kcal/(h $m^2$ °C) | 0.04892 Btu/(s $ft^2$ °F) |

| Measure | Metric | US |
|---|---|---|
| Viscosity (dynamic) | | |
| 1 cP | 0.001 Pa s | 2.886 (lbf s)/ft$^2$ |
| 1 Pa s | 1000 cP | 47.88 (lbf s)/ft$^2$ |
| Viscosity (kinematic) | | |
| 1 cSt (centistoke) | 1 mm$^2$/s | 0.0155 in$^2$/s |
| Energy | | |
| 1 joule (J) | 1 kg m$^2$/s$^2$ | 0.2389 cal |
| 1 kWh | 3,600 kJ | 3,414 Btu |
| 1 gigajoule (GJ) | 10$^9$ J | |
| 1 exajoule (EJ) | 10$^{18}$ J | 0.9488 quad |
| 1 British thermal unit (Btu) | 1,054 J | |
| 1 quadrillion Btu (quad) | 1.054 exajoule (EJ) | 10$^{15}$ Btu |
| 1 metric ton oil equivalent (toe) | 1270 m$^3$ natural gas | 7.4 bbl crude oil |
| 1 toe | 2.3 metric ton coal | |
| Power | | |
| 1 kilowatt (kW) | 3,600 kJ/h | 3,414 Btu/h |
| 1 kilowatt (kW) | 1 kJ/s | 1.341 hp |
| 1 megawatt (MW) | 10$^6$ W | 3.415 million (MM) Btu/h |
| 1 gigawatt (GW) | 1000 MW | 3415 MM Btu/h |
| 1 exawatt (EW) | 10$^{18}$ J/s | 3.415 quad/h |
| 1 horsepower (hp) (British) | 745.7 W | 550 ft-lb/s |
| 1 hp (British) | 1.0139 hp (metric) | 2545 Btu/h |
| Temperature | | |
| 0°C | 273.15 K (°C + 273.15) | 32°F (or 1.8 × °C + 32) |
| 1 K | 8.617 × 10$^{-5}$ eV (electron volts) | 1.8°R (or 1.8 × K) |
| Bioenergy | | |
| 1 MT biodiesel | 37.8 GJ | 300 US gal |
| 1 MT biodiesel | 1136 L (avg sg 0.88) | |
| 1 MT ethanol | 1262 L (avg sg 0.79) | 333 US gal |
| 1 MT ethanol | | 7.94 petroleum barrels |

# Conversion Factors and Constants

| Measure | Metric | US |
|---|---|---|
| **Constants** | | |
| R (gas constant) | 1.9872 cal/(gmol K) | 0.7302 atm ft$^3$/(lbmol°R) |
| R | 8.3145 J/(gmol K) | 10.73 psia ft$^3$/(lbmol°R) |
| g (gravitational constant) | 9.8055 m/s$^2$ | 32.173 ft/s$^2$ |
| N (Avogadro's number) | 6.022 × 10$^{23}$ molecules/gmol | |
| k (Boltzmann's constant) | 1.3807 × 10$^{-23}$ J/K | |
| ε (electron charge) | 1.6022 × 10$^{-19}$ C (coulombs) | |
| F (Faraday's constant) | 9.6485 × 10$^4$ C/gmol | |

# Index

2-hydroxy-1,4-napthoquinone (HNQ), 310
2,6-dimethyl-1,4-benzoquinone (DMBQ), 310
5-hydroxymethylfurfural (HMF), 41

## A

Abengoa Bioenergy Corporation ethanol plants, 182
Aberdeen Energy ethanol plant, 182
abiotic hydrogen production, 271
Absolute Energy, LLC ethanol plant, 182
ACE Ethanol, LLC ethanol plant, 182
acetate, 279–280
acetic acid, 108, 163–164
acetoclastic methanogens, 332
acetogenesis, 27–28, 330, 332
acetogenic bacteria, 332
acid catalysis, 202, 206–208
acid esterification, 240, 243–244, 246
acid transesterification, 202, 204, 206, 244
acidogenesis, 331
acids
  acetic, 108, 163–164
  lactic, 108
activation energy, 214
adenosine triphosphate (ATP)
  aerobic metabolism, 22, 24
  aerobic oxidation of organic compounds, 305
  catabolism, 18
Adkins Energy, LLC ethanol plant, 182
adsorbents, 124

Advanced Bioenergy ethanol plant, 182
aerobic environments, 19
aerobic respiration, 19–23, 33
AFEX (ammonia fiber/freeze explosion/expansion) process, 139–140
African ethanol production, 78
AGP company, 182
agricultural milling, 251
agricultural processing by-products, 84
agricultural residues, 82–84, 284–287
Agri-Energy, LLC, 182
*Alcaligenes faecalis*, 311
alcohols
  effect on biodiesel production, 204
  etherification, 207
  phases, 218
Al-Corn Clean Fuel ethanol plant, 182
algae, 236
algal biomass, 273
algal oils
  bioreactors, 230–233
  overview, 223–229
  raceway ponds, 229–230
  yields of, 198
alkali catalysis, 204–206, 240
alkali transesterification, 241–243, 245–246
alkaline peroxide treatment, 140–141
alkyl esters, 202
Al-Shyoukh, A.O., 207
alternative energies, 6, 12–13. *See also* biofuels

**351**

# Index

Al-Widyan, M.I., 207
Amaizing Energy, LLC ethanol plants, 182
American Oil Chemists Society (AOCS), 245
amino acid biosynthesis, 28–29
ammonia fiber/freeze explosion/expansion (AFEX) process, 139–140
ammonia, liquid, 115
Ammonia Recycle Percolation (ARP), 139
ammonia treatment, 139–140
ammonium sulfate, 115
amylase corn, 131
amylopectin, 69–70
amylose, 69–70
*Anabaena* cyanobacteria, 238, 308
anabolic reactions, 28–30, 33
anabolism, 18
anaerobic bacteria, 332
anaerobic digestion
  overview, 329
  systems for methane production from
    overview, 338–339
    process design, 340–343
    reactor conditions, 339–340
anaerobic environments, 19
anaerobic fermentation process, 273
anaerobic oxidation, 28
anaerobic respiration, 19, 23–24
anaerobic treatment processes, 329
*Anaerolineae* bacteria, 333
Andersons Albion Ethanol, LLC, The, 191
Andersons Clymers Ethanol, LLC, The, 191
Andersons Marathon Ethanol, LLC, The, 191
Andrews equation, 39
animals
  fats from, 88–91
  wastes, 91–93
anion exchange, 137–138
Ankistrodesmus microalgal oil, 93
annual capital charges, 258
annual production costs, 258
anodes, 305, 308–309, 320–321, 325
anoxic environments, 19
anoxygenic photosynthesis, 33
anthraquinone-2,6-disulfonate (AQDS), 310

antioxidants, 201
Antoine equation, 217, 292
AOCS (American Oil Chemists Society), 245
AQDS (anthraquinone-2,6-disulfonate), 310
ARA (arachidonic acid), 236–237
arabinose, 146–149
archae, 17, 27–28
Archer Daniels Midland ethanol plants, 182–183
Arkalon Energy, LLC ethanol plant, 183
Arkenol's process, 134–136
ARP (Ammonia Recycle Percolation), 139
Arrhenius equation, 48
artificial light, 231
ATP. *See* adenosine triphosphate
Aventine Renewable Energy, LLC ethanol plants, 183

## B

Bacillariophyceae algae, 227
*Bacillus stearothermophilus*, 151
bacteria. *See* individual types of bacteria by name
  defined, 17
  fermentation by
    of complex organic compounds, 27–28
    of sugars, 26–27
  hydrogen production by, 277–279
Badger State Ethanol, LLC ethanol plant, 183
bagasse, 80, 85–86
barley, 75–76, 85–87
base catalysis, 204–206, 208
base solutions for biomass pretreatment, 139
batch anaerobic digestion systems, 340–341
batch fermentation, 113–114, 139, 280–281
batch reactors
  fermentations in, 59–61
  mass balance equations for, 50
  overview, 57–58
  simulations of, 59–61
beef tallow, 92
Benemann, J.R., 274
Bennetto, H.P., 305

beta-glucan, 75–76
BHA (butylated hydroxyanisole), 201
BHT (butylated hydroxytoluene), 201
Big River Resources, LLC ethanol plant, 183
biobutanol, 201
biochemistry. *See also* fermentation
　chemical oxygen demand, 33–35
　lithotrophic growth, 30–31
　of microbial fuel cells, 303–305
　organoheterotrophic metabolism
　　aerobic respiration, 19–23
　　anaerobic respiration, 23–24
　　overview, 17–19
　phototrophic metabolism, 31–33
　*Thermotoga*, 276–277
bioconversion, 236. *See also* biochemistry; microbial modeling of biofuel production
biodiesel, 197–268. *See also* chemistry of biodiesel production; microbial oils
　commercial scale operations, 245–247
　coproducts, 236–238
　density, 210
　economics
　　capital cost, 255–257
　　feedstock cost, 252–254
　　manufacturing cost, 255
　　operating cost, 257–258
　　overview, 250–251
　energy density, 5
　environmental considerations, 199–200
　ethanol fuel additive, 201
　nitrogen oxide emissions, 200–201
　oil sources
　　biosynthesis, 234–236
　　modification of, 234–236
　　overview, 219
　　plant oils, 219–223
　　straight vegetable oil, 233–234
　　used cooking oils, 233
　overview, 197–199
　pilot scale operations, 245–247
　production procedures
　　acid esterification and transesterification, 244
　　acid esterification pretreatment, 243
　　alkali transesterification, 241–243
　　biodiesel water wash, 244–245
　　overview, 238–240
　　titration, 240–241
　quality control analytical technique, 247–250
　thermodynamics of
　　boiling point, 216–217
　　overview, 210
　　phase separation, 218
　　reaction kinetics, 211–215
　　solubility, 218
　　thermogravimetric kinetics, 215–216
　　vapor pressure, 216–217
　viscosity, 210
　zero-emissions production of, 4
Bioengineering Resources, Inc. (BRI) process, 169–170
BioFuel Energy-Buffalo Lake Energy, LLC ethanol plant, 183
BioFuel Energy-Pioneer Trail Energy, LLC ethanol plant, 183
BioFuel International ethanol plant, 183
biofuels. *See also* ethanol; feedstocks; microbial modeling of biofuel production
　overview, 5–6, 14–15
　production of
　　biorefineries, 3–5
　　environmental impact, 13–14
　　overview, 10–12
　use of, 10–12
biogas, 91, 93, 334, 338, 343–344
BioGasol biomass ethanol fermentation process, 155–156
biological acetate production, 280
Biological Modeling and Simulation Software. *See* BIOMASS
biomass
　calculating chemical oxygen demand, 35
　concentration of, 52–54, 57
　environmental impact, 13–14
　gasification of, 158–160
　lignocellulosics, 10–11
　measurements of, 314
　pretreatment
　　alkaline peroxide treatment, 140–141
　　ammonia treatment, 139–140
　　concentrated phosphoric acid fractionation, 141–142

# Index

biomass, pretreatment (*Cont.*):
   concentrated sulfuric acid hydrolysis, 134–136
   dilute sulfuric acid hydrolysis, 136–138
   ionic liquid fractionation, 142–143
   lime treatment, 140
   organosolv fractionation, 141
   overview, 134
   steam explosion, 138–139
   wet oxidation, 141
  rates of formation, 45–46
  residues, 81–82
  sources for methane generation, 334–338
  yields, 44–45
BIOMASS (Biological Modeling and Simulation Software)
  batch fermentations and simulations, 59–61
  CSTR fermentations and simulations, 61–64
  overview, 58–59
biophotolysis, 272–273, 324
bioreactors
  batch reactors, 50
  continuous stirred tank reactors
   for anaerobic digestion, 340
   with cell recycle, 52–54
   design strategies, 57–58
   fermentations, 61–64
   fermentors, 114
   overview, 50–52
   simulations, 61–64
  design strategies, 57–58
  fed-batch systems, 54–55
  overview, 48–50
  plug flow systems, 55–57
  sizing for PEM fuel cell use, 297–299
biorefineries, 3–5
biosynthesis of oils, 234–236
black liquors, 81
Blue Flint Ethanol plant, 183
BlueFire Ethanol company, 136
blue-green algae, 227, 230
boiling point of biodiesel, 216–217
Bonanza Energy, LLC ethanol plant, 184
*Botryococcus braunii* microalgal oil, 93
Brazilian ethanol production, 79, 115
BRI (Bioengineering Resources, Inc.) process, 169–170
Brix measurements, 112–113
brown grease, 233
Brown, R., 252
bubbling fluidized bed gasifiers, 159–163
Bushmills Ethanol, Inc. ethanol plant, 184
butylated hydroxyanisole (BHA), 201
butylated hydroxytoluene (BHT), 201

## C

CAFE (car fuel efficiency) standards, 10
Calgren ethanol plant, 184
Calvin cycle, 30–31
Canadian ethanol production, 73–75
Canakci, M., 204
*Candida antarctica* lipase B, 203, 208, 210
*Candida* yeast, 43
cane molasses, 111–114
canola, 219
capital cost of biodiesel, 255–257
car fuel efficiency (CAFE) standards, 10
carbohydrates, 87, 272
carbon dioxide. *See* $CO_2$
carbon sources for fermentation, 284–287
Cardinal Ethanol plant, 184
Cargill, Inc. ethanol plants, 184
Cascade Grain ethanol plant, 184
cassava, 78–79
Castle Rock Renewable Fuels, LLC ethanol plant, 184
castor oil, 216
catabolism, 18
catabolite repression, 107
catalysis
  acid, 202, 206–208
  alkali, 204–206, 240
  base, 204–206, 208
  enzyme, 203, 208, 215
catalysts, modified, 166–168
catalytic conversion of syngas, 166–168, 171, 173
catalytic tar cracking, 160
cathodes, 305, 311–312, 321
cation exchange membrane (CEM), 303
CBP (consolidated bioprocessing), 155
cellobiase enzyme, 143
cellobiose subunit of cellulose, 284
cellulase enzymes, 143, 331
cellulolytic microorganisms, 152

# Index

cellulose, 80, 150
cellulosic ethanol production, 208
cellulosic sugars, 232–233
CEM (cation exchange membrane), 303
Center Ethanol plant, 184
Central Indiana Ethanol, LLC plant, 184
central metabolic pathways, 331
Central MN Ethanol Coop plant, 184
cereal grains
　barley, 75–76
　corn, 71–73
　oat, 77
　overview, 69–71
　rice, 77
　sorghum, 76–77
　wheat, 73–75
cetane values, 249
CGF (corn gluten feed), 119
CGM (corn germ meal), 119
CGM (corn gluten meal), 119
char, 158
chemical oxygen demand (COD), 33–35, 314, 321, 335
chemistry of biodiesel production, 201–218
　alcohol effect, 204
　catalysis
　　acid, 206–208
　　alkali, 204–206
　　base, 204–206
　　enzyme, 208
　esterification, 202–203, 208–210
　hydrolysis, 203–204
　lipase-catalyzed interesterification, 203
　overview, 201–202
　saponification, 203–204
　transesterification
　　lipase-catalyzed, 203
　　overview, 202
　　supercritical, 208–210
chemolithotrophs, 17, 30
chemoorganotrophs, 17
chemotrophs, 17
Chief Ethanol plant, 184
Childers, S.E., 282
Chinese ethanol production, 71, 78
Chippewa Valley Ethanol Co. plant, 184
*Chlorella protothecoides* green algae, 227–228
*Chlorella* spp. green algae, 228

*Chlorella vulgaris*, 228
Chlorophyceae green algae, 227
CHP (combined heat and power) engine systems, 344
Christi, Y., 229, 233
chromatography methods, 247–248
Chrysophyceae golden algae, 227
Cilion Ethanol plant, 184
circuit resistance, power density as function of, 313
circulating fluidized bed gasifiers, 159–163
citric acid, 235
Clean Air Act, 14
closed reflux COD test, 34
closed-system pond method, 233
*Clostridium* bacteria
　Fe- and NiFe-hydrogenases, 270
　fermentation of cellulosic materials, 286
　methane production, 332–333
*Clostridium ljungdahlii* bacteria, 160, 163–164
*Clostridium phytofermentans* bacteria, 152
*Clostridium saccharolyticus* bacteria, 281
*Clostridium thermolacticum* bacteria, 278, 280
C:N:P ratio, 336
$CO_2$ (carbon dioxide)
　dry milling ethanol fermentation process, 122
　emissions
　　in biodiesel production, 4
　　in biofuel production, 13
　　in ethanol production, 4
　　from fossil fuels, 14
　theoretical ethanol yield, 124
cobalt, 164–165
COD. *See* chemical oxygen demand
COD:N:P ratio, 336
colorimetric method, 34
combined heat and power (CHP) engine systems, 344
combustible air, 329
commercial scale operations, 245–247
Commonwealth Agri-Energy, LLC ethanol plant, 184
competitive inhibition, 39
complementary substrates, 42
complex organic compound fermentation, 27–28

# Index

concentrated phosphoric acid fractionation, 141–142
concentrated sulfuric acid hydrolysis, 134–136
concentrations
  biomass, 52–54, 57
  of dissolved hydrogen, 294–296
  of hydrogen gas, 293
consolidated bioprocessing (CBP), 155
consumption, energy, 6–7, 10
contingency fees, 256
continuous dry-grind fermentation, 132
continuous fermentation process, 113–114, 119–120
continuous stirred tank reactors (CSTR)
  for anaerobic digestion, 340
  with cell recycle, 52–54
  design strategies, 57–58
  fermentations, 61–64
  fermentors, 114
  mass balance equations for, 50–52
  simulations, 61–64
cooking oils, 233
cooling sugarcane juice, 111
coproducts
  biodiesel production, 236–238
  hydrogen production, 279–280
corn
  ethanol production from, 13–14, 115, 117–118
  feedstocks, 71–73, 126–127, 131
  fiber, 84, 87
  fractionation of, 132
  stover, 11, 82–86
corn germ meal (CGM), 119
corn gluten feed (CGF), 119
corn gluten meal (CGM), 119
Corn, LP ethanol plant, 184
corn oil, 92
Corn Plus, LLP ethanol plant, 184
corn, stover, 11, 82–86
Cornhusker Energy Lexington, LLC ethanol plant, 184
Coshoctan Ethanol, OH plant, 184
costs
  biodiesel production
    capital, 255–257
    of feedstock, 252–254
    manufacturing, 255
    operating, 257–258
    overview, 250–251

biomass ethanol plants, 156–158
catalytic conversion of biomass-derived syngas, 171, 173
crude oil, 7
ethanol production
  in dry mill plants, 125–128, 130
  in wet mill plants, 125–127, 129–130
  gasifier, 170–171
  syngas fermentation process, 170–172
cosubstrates, 339
cottonseed hull, 87
cottonseed oil, 92, 218
Coulombic efficiency of MFCs, 315–318
Crabbe, E., 207
crambe oil, 92
crude glycerin, 251
crude oil, 7–8
*Crypthecodinium cohnii*, 227, 236
CSTR. *See* continuous stirred tank reactors
cull peaches, 286
cultures
  fermentation of complex organic compounds by mixed, 27–28
  MFC microbial, 309
cyanobacteria, 237–238, 270, 272–273, 308, 324–325
Cyanophyceae blue-green algae, 227
cysteine hydrochloride, 290

## D

DAG (diacylglycerol), 202
Dakota Ethanol, LLC ethanol plant, 185
dark reactions, 28–30, 33
DDGS (distillers dried grains with solubles), 122
DE (dextrose equivalent), 121
De Vrije, T., 286
decortication, 76
dedicated energy crops, 84–88
degrees of reduction, 35
dehydration, molecular sieve, 124–125
DENCO, LLC ethanol plant, 185
density, energy, 5–6, 269
Department of Energy (DOE)
  Energy Information Administration, 298
  NREL report, 222

Index    357

deproteinated milk serum (DPMS), 323
design
  bioreactor
    batch reactors, 50
    continuous stirred tank reactors, 50–54
    fed-batch systems, 54–55
    overview, 48–50
    plug flow systems, 55–57
    strategies, 57–58
  microbial fuel cell
    anode compartment, 308–309
    cathode compartment, 311–312
    exchange membrane, 312
    microbial cultures, 309
    power density as function of circuit resistance, 313
    redox mediators, 310–311
*Desulfovibrio* bacteria, 270
detection of hydrogen, 291
detoxification of dilute acid hydrolysates, 137–138
dextrose equivalent (DE), 121
DHA (docosahexaenoic acid), 227, 236
diacylglycerol (DAG), 202
diatoms, 227
Diesel cycle, 210
diesel engines, 8
Diesel, Rudolf, 197
digital spectrophotometers, 314
dilute acid hydrolysate detoxification, 137–138
dilute acid hydrolysis, 157
dilute sulfuric acid hydrolysis, 136–138
dinoflagellates, 227
direct biophotolysis, 272, 324
direct costs, bioenergy plant, 255
dissolved hydrogen concentration, 294–296
distillation
  biodiesel, 249
  ethanol, 124
distillers dried grains with solubles (DDGS), 122
DM (dry matter) content, 336
DMBQ (2,6-dimethyl-1,4-benzoquinone), 310
docosahexaenoic acid (DHA), 227, 236
DOE (Department of Energy)
  Energy Information Administration, 298
  NREL report, 222

downdraft gasifiers, 159–163
DPE (DuPont Performance Elastomers) Viton fluoroelastomer, 248
DPMS (deproteinated milk serum), 323
Drapcho, C., 290
dry fermentation systems, 338
dry matter (DM) content, 336
dry milling
  corn, 73
  cost of ethanol production from, 125–128, 130
  overview, 117–118
  process description, 119–122
*Dunaliella bardawil* microalgal oil, 93
*Dunaliella salina*, 227
DuPont Performance Elastomers (DPE) Viton fluoroelastomer, 248

## E

E Caruso ethanol plant, 185
E Energy Adams, LLC ethanol plant, 185
East Kansas Agri-Energy, LLC ethanol plant, 185
Eastern Regional Research Center (ERRC), 132
E-biodiesel, 198–199
economics
  of biodiesel production
    capital cost, 255–257
    feedstock cost, 252–254
    manufacturing cost, 255
    operating cost, 257–258
    overview, 250–251
  of ethanol production
    starch-based, 125–126
    sugar-based, 156–158
  of syngas process, 170–174
ED pathway. *See* Entner-Doudoroff pathway
efficiency of energy use, 8–10
eicosapentaenoic acid (EpA), 236–237
electrocatalytic conductive polymer, 320
electron transport system (ETS), 22–23
electrons, 272
Elkhorn Valley Ethanol, LLC ethanol plant, 185
ELSD (evaporative light scattering detector), 248

# Index

Embden-Meyerhof pathway (EMP), 20–21, 25, 276, 304, 331
E-milling (enzymatic milling) process, 132–133
emissions
  carbon dioxide
    in biodiesel production, 4
    in biofuel production, 13
    in ethanol production, 4
    from fossil fuels, 14
  nitrogen oxide, 200–201
  zero-emissions production of biodiesel, 4
EMP. *See* Embden-Meyerhof pathway
endothermic reactions of biomass gasification, 159
energetics of hydrogen production, 275
energy
  alternative, 12–13
  balance ratios, 198
  consumption, 6–7, 10
  density, 5–6, 269
  efficiency of use, 8–10
  geothermal, 8
  overview, 6–8
  solar
    biofuel production, 5
    efficiency of, 8
    phototrophic metabolism, 32–33
    technology for, 12–13
  wind, 8, 13
  yield, 5
engines
  diesel, 8
  hybrid diesel, 8, 10
  hybrid hydrogen fuel cell, 8, 10
  spark-ignition gasoline, 8
Entner-Doudoroff (ED) pathway, 20, 108–109, 276
environmental impact
  of biodiesel production, 199–201
  of biofuel production, 13–14
Environmental Protection Agency (EPA), 247
environments, biochemical pathway, 19
enzymatic milling (E-milling) process, 132–133
enzymes. *See also* individual enzymes by name
  catalysis of, 203, 208, 215
  for hydrogen production reactions, 269–270
  hydrolysis, 131–132, 143–144, 331

EPA (eicosapentaenoic acid), 236–237
EPA (Environmental Protection Agency), 247
ERRC (Eastern Regional Research Center), 132
erucic acid, 218
*Escherichia coli* microorganism, 149–150, 305
ESE Alcohol Inc. ethanol plant, 185
esterification
  acid transesterification and, 244
  homogeneous acid catalysts, 207
  overview, 202–203
  supercritical, 208–210
ETBE (ethyl tert butyl ether), 201
ethanol. *See also* feedstocks
  addition to gasoline, 14
  energy balance ratio, 198
  fermentation
    efficiency of, 123
    of glucose to, 25–26
    inhibition equations for, 107–108
  fuel additive, 201
  growth equations for, 107–108
  liquid volumes of, 123
  plant processes, 113–114
  production of
    from corn, 13–14
    dry mill plant economics, 125–128, 130
    stoichiometry of, 145–146, 164
    wet mill plant economics, 125–127, 129–130
    world sources, 10
    zero-emissions, 4
  recovery of, 124–125
  theoretical yield of, 122–124
Ethanol Grain Processors, LLC plant, 185
ethyl esters, 204, 216
ethyl tert butyl ether (ETBE), 201
ETS (electron transport system), 22–23
eucarya, 17, 227
European ethanol production, 80
evaporative light scattering detector (ELSD), 248
exchange membranes, MFC, 312
exogenous redox mediators, 310–311
exothermic reactions of biomass gasification, 159
external redox mediators, 311

## F

FAME (fatty acid methyl esters), 217
Faraday's constant, 304, 316
farm-gate price, 253
FAS (fatty acid synthase) system, 234
fatty acid ester composition, 233
fatty acid methyl esters (FAME), 217
fatty acid synthase (FAS) system, 234
fed-batch systems, 54–55, 232
feedstocks, 69–103. *See also* lignocellulosic feedstocks; starch feedstocks; sugar feedstocks
   animal fats, 88–91
   animal wastes, 91–93
   cost of, 252–254
   municipal solid waste, 94
   overview, 69
   plant oils, 88–91
Fe-hydrogenase enzymes, 270, 272
Fenimore mechanisms, 200
fermentation, 134–158
   anabolic reactions for heterotrophic organotrophic microorganisms, 28–30
   batch process, 113–114
   in batch reactors, 59–61
   in biological methane production, 330
   of complex organic compounds by mixed cultures, 27–28
   continuous dry-grind, 132
   continuous process, 113–114, 119–120
   continuous stirred tank reactors, 61–64
   efficiency of, 123
   of glucose, 30
   in hydrogen production
      by bacteria, 277–279
      batch, 280–281
      biochemical pathway for *Thermotoga*, 276–277
      carbon sources from agricultural residues, 284–287
      coproduct formation, 279–280
      energetics, 275
      inhibition, 281
      overview, 269, 274–275
      sulfidogenesis, 281–283
      Thermotogales, 276
   Melle-Boinot process, 113–114
   in methane production, 331
   overview, 19, 25
   rate of, 341
   of simple sugars
      by bacteria, 26–27
      by yeast, 25–26
   sugar platform
      Cellulolytic and Xylanolytic strains, 152
      mesophilic sugar-utilizing strains, 145–150
      overview, 144–145
      thermophilic sugar-utilizing strains, 151–152
   of syngas, 160–165
fermentation bioreactors, 297–299
fermentation gases, 291, 293
fermentative bacteria, 270, 331
ferredoxin, 274
ferricyanide, 311–312
FFAs. *See* free fatty acids
Fick's Law, 312
filamentous blue-green algae, 230
First United Ethanol, LLC (FUEL) plant, 185
first-order kinetics, 211
Fischer-Tropsch catalysts, 166–168
fish oil, 236–237
fixed-bed gasifiers, 159–163
flash point, 239–240
fluidized bed gasifiers, 159–163
forest products, 81–82
fossil fuels, 7–8, 14
fractionation
   concentrated phosphoric acid, 141–142
   of corn, 132
   ionic liquid, 142–143
   organosolv, 141
free fatty acids (FFAs), 202, 206, 233, 243
free glycerin, 250
Freedman, B., 211
Front Range Energy, LLC ethanol plant, 185
fructose polymers, 110
fruit and vegetable waste (FVW), 336, 340–341
FUEL (First United Ethanol, LLC) plant, 185
fuel cells, microbial, 5
fuels. *See also* biodiesel; biofuels; ethanol
   alternative, 6, 12–13
   fossil, 7–8, 14
   opportunity, 13
   transportation, 10

fungal lipids, 232
furfural, 42
FVW (fruit and vegetable waste), 336, 340–341

## G

G6P (glucose-6-phosphate), 271
GA (growth-associated) products, 52, 54
gas chromatography (GC), 247, 291
gasification, 134, 158–160
gasifiers, 159–163, 169–171
gasoline-ethanol mixtures, 201
Gateway Ethanol plant, 185
GC (gas chromatography), 247, 291
Genencor International company, 131–132
genetic-level expression systems, 236
*Geobacter* spp. cultures, 309
geothermal energy, 8
German production of ethanol, 116
Glacial Lakes Energy, LLC ethanol plant, 185
Global Ethanol/Midwest Grain Processor plants, 185
glucoamylase, 121
glucose
  combustion, 275
  energetics, 275
  fermentation of, 25, 30, 279
  growth rate of *Candida* based on, 43
  hydrogen production models
    batch fermentations and simulations, 59–61
    CSTR fermentations and simulations, 61–64
    overview, 58–59
  metabolized by *S. cerevisiae* yeast, 107
  oxidation of, 20–21, 29
  production of, 122
  theoretical ethanol yield, 122
glucose-6-phosphate (G6P), 271
gluten, 75, 119
glycerin, 239, 245–246, 250
glycerol, 202, 206
glycerolysis, 203–204, 247
glycolysis, 20–21, 25, 303
glycolytic pathway for ethanol production, 105–106
golden algae, 227
Golden Cheese Company of California ethanol plant, 185

Golden Grain Energy, LLC ethanol plant, 185
Golden Triangle Energy, LLC ethanol plant, 186
Goodland Energy Center ethanol plant, 185
Goodrum, J. W., 217
Graboski, M. S., 217
grain feedstock, 78. *See also* cereal grains
Grain Processing Corp. ethanol plant, 186
Grand River Distribution ethanol plant, 186
Granite Falls Energy, LLC ethanol plant, 186
graphite, 308
Greater Ohio Ethanol, LLC plant, 186
green algae, 227–228, 230, 272
Green Plains Renewable Energy ethanol plants, 186
growth models, microbial
  inhibition, 39–42
  for multiple limiting substrates, 42–44
  overview, 37–38
  unstructured, single limiting nutrient, 38–39
  yield parameters, 44–45
growth rate expressions, kinetic, 45–48
growth-associated (GA) products, 52, 54
Gunaseelan, N. V., 334
Guo, Y., 206

## H

$H_2S$ (hydrogen sulfide), 282
$H_2$-synthesizing hydrogenases, 288
Haas, M. J., 251
Hallenbeck, P. C., 274
Hawkeye Renewables, LLC ethanol plants, 186
headspace, hydrogen recovery from, 288, 291
Heartland Corn Products ethanol plant, 186
Heartland Grain Fuels, LP ethanol plants, 186
heavy steep water (HSW), 119
helical coil photobioreactors, 232
hemicellulase enzymes, 143, 331
hemicellulose, 80, 140–142
hemicellulosic hydrolysates, 286
Henry's law, 294–295
herbicides, 252

Heron Lake BioEnergy, LLC ethanol plant, 186
Hess, M. A., 201
heterotrophic growth of algae, 227
heterotrophic organotrophic microorganisms, 28–30
hexose monophosphate pathway (HMP), 20
hexose transporters, 105–107
HFC (high fermentable corn), 131
HFF (highly fermentable fraction), 131
high fermentable corn (HFC), 131
high oil fraction (HOF), 131
high pressure liquid chromatography (HPLC), 248
high total fermentables (HTF), 131
highly fermentable fraction (HFF), 131
HMF (5-hydroxymethylfurfural), 41
HMP (hexose monophosphate pathway), 20
HNQ (2-hydroxy-1,4-napthoquinone), 310
HOF (high oil fraction), 131
Holt County Ethanol plant, 186
Homeland Energy ethanol plant, 186
homoacetogenesis, 332
homogeneous acid catalysts, 207
homogeneous base catalysts, 205–206
HPLC (high pressure liquid chromatography), 248
HSW (heavy steep water), 119
HTF (high total fermentables), 131
Hubbert's curve, 197
Huber, R., 283
Hungate, Dr. Robert E., 330
Husker Ag, LLC ethanol plant, 186
hybrid diesel engines, 8, 10
hybrid hydrogen fuel cell engines, 8, 10
hybrids, corn, 131
hydraulic retention time, 51–52, 63–64, 321–322, 339, 341
hydrodynamic dispersion, 55–56
hydrogen
  biological production of, 269–302. *See also* fermentation; reporting hydrogen production
  abiotic, 271
  fermentation bioreactor sizing for PEM fuel cell use, 297–299
  important enzymes, 269–270
  overview, 269
  photobiological, 271–274
  process and culture parameters, 287–290
  combustion, 275
  detection of, 291
  product inhibition, 40–41
  production of
    batch fermentations and simulations, 59–61
    CSTR fermentations and simulations, 61–64
    overview, 5, 58–59
    potential for, 12
  solubility of, 288
hydrogen sulfide ($H_2S$), 282
hydrogenase enzymes, 270–271, 274
hydrogenation, 249
hydrolysis
  in biodiesel production, 203–204
  concentrated sulfuric acid, 134–136
  defined, 27
  dilute acid, 157
  dilute sulfuric acid, 136–138
  enzyme, 131–132, 143–144, 331
  in methane production, 330–331
  of particulate organic substrate, 47
  rate of, 341
  separate hydrolysis and fermentation, 154–155
  of starch, 122, 131–132
hydrotalcites, 206
hyperthermophilic microorganisms, 274, 282, 298

# I

Idaho Ethanol Processing plant, 187
Illinois River Energy, LLC ethanol plant, 187
immobilized systems, 208
impact, environmental, 13–14
Indiana Bio-Energy ethanol plant, 187
indirect biophotolysis, 273
indirect costs for bioenergy plants, 255
inert gases, 292
inflation, 255
inhibition
  competitive, 39
  of hydrogen, 281
  models for, 39–42
  noncompetitive, 39
  by-product, 40–41
  by substrate, 39–40
  uncompetitive, 39
  by xenobiotic compounds, 41–42
Innovase venture, 132

## Index

interactive Monod model, 42
interesterification, 203, 208
intermediate compounds, 28–29
International Critical Tables, 295
iodine values (IV), 90, 249–250
Iogen process, 153–154
ionic liquid fractionation, 142–143
iron-sulfur protein, 274
Iroquois Bio-Energy Company, LLC ethanol plant, 187
*Isochrysis* sp. microalgal oil, 93
IV (iodine values), 90, 249–250

### K

KAAPA Ethanol, LLC plant, 187
Kansas Ethanol, LLC plant, 187
Khan, A., 207
kinematic viscosity, 248
kinetics
  biodiesel reaction, 211–215
  rate expressions, 45–48
  thermogravimetric, 215–216
*Klebsiella oxytoca* microorganism, 150
Korus, R., 251
Krebs cycle, 21–22
Kusdiana, D., 210

### L

lactic acid, 108
Land O'Lakes ethanol plant, 187
LC-PUFA (long-chain polyunsaturated fatty acids), 223, 236
LDH (L-lactate dehydrogenase), 151
Leung, D. Y. C., 206
levan synthesis, 110
levansucrase enzyme, 110
Levelland/Hockley County Ethanol, LLC plant, 187
Lifeline Foods, LLC ethanol plant, 187
light reactions, 32–33
lignin, 140–142
lignocellulosic feedstocks, 80–88. *See also* sugar platform for ethanol production; syngas
  agricultural processing by-products, 84
  agricultural residues, 82–84
  dedicated energy crops, 84–88
  ethanol production from, 133–134
  forest products and residues, 81–82
  overview, 10–11, 80–81
lime treatment, 140

Lincolnland Agri-Energy, LLC ethanol plant, 187
Lincolnway Energy, LLC ethanol plant, 187
Lineweaver-Burk double reciprocal plot, 215
lipase-catalyzed interesterification, 203
lipase-catalyzed transesterification, 203
lipases, 203, 208
lipids, 234–236
liquefaction, starch, 119–120
liquid ammonia, 115
liquid volumes of ethanol, 123
lithotrophic growth, 30–31
Little Sioux Corn Processors, LP ethanol plant, 187
Liu, Y., 206–207, 317, 319–320
L-lactate dehydrogenase (LDH), 151
loading rate of methane reactors, 340
Logan, B. H., 305, 317, 319
logistic growth equation, 50
long-chain polyunsaturated fatty acids (LC-PUFA), 223, 236
longitudinal dispersion, 56
losses, vampire, 10
Lotero, E., 206
lymphoblastoid T-cells, 237

### M

MAG (monoacylglycerol), 202, 204
manioc, 78
manufacturing costs of biodiesel, 255
Marquis Energy, LLC ethanol plant, 187
Marysville Ethanol, LLC plant, 187
mass balance
  for batch reactors, 50
  in BioGasol process, 156–157
  for continuous stirred tank reactors, 50–54
  plug flow systems, 55–57
mass movement, one-dimensional, 56
Mavera corn hybrid, 131
McCormick, R. L., 217
measurements
  of biomass, 314
  of substrate, 314
media for *Thermotoga* spp., 288–289
Melle-Boinot process, 113–114
Merrick & Company ethanol plant, 187
mesophilic microorganisms, 145–150, 274

# Index

metabolism
  of glucose, 25
  lithotrophic growth, 30–31
  organoheterotrophic, 19–30. *See also* fermentation
    aerobic respiration, 19–23
    anaerobic respiration, 23–24
    phases of, 18
    phototrophic, 31–33
    pyruvate, 25
methane, 329–345
  biogas composition and use, 343–344
  biomass sources for, 334–338
  microbiology of, 329–334
  overview, 329
  production of, 28
  systems for anaerobic digestion
    overview, 338–339
    process design, 340–343
    reactor conditions, 339–340
methanogenesis, 27–28, 330, 332–333
methanogenic environments, 329–330
methanogens, 333
methanol, 201
methanol synthesis catalysts, modified, 166–168
methyl esters, 204, 216, 242
methyl tert-butyl ether (MTBE), 14, 201
MFCs. *See* microbial fuel cells
MGP Ingredients, Inc. ethanol plants, 187
Miao, X., 227
Michaelis-Menten kinetics, 215
microalgae, 222
microalgal oils, 91, 93. *See also* algal oils
microbial communities, 332–334
microbial conversion processes, 3–4, 13
microbial cultures, MFC, 309
microbial decay, 47
microbial fuel cells (MFCs), 303–327
  biochemical basis of, 303–305
  defined, 5
  design of
    anode compartment, 308–309
    cathode compartment, 311–312
    exchange membrane, 312
    microbial cultures, 309
    power density as function of circuit resistance, 313
    redox mediators, 310–311
    fabrication example, 322–323
  future directions for, 323–325
  overview, 303
  performance methods, 314–318
  performance of, 318–322
  research for, 305–308
microbial lipid production, 228
microbial modeling of biofuel production, 37–66
  bioreactor operation and design
    batch reactors, 50
    continuous stirred tank reactors, 50–54
    fed-batch systems, 54–55
    overview, 48–50
    plug flow systems, 55–57
    strategies for, 57–58
  glucose utilization and hydrogen production
    batch fermentations and simulations, 59–61
    CSTR fermentations and simulations, 61–64
    overview, 58–59
  growth models
    inhibition models, 39–42
    for multiple limiting substrates, 42–44
    overview, 37–38
    unstructured, single limiting nutrient, 38–39
    yield parameters, 44–45
  kinetic rate expressions, 45–48
  overview, 37
microbial oils
  algal oils, 227–229
  bioreactors, 230–233
  overview, 223–226
  raceway ponds, 229–230
microbiology of methane production
  methanogenic environments, 329–330
  microbial communities, 332–334
  process description, 330–332
microorganisms. *See also* individual microorganisms by name
  classifications of, 17
  for ethanol production
    *Saccharomyces cerevisiae*, 105–108
    *Zymomonas mobilis*, 108–111
Mid America Agri Products ethanol plants, 188

Mid-Missouri Energy, Inc. ethanol plant, 188
Midwest Grain Processor ethanol plants, 185
Midwest Renewable Energy, LLC ethanol plant, 188
milling, enzymatic, 132–133. *See also* dry milling; wet milling
Minnesota Energy ethanol plant, 188
*Miscanthus* biomass, 286–287
mixed cultures, 27–28
mixing, single-chamber MFC, 321
mixotrophic growth of algae, 228
Mo-based catalysts, modified, 166–168
Modi, M. K., 208
modified catalysts, 166–168
moisture content in methane reactors, 339
mol $H_2$/L media, 294
molar ratios, 207
molasses, 80, 111–114
molecular sieve dehydration, 124–125
monoacylglycerol (MAG), 202, 204
Monod model
  for multiple complementary nutrients, 42
  power density as function of substrate, 318–319
  for single limiting nutrients, 38–39
Monsanto company, 131
*Mortierella alpina*, 236
MSW (municipal solid waste), 94
MTBE (methyl tert-butyl ether), 14, 201
multiple limiting substrates, 42–44
multiple substitutable substrates, 42–44
municipal solid waste (MSW), 94

### N

NADPH, 18, 22–24, 30–33, 235, 272, 282
Nafion membranes, 312
*Nannochloris* sp. microalgal oil, 93
National Biodiesel Board, 247
National Renewable Energy Laboratory (NREL), 136–137, 222
natural gas, 8
NEDAK Ethanol plant, 188
net present value (NPV), 253
New Energy Corp. ethanol plant, 188
NGA (nongrowth-associated) products, 46, 52, 54
NiFe-hydrogenases, 270
Nigerian ethanol production, 78
nitrification, 30
nitrogen, 234–235

nitrogen oxide, 200–201
nitrogenase enzymes, 270, 274
nitrogen-intensive crops, 219
*Nitzschia* sp. microalgal oil, 93
noncatalyzed reactions, 239
noncompetitive inhibition, 39
nongrowth-associated (NGA) products, 46, 52, 54
noninteractive Monod model, 42
North Country Ethanol, LLC plant, 188
Northeast Biofuels ethanol plant, 188
Northwest Renewable, LLC ethanol plant, 188
Noureddini, H., 211
Novozymes, 132
NPV (net present value), 253
NREL (National Renewable Energy Laboratory), 136–137, 222
nutrients
  complementary, 42
  growth models for single limiting, 38–39
  hydrolysis modeling for, 47
  inhibition by, 39–40
  models for multiple limiting, 42–44
  Monod model, 38–39, 42
  rates of utilization, 47
  substitutable, 42–44

### O

oat straw, 85–86
oats, 77
OD (optical density), 314
Official Method Ca 14-56 testing, AOCS, 245
Ohm's law, 315
oils. *See also* microbial oils
  corn, 92
  cottonseed, 92
  crambe, 92
  crude, 7–8
  microalgal, 91, 93
  peanut, 92
  rapeseed, 88, 90, 92
  as sources for biodiesel production
    biosynthesis and modification of, 234–236
    overview, 219
    plant oils, 219–223
    straight vegetable oil, 233–234
    used cooking oils, 233
  soybean, 88–90, 92
  sunflower, 92

Index 365

OM (organic matter), 336
one-dimensional mass movement, 56
operating costs of biodiesel
    production, 257–258
opportunity fuels, 13
optical density (OD), 314
organic compound fermentation,
    27–28
organic intermediate compounds,
    28–29
organic matter (OM), 336
organoheterotrophic metabolism. *See
    also* fermentation
    aerobic respiration, 19–23
    anaerobic respiration, 23–24
organosolv fractionation, 141
Otter Tail Ag Enterprises ethanol
    plant, 188
Otto cycle, 197
output power of fuel cells, 297
overhead expenses for bioenergy
    plants, 255
overliming, 137–138
oxidation
    anaerobic, 28
    of glucose, 20–21, 29
    iodine values, 249–250
    in MFCs, 303
    of organic substrate, 305
    of pyruvate, 21–22
    states for common elements, 17–18
    wet, 141
oxidative stability, 219
oxygen
    calculating diffusion rate, 318
    effect on *S. cerevisiae* metabolism,
        108
    rates of utilization, 47
oxygenic photosynthesis, 32–33, 271

### P

Pacific Ethanol plants, 188
palm oil, 222
Panda Energy ethanol plant, 188
*Panicum virgatum* grass, 335–336
pantothenic acid, 164–165
PAR (Photosynthetically Active
    Radiation), 32
Parallel Products ethanol plants,
    188–189
partial pressure, 292–293
partitioned aquaculture systems (PAS),
    229

pasteurization of sugarcane juice, 111
Patriot Renewable Fuels, LLC ethanol
    plant, 189
PDC (pyruvate decarboxylase), 151
PDH (pyruvate dehydrogenase), 151
peaches, 286
peak power density, 317, 321
peanut oil, 92
pearl millet, 78
*Pelotomaculum* bacteria, 333
PEMs. *See* proton exchange
    membranes
Penford Products ethanol plant, 189
performance of microbial fuel cells
    methods for, 314–318
    overview, 318–322
pesticides, 252
PET (production of ethanol) operon,
    149–150
PFA (polyunsaturated fatty acids),
    90–91
PFK enzyme, 108–109
PFL (pyruvate formate lyase), 151
pH
    of methane reactors, 340
    optimum, 290
phase separation, 218
Phoenix Biofuels ethanol plant, 189
phosphoric acid fractionation,
    concentrated, 141–142
photoautotrophic growth, 272
photoautotrophic microorganisms, 33
photobiological hydrogen production,
    269, 271–274
photobioreactors, 231, 233, 273
photofermentation, 273–274
photosynthesis
    anoxygenic, 33
    biofuel production, 5
    direct biophotolysis, 272
    microbial fuel cell, 308
    oxygenic, 32–33
photosynthetic cyanobacteria, 324
photosynthetic pigments, 271
Photosynthetically Active Radiation
    (PAR), 32
phototrophic metabolism, 31–33
phototrophs, 17
photovoltaic technology, 12–13
Piedmont Biofuels Coop, 199
pigments, photosynthetic, 271
pilot scale operations, 245–247
Pinal Energy, LLC ethanol plant, 189

Pine Lake Corn Processors, LLC ethanol plant, 189
Ping Pong Bi Bi inhibition model, 215
Pioneer Hi-Bred International, Inc., 131
plant oils, 88–91, 219–223
plasmid-bearing yeast strains, 146–147
platinum carbon anodes, 320
Platinum Ethanol, LLC plant, 189
plug flow systems, 55–57
Plymouth Ethanol, LLC plant, 189
Poet company, 132, 189–190
polar lipids, 237
polarization, 313
polymerization, 250
polysaccharides, 331
polysulfides, 282
polyunsaturated fatty acids (PFA), 90–91
poplars, 87
potassium acetate, 280
potassium hexacyanoferrate, 308
potassium hydroxide, 204–206
potassium soap, 206
power calculations, MFC, 315–317
power density
   calculating for MFCs, 315, 317
   as function of circuit resistance, 313
   as function of substrate, 318–320
power output of MFCs, 315
Prarie Horizon Agri-Energy, LLC ethanol plant, 190
preservation of *Thermotoga* spp., 290
pressure
   partial, 292–293
   of total gas, 292
   vapor, 216–217
   of water vapor, 292
pressure swing adsorption (PSA), 124–125
pretreatment, acid esterification, 243. *See also* biomass
processor preferred high fermentable corn (HFC), 131
product inhibition, 40–41
product yields, 44–45
production. *See* biodiesel; ethanol
production of ethanol (PET) operon, 149–150
products
   classification of, 45
   growth-associated, 52, 54
   nongrowth-associated, 52, 54

profit fees, 256
protein levels in wheat, 74–75
proton exchange membranes (PEMs)
   calculating oxygen diffusion rates through, 318
   fermentation bioreactor sizing for fuel cells, 297–299
   fuel cells, 344
   MFCs with
      calculating Coulombic efficiency, 317–318
      calculating peak power density for, 317
      defined, 303
      single-chamber design, 320–321
protonation, 206
protons, 272, 312
PSA (pressure swing adsorption), 124–125
purification of sugarcane juice, 111
*Pyrococcus furiosis*, 270, 281–282
pyrolysis, 158–159
pyruvate decarboxylase (PDC), 151
pyruvate dehydrogenase (PDH), 151
pyruvate formate lyase (PFL), 151
pyruvate metabolism, 25
pyruvate oxidation, 21–22
*Pythium irregulare*, 236–237

## Q

Quad-County Corn Processors ethanol plant, 190
quality control analytical technique, 247–250
quantification of hydrogen fermentation
   dissolved concentration in liquids, 294–296
   gas concentration, 293
   hydrogen detection, 291
   mol $H_2$/L media, 294
   overview, 290–291
   partial pressure, 292–293
   production rates, 294
   total gas pressure, 292
   water vapor pressure, 292

## R

Rabaey, K., 308
rapeseed oil, 13–14, 88, 90, 92, 210, 218
Ratledge, C., 234
raw material costs, 257

reactions
  anabolic, 28–30, 33
  endothermic, 159
  exothermic, 159
  hydrogen production, 269–270
  kinetics of, 211–215
  light, 32–33
  noncatalyzed, 239
  soap, 240
reactors, 287–288. *See also* bioreactors
recovery of ethanol, 124–125
Red Trail Energy, LLC ethanol plant, 190
Redfield Energy, LLC ethanol plant, 190
redox mediators, MFC, 310–311
redox potential, 22–23
reduction, degrees of, 35
Reeve Agri-Energy ethanol plant, 190
Reid, R. C., 216
Renessen venture, 131
Renew Energy ethanol plant, 191
Renova Energy ethanol plant, 193
reporting hydrogen production
  dissolved concentration in liquids, 294–296
  gas concentration, 293
  hydrogen detection, 291
  mol $H_2$/L media, 294
  overview, 290–291
  partial pressure, 292–293
  production rates, 294
  total gas pressure, 292
  water vapor pressure, 292
residues
  agricultural, 82–84, 284–287
  biomass, 81–82
resistors, 323
respiration, aerobic, 19–23, 33. *See also* anaerobic respiration
retention time of methane reactors, 339
Rh-based catalysts, 166–168
rice, 77
rice hulls, 84, 87
rice straw, 85–86
Richmond, A., 229
ROI comparisons, 130
roots, 78–79

## S

SAA (soaking in aqueous ammonia) process, 139
*Saccharomyces cerevisiae*, 144, 147–149, 152, 334

sagu, 78
Saka, S., 210
salt content, 290
Sander, R., 295
saponification, 203–204
saturated fats, 218
saturation of feedstock, 90
*Schizochytrium* spp. microalgae, 236
Schroder, C., 275, 281, 283, 290
SCO (single-cell oils), 227
second-order kinetics, 211
separate hydrolysis and fermentation (SHF), 154–155
*Shewanella* bacteria, 309
SHF (separate hydrolysis and fermentation), 154–155
short-chain unsaturated fatty acids, 234
SI (spark-ignition) gasoline engines, 8
sieves, molecular, 124
silage, 338
silica, 291
simulations
  of batch reactors, 59–61
  continuous stirred tank reactors, 61–64
simultaneous saccharification and cofermentation (SSCF), 153–154
simultaneous saccharification and fermentation (SSF), 121–122, 150, 153–154
single limiting nutrients, 38–39
single-cell oils (SCO), 227
single-chamber MFC designs, 320–321
Siouxland Energy & Livestock Coop ethanol plant, 191
Siouxland Ethanol, LLC ethanol plant, 191
slime molds, 227
slurry, corn, 119–121
soaking in aqueous ammonia (SAA) process, 139
soap reaction, 240
sodium acetate, 280
sodium hydroxide, 204–206
sodium methoxide, 205–206
sodium soap, 206
SOI (starts of injection), 200
solar energy
  biofuel production, 5
  efficiency of, 8
  phototrophic metabolism, 32–33
  technology for, 12–13

solar radiation, 32–33
solid acid catalysts, 207–208, 239
solid base catalysts, 206
solubility
  of biodiesel, 218
  of hydrogen gas, 288
sorbitol, 110
sorghum, 76–77
sorghum straw, 85–86
Southwest Iowa Renewable Energy, LLC ethanol plant, 191
soy protein isolate, 253
soybean hull, 87
soybean oil, 88–90, 92
soybeans, 198, 219
spark-ignition (SI) gasoline engines, 8
spectrophotometers, digital, 314
SQDG (sulfoquinovosyl diacylglycerols), 237
SSCF (simultaneous saccharification and cofermentation), 153–154
SSF (simultaneous saccharification and fermentation), 121–122, 150, 153–154
starch feedstocks, 69–79, 105–133. *See also* cereal grains
  microorganisms
    *Saccharomyces cerevisiae*, 105–108
    *Zymomonas mobilis*, 108–111
  other grains, 78
  process technology
    dry milling, 119–122
    economics of, 125–126
    overview, 115–118
    recent developments in, 126–133
    recovery of ethanol in, 124–125
    sugar feedstocks, 111–115
    theoretical ethanol yield, 122–124
    wet milling, 118–119
  tubers and roots, 78–79
starch hydrolysis, 122, 131–132
starch liquefaction, 119–120
starch processing, 118
Stargen products, 131–132
starts of injection (SOI), 200
steady-state concentrations, 52–54
steam explosions, 138–139
Steele's model, 40
steeping corn, 118
Sterling Ethanol, LLC ethanol plant, 191
stillage, 122
stoichiometry of ethanol production, 145–146, 164
storage of *Thermotoga*, 290
straight vegetable oil (SVO), 233–234
structured growth models, 38
substitutable substrates, 42–44
substrates
  complementary, 42
  hydrolysis modeling for, 47
  inhibition by, 39–40
  measurements of, 314
  in MFCs, 308
  Monod model, 38–39
  multiple limiting, 42–44
  photofermentation, 274
  power density as function of, 318–320
  rates of utilization, 47
  substitutable, 42–44
sucrose, 107, 279
sugar
  from cull peaches, 286
  yeast strains utilizing
    mesophilic, 145–150
    thermophilic, 151–152
sugar beets
  ethanol production from, 116–117
  juice from, 111–113, 115
  molasses from, 111–113
  overview, 80
  production costs of ethanol from, 116
sugar feedstocks
  for ethanol production. *See also* microorganisms; starch feedstocks
    process technology, 111–115
    sugarbeets, 116–117
  fermentation of
    by bacteria, 26–27
    by yeast, 25–26
  overview, 79–80
sugar platform for ethanol production. *See also* biomass
  enzyme hydrolysis, 143–144
  fermentation
    Cellulolytic and Xylanolytic strains, 152
    mesophilic sugar-utilizing strains, 145–150
    overview, 144–145
    thermophilic sugar-utilizing strains, 151–152
  process economics, 156–158
  process integration, 152–156

sugar transport, 105–107
sugarcane, 79–80, 111–114
sulfhydrogenase, 282
sulfide dehydrogenase, 282
sulfidogenesis, 281–283
sulfonic acid, 207
sulfoquinovosyl diacylglycerols (SQDG), 237
sulfuric acid, 134–136, 138, 207
sunflower oil, 92
SunOpta company, 138–139
supercritical esterification, 208–210
supercritical transesterification, 208–210
SVO (straight vegetable oil), 233–234
sweet potatoes, 79
switchgrass, 87–88, 335–336
syngas
　overview, 80
　platform for ethanol production
　　biomass gasification, 158–160
　　catalytic conversion of, 166–168
　　fermentation, 160–165
　　process economics, 170–174
　　process integration, 169–170
Syngenta company, 131
Syntec process, 170
synthetic zeolite adsorbents, 124
syntrophic bacteria, 279, 332–333

## T

TAG. *See* triacylglycerol
TAN (total acid number), 249
tapioca, 78
Tate & Lyle ethanol plant, 191
TCA (tricarboxylic acid), 21–22, 304
TCD (thermal conductivity detector), 291
technology
　corn ethanol, 126–127
　for ethanol production from starch
　　dry milling, 119–122
　　economics of, 125–126
　　overview, 115–118
　　recent developments in, 126–133
　　recovery of ethanol in, 124–125
　　theoretical ethanol yield, 122–124
　　wet milling, 118–119
　for ethanol production from sugar, 111–115
　photovoltaic, 12–13
temperatures
　effect on kinetic constant values, 47–48

of methane reactors, 339
in steam explosion, 138
testing, viscosity, 248
tetracycline, 146–147
TGA (thermogravimetric analysis), 216
Thailand ethanol production, 78
Tharaldson Ethanol plant, 191
Thauer, R., 275
theoretical ethanol yield, 122–124
thermal conductivity detector (TCD), 291
thermal conversion processes, 11
*Thermoanaerobacter mathranii*, 151–152
thermodynamics of biodiesel production
　boiling point, 216–217
　overview, 210
　phase separation, 218
　reaction kinetics, 211–215
　solubility, 218
　thermogravimetric kinetics, 215–216
　vapor pressure, 216–217
thermogravimetric analysis (TGA), 216
thermogravimetric kinetics, 215–216
thermophilic bacteria, 151–152, 274, 277
thermophilic digestion, 339
thermophilic fermentations, 288
*Thermotoga* bacteria
　from agricultural residues, 284–287
　batch curve growth for, 59–60
　batch fermentation, 280
　biohydrogen production, 298
　biomass and product yields for, 45
　fermentative hydrogen production with, 276–277
　growth model for, 44
　hydrogen inhibition, 281
　hydrogenase activities, 270
　media for, 288–289
　preservation of, 290
　storage of, 290
　sulfidogenesis, 281–283
thin stillage, 122
thionine, 305
titration, 240–241
tocopherols, 201
toga, 276
total acid number (TAN), 249
total gas pressure, 292
total glycerin, 250
total solids (TS), 338, 340

transesterification
  acid esterification and, 244
  lipase-catalyzed, 203
  overview, 202
  supercritical, 207–210
  titration, 240
transportation fuels, 10
Trenton Agri Products, LLC ethanol plant, 191
triacylglycerol (TAG), 202, 204, 228
tricarboxylic acid (TCA), 21–22, 304
triglycerides, rapeseed oil, 210
TS (total solids), 338, 340
tubers, 78–79
two-chamber MFC designs, 320

## U

UASB (upflow anaerobic sludge blanket), 341
ultraviolet (UV), 228
uncompetitive inhibition, 39
unicellular green algae, 230
United Ethanol plant, 191
United States
  corn production in, 96–97
  distribution of crop residues in, 100–101
  distribution of wood residues in, 98–99
  ethanol plants in, 182–195
  ethanol production
    barley-based, 75–76
    corn-based, 71–73
    sorghum-based, 76
  forest-based biofuel production, 81–82
  projected production of switchgrass, 102–103
United WI Grain Producers, LLC ethanol plant, 191
unmethylated esters, 211
unsaturated hydrocarbons, 219
unsaturation, 250
unstructured growth models, 38–39
updraft gasifiers, 159–163
upflow anaerobic sludge blanket (UASB), 341
U.S. Department of Energy (DOE)
  Energy Information Administration, 298
  NREL report, 222
used cooking oils, 233

Utica Energy, LLC ethanol plant, 191
UV (ultraviolet), 228

## V

vampire losses, 10
Van Gerpen, J., 204
Van Niel, E. W. J., 290
Van Ooteghem, S. A., 288
vapor pressure, 216–217
vegetable-based fuels, 197
VeraSun Energy Corporation ethanol plant, 192
virgin canola feedstock, 250–251
viscosity, biodiesel, 210, 248
Viton tubing, 248
volatile solids (VS), 338, 340
Volta, Alessandro, 329
VS (volatile solids), 338, 340

## W

Wang, Xi, 202
washes, biodiesel, 244–245
washout retention time, 63–64
waste products
  cooking oils, 250–251
  fuels from, 13
wastewater treatment, 321–322
water scrubbing, 160
water vapor, 291–292
water wash, biodiesel, 244–245
WDG (wet distillers grains), 122
WDGS (wet distillers grains with solubles), 122
Western New York Energy, LLC ethanol plant, 192
Western Plains Energy, LLC ethanol plant, 192
Western Wisconsin Renewable Energy, LLC ethanol plant, 192
wet distillers grains (WDG), 122
wet distillers grains with solubles (WDGS), 122
wet fermentation systems, 338
wet milling, 73, 118–119, 125–127, 129–130
wet oxidation, 141
wheat, 73–75, 116
wheat straw, 85–86
wheat straw xylan hydrolysis, 143–144
White Energy ethanol plant, 193
willows, 87
wind energy, 8, 13
Wind Gap Farms ethanol plant, 193

wood chip-derived syngas fermentation, 170–172
wood-based biofuel production, 81
Wu, Q., 227

### X

xenobiotic compounds, 41–42
Xethanol BioFuels, LLC ethanol plant, 193
Xylanolytic strains, 152
xylose, 43, 331
xylose isomerase, 147
xylose metabolism, 145–146, 148, 152

### Y

yeast, 25–26, 290. *See also* individual yeasts by name

yellow grease, 233
yield, energy, 5
yield parameters, 44–45
Yu, X., 288
Yuan, W., 216, 217
yucca, 78
Yuma Ethanol plant, 193

### Z

zero-emissions goal, 4
Zhang, Y., 200, 233, 245, 250, 251
Zhu, D., 211
*Zymomonas mobilis*, 145–147

CPSIA information can be obtained at www.ICGtesting.com
Printed in the USA